Advances in Earthworm Ecotoxicology

Other titles from the
Society of Environmental Toxicology and Chemistry (SETAC)

ADVANCES IN EARTHWORM ECOTOXICOLOGY

Edited by

Stephen C. Sheppard
ECOMatters Inc.
Pinawa, Manitoba, Canada

John D. Bembridge
ZENECA Agrochemical
Berkshire, Braknell, England

Martin Holmstrup
National Environmental Research Institute
Silkeborg, Denmark

Leo Posthuma
RIVM
Bilthoven, The Netherlands

Proceedings from the Second International Workshop on Earthworm Ecotoxicology
April 1997
Amsterdam, The Netherlands

SETAC Technical Publications Series

SETAC Liaison
Roman Lanno
Department of Zoology, Oklahoma State University, Stillwater, Oklahoma, USA

Current Coordinating Editor of SETAC Books
C.G. Ingersoll
U.S. Geological Survey, Midwest Science Center

Publication sponsored by the Society of Enviromental Toxicology and Chemistry
(SETAC)

Cover by Michael Kenney Graphic Design and Advertising
Editing by Wordsmiths Unlimited

Library of Congress Cataloging-in-Publication Data

International Workshop on Earthworm Ecotoxicology (2nd:1997 Amsterdam, Netherlands)
Advances in earthworm ecotoxicology/edited by Stephen Sheppard ... [et al.].
  p. cm. -- (SETAC technical publications series)
"Proceedings from the Second International Workshop on Earthworm Ecotoxicology, 2–5 April 1997, Amsterdam, Netherlands."
"Publication sponsored by the Society of Environmental Toxicology and Chemistry (SETAC) and the SETAC Foundation for Environmental Education."
Includes bibliographical references.
ISBN 1-880611-25-2
1. Earthworms–Effect of chemicals on–Congresses.
2. Agricultural chemicals–Toxicology–Congresses. 3. Ecological risk assessment–Congresses. I. Sheppard, Stephen, 1951–. II. SETAC (Society) III. SETAC Foundation for Environmental Education. IV. Title. V. Series.

QL391.A6I625 1997   98-16409
592'.6417–dc21   CIP

International Standard Book Number 1-880611-25-2
Printed in the United States of America
05 04 03 02 01 00 99 98   10 9 8 7 6 5 4 3 2 1

∞The paper used in this publication meets the minimum requirements of the American National Standard for Information Sciences—Permanence of Paper for Printed Library Materials, ANSI Z39.48-1984.

Reference Listing: Sheppard SC, Bembridge JD, Holmstrup M, Posthuma L. 1998. Advances in earthworm ecotoxicology. Proceedings from the Second International Workshop on Earthworm Ecotoxicology. 2–5 April 1997. Amsterdam, The Netherlands. Pensacola FL: Society of Environmental Toxicology and Chemistry (SETAC) 472 pp.

The SETAC Technical Publications Series

The SETAC Technical Publications Series was established by the Society of Environmental Toxicology and Chemistry (SETAC) to provide in-depth reviews and critical appraisals on scientific subjects relevant to understanding the impacts of chemicals and technology on the environment. The series consists of single- and multiple-authored or edited books on topics reviewed and recommended by the Publications Advisory Council and approved by the SETAC Board of Directors for their importance, timeliness, and contribution to multidisciplinary approaches to solving environmental problems. The diversity and breadth of subjects covered in the series reflect the wide range of disciplines encompassed by environmental toxicology, environmental chemistry, and hazard and risk assessment. Despite this diversity, the goals of these volumes are similar: they are to present the reader with authoritative coverage of the literature, as well as paradigms, methodologies, controversies, research needs, and new developments specific to the featured topics.

The SETAC Technical Publications Series is useful to environmental scientists in research, research management, chemical manufacturing, risk assessment, regulation, and education, as well as to students considering or preparing for careers in these areas. The series provides information for keeping abreast of recent developments in familiar subject areas and for rapid introduction to principles and approaches in new subject areas.

Preface

Progress in earthworm ecotoxicology

Earthworms are widely considered as representative terrestrial organisms closely associated with the soil; therefore, they are good indicators of soil quality. They have been recognized for almost 2 decades as keystone species for regulatory testing and environmental risk assessment of agricultural chemicals. To summarize and discuss progress in the field of applied earthworm ecotoxicology, a very successful workshop was held in Sheffield, United Kingdom, in 1991. As a follow-up, a second workshop was held 2–5 April 1997 by scientists involved in earthworm ecotoxicology, representing universities, governmental agencies, and industrial research groups. The aims of this second workshop, held as a satellite meeting to the 1997 SETAC-Europe conference in Amsterdam, were to assess progress made with the recommendations since 1991, to advance collaboration, and to optimize assessment methods to promote sound regulation policy-setting.

The workshop organizers intended to create a forum for scientists from different backgrounds and to generate scientific knowledge firmly based in ecological theory with applied questions posed by the speed of societal change. In this setting, researchers in the field of earthworm ecotoxicology were asked to report on actual progress made since the 1991 meeting, with emphasis on new developments and applicability.

In order to achieve this goal, the workshop program was developed on the basis of a formal opinion survey of scientists from industry, governments, and scientific organizations. Based on the answers to the questionnaire, the workshop program consisted not only of oral and poster presentations but also of ample time for discussions, with the aim of producing comprehensive conclusions and recommendations. This procedure was implemented by assigning a group of rapporteurs from different origins. This group, chaired by John Bembridge, scrupulously noted the main lines of the discussion and formulated a series of draft recommendations. The draft recommendations were circulated among the members of the organizing committee, the scientific committee, and all participants. Through the circulation, it was ensured that the conclusions reached on site not only were supported by the participants but also were accurately phrased and operational.

It is a pleasure to see that the aims of the workshop and the aims of both the scientific and the organizing committees have resulted in this book. Earthworm ecotoxicology receives scientific attention at many levels of biological organization, so achieving consensus among the different workshop contributions seldom succeeds during the course of such a meeting in spite of all good intentions. However, authors, reviewers, and editors alike attempted to present workshop contributions that were relevant and challenging for science, product, and policy development. In our opinion, a coherent picture of the state-of-the-art has emerged. The extent to which this and the broader aims of the workshop have been achieved is here for the reader to assess. Ecotoxicologists aim to understand and protect earthworm populations, communities, and ecosystem function and structure by using the predictive value of responses at different levels of organization, be they biomarkers, bioindicators, or laboratory tests. Each of these approaches can result in real scientific progress, but only if scientists are aware of the need to synthesize the research outcomes at these different levels of biological organization. Such synthesis will not only benefit regulatory practice but also will identify research needs.

Adriaan J. Reinecke, Chair, Scientific Committee

Leo Posthuma, Secretary, Organizing Committee

Foreword
Earthworm ecotoxicology: a squirming quest

Introduction to the proceedings of the Second
International Workshop on Earthworm Ecotoxicology,
Amsterdam, April 1997

H. Eijsackers

Despite the squirming character of earthworms, it is a straightforward matter to explain to people the earthworm's benefit, utility, and potential for profit. Earthworms improve soil quality, enhance nutrient release and uptake, and maintain and increase soil aeration and drainage. They are a major food source for mammalian and bird predators that are important for nature conservancy. For both recreation (angling) and waste processing, they supply increasing markets. Is it a wonder that they are widely used for assessment of the potential environmental effects of human activities, as important representatives of soil animals, and the impact of soil animals on soil quality?

The use of earthworms in laboratory and field testing has been central to both soil ecology and ecotoxicological risk assessment and has received increasing attention in the last 10 to 20 years. Most recently, the use of earthworms in field bioassays and the application of earthworm biomarkers has gained growing interest.

In 1991, researchers from universities, government research institutes, pesticide and chemical industries, and regulatory bodies met for the first time in the framework of the First International Workshop on Earthworm Ecotoxicology (IWEE-1) in Sheffield to discuss laboratory and field testing [1]. The informal setting and intense discussions greatly contributed to the success of that meeting.

During this second workshop (IWEE-2) in Amsterdam in April 1997, these discussions continued. Experiences in the regulatory use of earthworm testing were described; new insights about bioavailability and internal load as discriminating factors for a proper assessment of environmental impacts were described as well. Other major developments were discussed, including the use of modeling to estimate long-term implications of impacts of toxic compounds on individuals for population survival as well as on food chain transfer and ecosystem structure and functioning. The substantial increase in interest in earthworm biomarkers was also reported.

This introduction does not aim to give a full overview of all the oral and poster presentations given: the intention is to sketch a coherent line in the workshop presentations, discussions, and recommendations. The line runs from first-tier standardized laboratory testing to flexible and specific field-testing schemes, underpinned both by basic research and by well-organized, intergroup ring testing. The recent interest in biomarkers, for instance, is reflected more extensively in these proceedings than in the workshop program, with the underlying question of whether biomarkers are indicators of species, or even system, health, or merely generally applicable physiological or biochemical-effect parameters.

The first section, "Laboratory testing: experiences, improvements, and advances," covers the latest observations in first-tier testing with earthworms, starting with an overview by Kula on the experiences within the BBA on acceptance of pesticides. The application of the 2 available earthworm tests (mortality and reproduction of *Eisenia fetida* or *E. andrei*) shows that they are suitable for first-tier testing but need improvement and extension with respect to the use of reference compounds and the comparability to or application of other earthworm species (*Eisenia* is a compost worm, not an earthworm). Spurgeon and Weeks address this question from a research point of view, evaluating whether these types of laboratory-applied tests with compost worms are really predictive of effects of chemicals on earthworms exposed in the field. Their comparative studies show that soil type and earthworm species should be taken into account. With respect to Zn, soil pH, organic matter, and clay type and content have a strong impact on the toxicity of Zn (see also Posthuma et al., Chapter 10). Further suggestions come from McIndoe et al. on improving accuracy of testing by including pretreatment measurements, especially on earthworm bodyweight as an important discriminative indicator of potential chronic effects. Litzel and Kula concentrate on the impact of pesticide application methods: the usual practice of applying the full dosage in the beginning of the test gives comparable results to several applications at lower rates typical of agronomy practice.

New topics in the area of first-tier testing are addressed by Lanno et al. with a contribution on the use of body residues to assess ecotoxicity. They found similar critical body residues (CBRs) for pentachlorophenol over a 10-fold concentration gradient with 3 earthworm species differing 1 order of magnitude in bodyweight. The relevance of CBRs is further supported by Posthuma et al., who studied the relationship between Zn partitioning and Zn uptake (internal load) in assessment procedures.

Another new topic is the introduction of behavioral responses to toxicity assessment. Stephenson et al. evaluated a series of pesticides with an avoidance response test that proved useful as an initial screen with both the compost worm *E. fetida* and the earthworm *Lumbricus terrestris*. In a later session on understanding field earthworm ecotoxicology, the contribution by Mather and Christensen further stresses this aspect of behavioral responses by describing the significance of earthworm surface migration as an important part of the reduction of earthworm numbers in the field under a pesticide regime: lower numbers after pesticide application could be partly the result of earthworms disappearing from the treated plot, hence mortality can be overestimated if migration is not accounted for. Other major groups of annelids like the enchytraeids (dominant in organic, rich, acidic forest soils) received increased attention from Römbke et al. so as to broaden the array of species on which ecological risk assessments can be based. Römbke et al. sketch the present status of the test procedures with enchytraeids. In a special workshop, succeeding IWEE-2, further technical aspects of testing with enchytraeids were covered that are not addressed in this book.

The second section, "Processes important to earthworm ecotoxicology," covers 2 major items that both deal with multiple stresses. Both Holmstrup and Petersen and Posthuma et al. show that we cannot assess contaminant effects isolated from other stress factors. Although this may seem usual and natural in experimental testing of effects of pesticides with standardized agronomy conditions, the role of variation in soil and exposure characteristics is important. For all types of environmental contamination, chemical impacts interfere with other environmental stress factors to produce a response. Holmstrup and Petersen studied desiccation and frost in relation to Cu and observed distinct synergism: the no-observed-effect concentration (NOEC) for Cu decreased 50% when relative humidity in air decreased from 100% to 96.5%. Clearly, there was less tolerance for Cu contamination when other stresses were present. Posthuma et al. examined soil acidity in relation to Zn availability in the pore water (measured as the solid/liquid partitioning, K_p or K_d) as well as Zn bioavailability (measured as a biota-to-soil-accumulation factor [BSAF] for 2 annelid species). Soil acidity was the primary descriptor for both K_p and BSAF, with Al-oxyhydroxides and clay content as secondary descriptors for BSAF.

Posthuma's suggestion that future work on uptake quantification should concentrate on toxicokinetic modeling is also acknowledged in the papers of Janssen et al., Sheppard et al., and Lanno et al. Janssen et al. studied the impact of environmental factors on the availability of radiocesium, finding that potassium and organic matter are important modulators for uptake. On the other hand, they found that the presence of earthworms may influence the partitioning of cesium. The advantages of working with radiochemicals are clearly illustrated by Sheppard et al.; by studying depuration and uptake kinetics of I, Cs, Mn, Zn, and Cd in radiotracer-spiked litter, they estimated the change of internal load in a realistic, nondestructive way for the same animals over a time series. They suggest that depuration could be a good indication of bioavailability and, moreover, conclude that most bioassays are executed in the first steep uptake phase. This may mean that impacts are assessed on the basis of laboratory tests in which body burdens are far below those at steady state.

In the third section, "Earthworm biomarkers: advances relevant to earthworm ecotoxicology," Scott-Fordsman and Weeks review 5 selected biomarkers with special regard to their use in risk assessment and conclude that only a few biomarkers show a dose-related response. Furthermore, very little is known about the persistence of a biomarker response once exposure ceases or about the relationships between biomarkers and natural variability. So far, no links between earthworm biomarker responses and adverse effects at higher levels of biological organization (population, community) have been established. Examples by Stürzenbaum and Svendsen et al. do show prospects for practical application of biomarker responses with heavy-metal-adapted populations and in assessment of contaminated areas.

The fourth section, "Experience with field earthworm ecotoxicology," in combination with the sixth and last section, "Progress toward understanding field earthworm ecotoxicology," gives an overview of the state-of-the-art in field assessment. Two topics are discussed most intensively:

1) to what extent are laboratory tests comparable to field assessments with respect to both under- or overestimation and false positives or negatives; and

2) to what extent can earthworm populations be reduced and over what period of time be able to recover fully?

Heimbach used the regular tier-2 reproduction test with field studies for 21 pesticides for laboratory studies and 15 pesticides for field studies; he concludes that field results are at least 5 to 10 X less sensitive than the laboratory tests, when expressed as the ratio between the effect concentration (EC50), based on percentage deviation from control, and the predicted environmental concentration (PEC). In a joint presentation by the various pesticide industries, McIndoe et al., conclude that regulatory short-term tests can be effectively used to screen out low-risk compounds, which leaves room for longer-term (e.g., field) tests to be applied to assess products for which risks are less certain. Triggers for further tiered testing should avoid false-negative assessments and at the same time minimize false-positive assessments.

For this kind of flexible, tailor-made field testing, Edwards underlines the importance of testing procedures that specify acceptable site variables, well-defined treatment variables, formulation and methods of chemical application (also special applications like granules and seed dressings), and endpoints for both final and intermediate assessment. Earthworm mobility may be an underestimated aspect in tuning field conditions to a specific situation, as illustrated by Mather and Christensen's observations.

The second item on the recovery potential of earthworm populations focuses on the issue of what degree of reduction in population size is permissible. Jones and Hart conclude from a literature review of laboratory toxicity tests and field studies on pesticides that acute effects in the field can occur at much lower concentrations than in laboratory studies. For nonpersistent chemicals with an environmental half-life (DT50) of less than 50 d, reductions of over 90% may allow full recovery within 1 y (one of the key issues according to Edwards' assessment procedure). For persistent chemicals with DT50s of more than 50 d, this level of population reduction may not be acceptable. Bembridge et al. used redundancy analysis to assess different kinds of perturbations of earthworm populations, discriminating the effects of pesticide treatments from other effects of agronomy practice. They report maximum increases of 130% and decreases of 50% from pesticide application, increases of 300 to 400% from manuring, and decreases of 70% from cultivation. Food was revealed to be the main limiting factor in this approach.

These observations are supported by the outcomes of 2 model studies presented in the fifth section, "Modeling earthworm population: an approach to field ecotoxicology." In models by both Axelsen and Holmstrup and Klok and de Roos, recovery potential of earthworm populations was sufficient to overcome population reductions up to 50%. In Axelsen and Holmstrup's model, available food was the most important limiting factor. Consequently, population growth retarded by food limitation was a key item in population recovery, although reduced competition from the reduced number of mature earthworms optimized growth potential of the hatching juveniles. Also, Klok and de Roos conclude that earthworm populations are most sensitive to toxicants that retard maturation.

From these field observations and the models based thereon, the conclusion arises that behavior (mobility and avoidance) together with maturation could be interesting descriptors for an improved, flexible field assessment. Moreover, as Verhoef stresses, from an ecological point of view, it should be worthwhile to use multi-species studies, as these allow for longer exposure periods, greater levels of biological organization, more toxicity endpoints, and better integration with ecosystem structure and function than exist in current field practices. To be sure, food-web models deserve further attention.

One of the major challenges for the forthcoming years will be to combine the positive elements of generic, preventive, standardized first-tier test systems with the site-specific (and frequently suggested tailor-made), flexible field testing. Especially in the field, the concepts of multi-species impacts and complex food-web modeling are relevant, but their proper application demands a large and detailed dataset. Such a dataset can be partly filled with the results of laboratory studies, but it is clear that a good deal of artisan-like field work is still needed.

From the lively discussions among representatives from research institutes, industries, and regulatory bodies as presented in the well-balanced and meticulously edited recommendations by Bembridge, it is clear that much still must be done to achieve an ecologically relevant test procedure. IWEE-3 certainly will provide some results.

Acknowledgment - The help of the editors, in particular Steve Sheppard and Leo Posthuma, in the preparation of this introduction is gratefully acknowledged.

References
1. Greig-Smith PW, Becker H, Edwards PJ, Heimbach F. 1992. Ecotoxicology of earthworms. Andover UK: Intercept.

Contents

List of tables (abbreviated titles)

List of figures (abbreviated titles)

Acronmyms and abbreviations

1CFOK	1-compartment first-order kinetics
AAS	flame-atomic absorption spectrophotometry
AchE	acetylcholinesterase
ANOVA	analysis of variance
AS	artificial soil
BaP	benzo[a]pyrene
BBA	Biologische Bundesanstalt Für Land-Und Forstwirtschaft
BSA	bovine serum albumin
BSAF	biota-to-soil accumulation factor
BSAR	biota-to-soil accumulation rate
CBR	critical body residue
CEC	cation exchange capacity
ChE	cholinesterase
CoE	Council of Europe
CR	concentration ratios
DAB	di-aminobenzidine dihydrochloride
DEB	dynamic energy budget
DNA	deoxyribonucleic acid
DOC	dissolved organic carbon
DT50	disappearance half-life
EC	European Council/Emulsifiable concentrate
ECB	European Chemical Bureau
ECPA	European Crop Protection Association
EDTA	ethylenediaminetetra-acetic acid
EEC	European Economic Community
ELISA	enzyme-linked immunosorbent assay
EPPO	European Plant Protection Organization
ER	erythrocyte rosette
ERT	Enchytraeid reproduction test
EU	European Union
EW	emulsion in water
FIAM	free-ion activity model
GST	glutathione S-transferase

HF	hydrogen fluoride
HPLC	high press liquid chromotography
HSP	heat-shock protein
ILL	incipient lethal level
ISO	International Standards Organization
IWEE-1	First International Workshop on Earthworm Ecotoxicology
IWEE-2	Second International Workshop on Earthworm Ecotoxicology
Kow	octanol/water partition coefficient
K_p	solid liquid partition coefficient
LC50	lethal concentration to 50% of test population
LOAEL	lowest-observed-adverse-effects level
LOEC	lowest-observed-effects concentration
LSC	liquid scintillation counting
MBC	methyl 2-benzimidazole carbamic acid
MBP	metal-binding protein
MDPE	medium density polyethylene
mRNA	messenger ribonucleic acid
MT	metallothionein
NADPH	nicotinamide-adenine dinucleotide phosphate
NBT	nitroblue tetrazolium dye
NEC	no-effect concentration
NOAEL	no-observed-adverse-effects level
NOEC	no-observed-effects concentration
NRR	neutral red retention time
OECD	Organization for Economic Cooperation and Development
OM	organic matter
OP	organophosphorus
PAH	polycyclic aromatic hydrocarbon
PBS	phosphate-buffered saline
PCA	principal component analysis
PCB	polychlorinated biphenyl
PCDD	polychlorinated dibenzo dioxin
PCDF	polychlorinated dibenzo furan
PCP	pentachlorophenol
PCR	polymerase chain reaction

PEC	predicted environmental concentration
PEC$_s$	predicted environmental concentration of compound in soil
pF	soil water potential
PRC	principal response curve
RDA	redundancy analysis
RH	relative humidities
RNA	ribonuclic acid
RS	reference soil
SC	suspension concentrate
SCARAB	Seeking Confirmation about Results at Boxworth project
SD	standard deviation
SE	standard error
SL	soluble concentrate
SR	secretory rosette
STPF	stabilized temperature platform furnace
TCTP	translationally controlled tumor protein
TER	toxicity/exposure ratio
TER$_a$	acute toxicity/exposure ratio
TNT	trinitrotoluene
[tox]$_{thr}$	toxic threshold level
TPTH	triphenyltin hydroxide
UBA	German Federal Environment Agency
UK	United Kingdom
WG	water dispersible granules
WP	wettable powder

Conference participants*

Anna-Karin Augustsson
Lund University
Lund, Sweden

Jorgen Axelsen
National Environmental Research
Institute
Silkeborg, Denmark

Hans Georg Back
Getave GmbH

Nussloch, Germany

Ian Barber
AgrEvo UK Ltd.
Essex, England

John D. Bembridge
ZENECA Agrochemicals
Berkshire, England

Marco Candolfi
Springborn Laboratories Europe
Horn, Switzerland

Ole M. Christensen
Aarhus University
Aarhus, Denmark

Rut Maria Callado de la Pena
Universitaet Osnabrueck
Osnabrueck, Germany

Trudy Crommentuijn
RIVM/CSR
Bilthoven, The Netherlands

Peter Dohmen
BASF Aktiengesellschaft
Limburgerhof, Germany

Charles T. Eason
Manaaki Whenua Landcare Research
Lincoln, New Zealand

Clive Edwards
Ohio State University
Columbus, OH

Peter J. Edwards
ZENECA Agrochemicals
Bracknell, England

Herman Eijsackers
RIVM
Bilthoven, The Netherlands

Torunn Gauteplass
University of Oslo
Oslo, Norway

Angela Gossmann
IBACON GmbH
Rossdorf, Germany

Jari Haimi
University of Jyvaskyla
Jyvaskyla, Finland

Fred Heimbach
Bayer AG
Leverkusen, Germany

Ranier Huesel
AgrEvo
Frankfurt am Main, Germany

Martin Holmstrup
National Environmental Research Institute
Silkeborg, Denmark

Simon Hoy
Pesticides Safety Directorate
York, England

Tjalling Jager
RIVM
Bilthoven, The Netherlands

Martien Janssen
RIVM
Bilthoven, The Netherlands

Alison Johnson
Huntingdon Life Sciences
York, England

A. Jones
Central Science Laboratory
York, England

Jürgen C. Kühle
ITEC
Kubschutz, Germany

Chris Klok
IBN-DLO
Wageningen, The Netherlands

Dmitri Krivolutsky
Institute of Ecology and Evolution
Moscow, Russia

Christine Kula
Biologische Bundesanstalt für Land-
und Forstwirtschaft
Braunschweig, Germany

Hartutm Kula
Biologische Bundesanstalt für Land-
und Forstwirtschaft
Braunschweig, Germany

Sherry Le Blanc
Oklahoma State University
Stillwater, OK

Fuencisla Marino Callejo
Universidad de Vigo
Vigo, Spain

Nicholas Martin
Mount Albert Research Centre
Auckland, New Zealand

Sabine Martin
Federal Environmental Agency
Berlin, Germany

Janice Mather
University of Aarhus
Aarhus, Denmark

Eddie McIndoe
ZENECA Agrochemicals
Bracknell, England

A. John Morgan
University of Wales, Cardiff
Cardiff, Wales, United Kingdom

Jos Notenboom
RIVM
Bilthoven, The Netherlands

Odin-Fevrtet
Rhone Poulenc Agro
Sophia Antipolis, France

Tonny Olesen
Institute of Terrestrial Ecology
Huntingdon, England

Ralf Petto
IBACON GmbH
Rossdorf, Germany

Vaclav Pizl
Institute for Soil
Ceske Budejovice, Czech Republic

Andrei Pokarzhevskii
Russian Academy of Sciences
Moscow, Russia

Leo Posthuma
RIVM
Bilthoven, The Netherlands

Nick F. Protopopov
Tomsk Polytechnical University
Tomsk, Russia

Jörg Römbke
ECT Oekotoxikologie Gmb
Flörsheim, Germany

Adriaan Reinecke
University of Stellenbosch
Matieland, South Africa

Sophié Reinecke
University of Stellenbosch
Matieland, South Africa

A. Rozen
Jagiellonian University
Krakow, Poland

Hans Rufli
Novartis Crop Protection A
Basel, Switzerland

Sten Rundgren
Lund Universit
Lund, Sweden

Janne Salminen
University of Helsinki
Lahti, Finland

Ruediger Schmelz
Universitaet Osnabrueck
Osnabrueck, Germany

Janeck Scott-Fordsmand
National Environmental Research
Institute
Silkeborg, Denmark

Steve Sheppard
AECL Whiteshell Laboratories
Pinawa, Manitoba, Canada

David Spurgeon
Reading University
Reading, England

Jorgen Stenersen
University of Oslo
Oslo, Norway

Gladys Stephenson
University of Guelph
Guelph, Ontario, Canada

Sven Strand
American Cyanamid Company
Princeton, NJ

S. Stuerzenbaum
University of Wales, Cardiff
Cardiff, Wales, United Kingdom

Zhenjun Sun
China Agricultural University
Beijing, China

Geoffrey Sunahara
National Research Council Canada
Montreal, Quebec, Canada

Claus Svendsen
Institute of Terrestrial Ecology
Huntingdon, England

Tom Tooby
Mallard House
York, England

Yvette van Erp
NOTOX
's-Hertogenbosch, The Netherlands

Kees van Gestel
Vrije Universiteit
Amsterdam, The Netherlands

Herman Verhoef
Vrije Universiteit
Amsterdam, The Netherlands

Alexander Viktorov
Russia Academy of Sciences
Moscow, Russia

Nadja Wagman
Umeå University
Umeå, Sweden

Jason Weeks
Institute of Terrestrial Ecology
Huntingdon, England

*These affiliations were accurate at the time of the workshop.

Laboratory testing: experiences, improvements, and advances

Chapter 1

Endpoints in laboratory testing with earthworms: experience with regard to regulatory decisions for plant protection products

Christine Kula

Testing side effects of chemicals and plant protection products on earthworms in the laboratory should produce as much relevant information on effects of a substance as possible. Two standard tests are available for laboratory testing on earthworms. The main endpoint of the Organization for Economic Cooperation and Development (OECD) acute test [1] is mortality. However, a new laboratory test for sublethal effects, available in a draft international guideline and a national German guideline, uses reproductive success, i.e., the number of juveniles, as its main endpoint. In this chapter, the different endpoints are reviewed and discussed using confidential data from tests submitted to the German authority responsible for the authorization procedure for plant protection products. Reproducibility, practicability, validity criteria, and relevance of the tested parameters for field populations are discussed, using examples from the stepwise risk-assessment procedures according to the European Union (EU) and the European Plant Protection Organization (EPPO).

Laboratory tests play an important role in risk assessment of chemicals and pesticides for earthworms. Two standard laboratory tests are available. A test of acute toxicity is the internationally standardized OECD Guideline 207 [1], and a test of sublethal effects is in the international standardization procedure [2]. Both methods play a central role in risk assessment for legal purposes within the EU [3,4]. The questions are whether gaps in knowledge still exist and whether there are problems using the tests in a fixed risk-assessment framework.

The following endpoints are standardized and well described in the guidelines [2,5,6]:

mortality,

reproduction, and

bodyweight change.

Other endpoints like behavioral, morphological, and physiological changes are reported occasionally, but they are not evaluated in a standardized way.

Endpoints used in laboratory tests are discussed here with respect to the parameters of reproducibility, practicability, validity criteria, and relevance for field conditions. "Reproducibility" means that repeated tests yield similar results. This can be proven in ring tests, e.g., where identical substances are tested in different facilities according to the same guideline. "Practicability" means that the test design is not too difficult to implement, thus minimizing the sources of mistakes. "Validity criteria," e.g., for control boxes, prescribe certain ranges for mortality or bodyweight changes to ensure that the results of a test meet a certain standard. Testing a reference compound is another way to reach this aim. "Relevance" means that the endpoints chosen for laboratory evaluation are pertinent to the earthworm population in the field.

Test methods

The acute test is the most relevant for standard testing in the laboratory [1]. A nearly identical test is described in an International Standards Organization (ISO) guideline [5]. Concerning sublethal effects, a reproduction test is in the international standardization procedure (ISO DIS 11268-2), which is very similar to the national guideline on reproduction tests of earthworms in Germany [6]. Table 1-1 summarizes the key characteristics of the acute and reproduction tests, and endpoints are listed in order of importance within the test. One of the main differences between the tests conducted for risk assessment of plant protection products is exposure. In the acute test, all test substances are mixed into the soil homogeneously, whereas in the reproduction test, plant protection products are applied in a manner as close to field conditions as possible. For most plant protection products, this means that an application on the soil surface is the relevant application mode.

Table 1-1   Key features of guidelines for laboratory tests on earthworms

Feature	Acute test[1]	Reproduction test[2]
Species	*Eisenia fetida/Eisenia andrei*	*Eisenia fetida/Eisenia andrei*
Test substrate	Artificial soil	Artificial soil
Application of test substance	Mixed into the soil	for chemicals other than pesticides, mixed into the soil for pesticides, as close to field conditions as possible (i.e., in most cases on soil surface)
Application rate	Up to 1000 mg/kg	for chemicals other than pesticides, not more than 1000 mg/kg for pesticides, at least one relevant
Test duration	2 weeks	8 weeks
Food	—	Cow manure
Validity criteria (control)	Mortality ≤ 10%	Mortality ≤10%; Number of juveniles ≥ 30; Coefficient of variation < 30
Endpoints	Mortality Body weight of adults Others (e.g., behavior)	Number of juveniles Body weight of adults Mortality Others (e.g., behavior)

[1] Guidelines: OECD 207, ISO 11268-1
[2] Guidelines: ISO DIS 11268-2, BBA VI 2-2

In the following subsections, the main endpoints, the main parameters, and the test-method guidelines are discussed.

Mortality

Mortality expressed by means of LC50 after 7 and 14 d of exposure is the main endpoint in the acute toxicity test [1]. In Table 1-2, the key characteristics of this endpoint are summarized.

Table 1-2   Characteristics of parameters for endpoints in acute and reproduction tests

Endpoint	Reproducibility	Practicability	Validity criteria (for control)
Mortality	Sufficient	Good (disappeared = dead)	≤ 10%
Reproduction	Sufficient	Good (juveniles allowed to hatch and grow up; hand sorting or heat extraction)	I. ≥30 juveniles/test box *Reached in 85% of tests (n = 40)* II. Coefficient of variation ≤ 30% *Reached in 97% of those tests where criterion I also fulfilled (n = 30)*
Bodyweight change	Sufficient	Good	Acute test: OECD: none ISO: weight decrease ≤ 20% Reproduction test: ISO: none BBA: weight decrease ≤ 20%
Behavior	No scale for observations and reporting	No scale	None

Reproducibility of mortality is considered sufficient for legislative purposes, in other words, to control a compound's use through legislation. Ring test results available for the acute test [7,8] and the reproduction test [9] demonstrate that there should be no essential reproducibility problems.

Practicability of this endpoint is considered sufficient. The OECD Guideline 207 definition "reaction to a mechanical stimulus" is practicable but may cause problems with severely injured test organisms that could die hours or days later. Because remains of dead animals in most cases disappear quickly and may only be found by chance, there is no need to include searching for the remains of dead animals into the test evaluation. Searching for and counting the live animals is sufficient.

For both the acute test and the reproduction test, a validity criterion for mortality in the control is included in the guidelines: mortality should not exceed 10%. In an evaluation of 170 acute tests available from the authorization procedure of plant protection products in Germany, there was no control mortality in 77% of the studies, up to 5% control mortality in 16% of the studies, and between 5 and 10% control mortality in 7% of the studies. Even better results were achieved for the reproduction test control boxes where the animals were fed. No mortality was found in 85% of the controls of the studies (n = 40 tests). In the remaining 15% of the controls, mortality was below 5%.

Relevance of mortality as an endpoint for the field is considered high. For plant protection products, exposure in the acute test differs from the reproduction test. In the acute test, products are homogeneously mixed into the soil, which may not correspond to normal exposure in the field. In the reproduction test, relevance for the field is higher than in the acute test because application of the plant protection product in the laboratory more closely mimics application in the field. Exposures in a reproduction test and in a field test are nevertheless not completely identical because the test vessels in the laboratory have a specific depth, whereas field "vessels" do not. In the field, distribution of the substance in the soil depends on several factors, e.g., precipitation, which do not apply in the laboratory.

The reference compound for the mortality endpoint, chloracetamide, should be tested occasionally (either 2 to 3 times a year or every time when only a few tests are conducted). The OECD guideline offers no advice concerning a target range for the results of the reference substance. The corresponding ISO guideline (ISO 11268-1) advises that the LC50 should be within 20 to 80 mg/kg. An evaluation of 40 chloracetamide tests, available from the authorization procedure of plant protection products in Germany shows that only 65% of the studies produced results within this range. Of the studies that failed, most (27.5% of the total) produced LC50s below 20 mg/kg. Because the upper limit is the most critical, it might be sufficient to require only that LC50 for chloracetamide should be below 80 mg/kg. In the subject studies, this criterion was reached 92.5% of the time. A recommendation to change the LC50 range for the reference substance accordingly should be taken into account when OECD Guideline 207 is revised.

Reproduction
Reproduction is the main endpoint in the sublethal test according to ISO DIS 11268-2. The test was originally designed for evaluating the effects of plant protection products on earthworms [6]. The number of juveniles was chosen as a parameter for reproduction (see Table 1-2 for characteristics). This test design is new, and a limited number of tests are available.

Two ring tests [9] concluded that test reproducibility is sufficient for legislative purposes. However, an inverse relationship between reproduction and bodyweight change of the adults might reduce reproducibility (see section on bodyweight change).

Practicability is considered sufficient. For reasons of practicability, juveniles are allowed to hatch and grow; they then either can be counted directly by examining the whole soil, or they can be extracted in a water bath, picked from the soil surface, and then counted.

There are 2 validity criteria for juvenile numbers in the control. The first criterion is that the number of juveniles per test box should exceed 30 after 4 weeks of exposure. This is a rather low value, keeping in mind the high reproductive potential of *Eisenia fetida* and *Eisenia andrei* [10]. However, in the ring tests, problems occurred with this validity criterion: some testing facilities yielded low numbers of juveniles. By systematically querying all German testing facilities, it was found that commercial manure used as food was the most likely cause of the observed problems, probably due to heating, sterilizing, or a high ammonia content. When cattle manure is used, it should be self-collected, rather than purchased commercially, and dried carefully. However, if commercial fertilizer is used, a quality check of this food is essential.

The second validity criterion is that the coefficient of variation for juvenile numbers in the control should not exceed 30. This criterion was reached in 80% of the tests (n = 40) submitted to the German authorization agency for plant protection products. Among those tests in which the number of juveniles $\geq$ 30, the variation coefficient did not exceed 30 in 97% of the tests (Table 1-2).

Relevance is considered to be high for the field. In risk-assessments of plant protection products in the laboratory, the product is always applied to the bare soil surface, whereas in the field, a crop cover may reduce the exposure. On the other hand, the use in the laboratory of *Eisenia fetida* and *Eisenia andrei*, which normally are not found in agricultural soils, and of an artificial soil instead of a natural soil require the application of a safety factor for assessing the risk for the field. However, Chapter 19 concludes from a comparison of several reproduction tests and corresponding field data on the same plant protection product that the reproduction test is more sensitive than the field test.

In order to evaluate the effects from plant protection products, at least a single and a 5-fold dose should be tested because a single-dose test does not give sufficient information about a dose–response relationship. The most informative way to do this test would be a full dose–response test and an evaluation of no-observed-effects concentration (NOEC) or EC_x.

If several tests are done each year, a reference compound should be tested at least twice a year. If only a few tests are conducted each year, the reference substance should be tested each time. From the ring test conducted [9], Carbendazim and Benomyl have proven to be suitable reference substances. At least 250 g active substance/ha is recommended to achieve a decrease in reproduction $\geq$ 30%.

The mode of application outlined in the reproduction test guideline for pesticides is a field-oriented application, which in most cases means an application to the soil surface. Table 1-3 shows a comparison of reproduction results for different application types in 3 tests. Both spray and mixed applications yielded a significant decrease of reproduction compared to the control, but the spray application tended to show a weaker effect than the mixed. As this spray application is the most relevant one for the field, there is no need to change the test design. The only advantage of mixing a substance into the soil is that doing so can be more standardized than can spraying.

Table 1-4 summarizes the other parameters of the reproduction process. The number of juveniles might be the most relevant endpoint but may not explain which part of the reproductive process is affected, e.g., the number of cocoons or the number of cocoons

hatching. If there is a need to determine which reproductive mechanism is affected, the reproductive
process must be split into distinct
parts. Doing so is disadvantageous
because the substrate is destroyed or
the soil is mixed, thus disturbing the
test organisms. Cocoon production
gives good and practical information on fertility, but the substrate is
destroyed when the cocoons are
washed out from it. If hatching juveniles must be counted, fresh substrate is necessary, and it must be
either uncontaminated or freshly contaminated.

Table 1-3   Results from different modes of application in reproduction test

Modes of application	# Juveniles produced		
	Test A	Test B	Test C
Control	109.5	118.0	189.0
Product sprayed onto soil surface[a]	6.3[b]	86.5[b]	95.0[b]
Product mixed into soil homogenously (5 cm soil depth)[a]	0[b]	69.3[b]	67.3[b]

[a] Identical amounts of product were applied.
[b] Significant $P \leq 0.05$ (compared to control boxes)

Table 1-4   Advantages and disadvantages of parameters for evaluating reproduction in the laboratory

Parameter	Advantage	Disadvantage
Cocoon production	Well-defined, sensitive	Substrate destroyed Hatching in uncontaminated or freshly contaminated substrate
Number of juveniles	Realistic exposure (integrating different exposure pathways and life stages)	No information on cocoon number and mode of action because only juvenile number is evaluated
Weight of juveniles	—	Difficult to handle Not sensitive
Sexual maturation	—	Time-consuming (no fixed evaluation date)
Life cycle (chronic test)	High relevance	Time-consuming (at least 3 months)

Determining the weight of juveniles could provide important information about reproduction, but juveniles with a mean weight of 20 mg or less are difficult to handle, and results may be highly variable. The tests done with the additional determination of juvenile
weight reveal that juvenile weight is generally a less sensitive measure than is the number of juveniles.

Compared to other endpoints, sexual maturation is more time consuming to determine
because development time must be evaluated. Even if time to sexual maturation is measured, the number of juveniles (including those hatching in contaminated substrate)
should be maintained as an endpoint, especially when hatched juveniles are exposed to
pesticides in the upper soil layer, a very relevant exposure pathway.

Using species other than *Eisenia fetida* and *Eisenia andrei* in the laboratory is complicated
because rearing of all the relevant species like *Lumbricus terrestris* and *Aporrectodea*

caliginosa is difficult. Some laboratories have had success in rearing indigenous earthworms like *A. caliginosa*, but at the moment reproducibility is in question.

Bodyweight change

Bodyweight change in adults is used in both the acute and the reproduction test guidelines, although it is not the main endpoint in either of them. Because this endpoint is not defined clearly and is interpreted in very different ways, some clarification is needed. Table 1-2 summarizes the main characteristics of this endpoint.

From different ring tests [7,8,9] it can be concluded that reproducibility of bodyweight change is sufficient. However, an inverse relationship between reproduction and bodyweight change might influence reproducibility. Animals that rapidly gain weight normally do not reproduce at the same time [11]. The mechanisms influencing this process (e.g., individual starting weight, condition of animals before test) are not yet fully understood. However, the data evaluated in this context, available from different tests (n = 33) and testing facilities, do not show such an inverse relationship. This relationship, therefore, needs further clarification.

The practicability of this endpoint is considered to be good, with no known major problems. Weighing individuals separately is possible, but weighing groups of animals is sufficient. Gut content is not removed when worm body weight is evaluated, and it is not expected that taking gut content into account will improve the information on bodyweight change.

There are no validity criteria concerning bodyweight change in control animals in either OECD Guideline 207 or the ISO reproduction test guideline. Test animals are fed in the reproduction test, and 56% of the tests suitable for evaluation (n = 34) revealed a weight increase of more than 10%, according to evaluation of the confidential data. Weight decreases of more than 10% occurred in only 3% of the tests. Decreases in weight of more than 20% were not found in the reproduction tests. For this author's evaluation, only tests with no mortality were chosen because, otherwise, mean bodyweight change may reflect fewer animals and therefore could lead to misinterpretation.

In the acute test, in which test animals are not fed, about 5% of the tests showed a weight decrease of > 20% in the control boxes, a loss that might have been caused by desiccation and reduced burrowing activity. Because of the reduced activity, and thus reduced exposure, it must be assumed that exposure in the treated boxes may have been suboptimal. Therefore, it must be questioned whether relevant information on bodyweight change in treated boxes can be derived when control bodyweight loss is high. Although bodyweight change normally occurs with a low variation (in a survey of 90 acute test controls, the variation coefficient was < 10% in 97% of the tests), there are some tests with a higher variation coefficient when a high bodyweight loss in the test occurred.

For this reason, a validity criterion for bodyweight change in the control should be included when the OECD guideline is revised. As specified in ISO Guideline 11268-1, body weight in control boxes should not decrease by more than 20% of initial weight (see Table

1-2). In the reproduction test, this problem normally does not occur because the animals are fed during the test; this is the reason that ISO DIS 11268-2 does not have such a validity criterion for body weight, whereas the older BBA guideline specified the 20% limit (Table 1-2).

Sometimes an increase in body weight in treated boxes compared to control boxes is found. Such an increase is not defined as an adverse effect of a substance. However, there is a lack of knowledge about the importance of this weight gain and about whether and how this parameter is relevant for field populations.

When mortality occurs in a test, care should be taken in evaluating bodyweight change. The surviving adults might by chance be the biggest ones, so mean bodyweight change per adult can be skewed. For this reason, bodyweight change should be evaluated only when < 15% mortality occurs in the test boxes. Another option is to express bodyweight change in terms of overall biomass, thus including the mortality endpoint. The connection between bodyweight change and mortality is a more technical problem, as bodyweight change is a sublethal endpoint and the effect concentrations of both endpoints normally are different.

One problem in practice is the evaluation of acute limit tests (with only one concentration tested) with regard to bodyweight change. Limit tests are often conducted when no effect on mortality was found in the pretest. Nevertheless, an evaluation of 60 randomly selected limit tests showed that about 50% of these tests revealed a significant reduction in body weight. In 25% of these tests, the reduction amounted to more than 15% of initial body weight. A reduction in body weight up to about 50% of initial body weight has been observed by this author. This means that no NOEC for body weight can be derived from the acute test, as it is designed for the evaluation of mortality. In the risk assessment procedure in Germany, such cases require an additional full-dose test, or the product is labeled as toxic to earthworms because there is insufficient information to evaluate sublethal effects.

Another point must be clarified concerning bodyweight change: the test quality is improving, and there are tests that result in a very small but statistically significant weight difference between control and treated boxes. It may happen that a weight loss of ~ 3 to 5% compared to control is significantly different from control values. This demonstrates again that, from the evaluation point of view, bodyweight change is a precise endpoint. However, the question arises whether a difference of 5%, for example is a biologically significant effect. Taking into consideration the natural variation of body weight in control boxes (variation coefficient < 10% in 97% of controls), it is suggested that a 10% decrease relative to the control be defined as an adverse effect.

Additional endpoints

According to the acute and reproduction test guidelines, any additional effects (e.g., prolonged burrowing time, crawling on the soil surface, flaccidity, hardened test animals, and color changes) observed should be reported. In most cases, it is not clear whether these effects are relevant for field populations and whether these effects will be detectable in higher tier studies, so further research is needed.

Table 1-2 shows characteristics concerning behavior endpoints. For both the acute and the reproduction tests, knowledge about behavior reproducibility is limited. Concerning practicability, a scale for reporting and interpreting behavior would be helpful. Because no standardized evaluation scheme has been developed, no validity criteria are available. The relevance for the field might be different for both test types. In the acute test, the test animals are added to the substrate when the test substance is already in the soil. A repellent or avoidance reaction measured by burrowing time can be derived from the behavior of the animals that crawl on the soil surface. This is a very unusual stress situation for the test animals because they cannot hide in the soil and are exposed to light. In the reproduction test, the situation is much more comparable to the field, where the animals are already in the soil; therefore, avoidance by seeking deeper soil layers is possible. Depending on the time scale used, a repellent effect might be advantageous when its consequence is that the toxic substance and the organism do not interact. On a longer time scale, some earthworm species would not reach their food on the surface when being repelled; therefore, the effect would be regarded as disadvantageous.

Additional species

Eisenia fetida and *Eisenia andrei* are the standard species for routine testing, although these species do not occur in agricultural soils. Because mass rearing of indigenous species is still difficult, it is likely that no additional species will be available in the future for routine testing. Therefore, special cases should be identified when risk assessment based on *Eisenia fetida* and *Eisenia andrei* is not sufficient. It might be possible to find general rules for the failure of risk assessment with *Eisenia fetida* and *Eisenia andrei*. For example, for special types of chemical classes like carbamate insecticides [12], different enzymes in a species lead to different reactions to toxicants. As an alternative to laborious field tests, which also are difficult to interpret, special laboratory tests (such as those on growth rates of species relevant for the field) are believed to be helpful. Up to now, very few proposed standards are available on this subject. Because exposure routes also may differ from species to species and from substance to substance depending on chemical properties, e.g., uptake route via soil pore water or uptake route via feeding of contaminated material, a special design could take these routes into account.

Risk assessment

Both the acute and the reproduction tests are used in the EU and EPPO risk assessment procedures for plant protection products. The acute test currently serves as a standard test, whereas the reproduction test is required only under certain circumstances.

Since the reproduction test was introduced into the German authorization procedure for plant protection products in 1994, experience with this test might be useful for further discussions (e.g., triggers for testing). Triggers for the reproduction test in the German risk assessment procedure include the degradation time of an active substance in soil, the number of applications, the ratio LC50/PEC (PEC = predicted environmental concentration), the ratio NOEC/PEC, and the comparison of LC50s after 7 and 14 d. Triggers are used on a flexible, case-by-case basis. This has led to some uncertainty, as not all plant protection producer-applicants in all cases could foresee this requirement. Fixed triggers like a $DT90_{field}$ of > 100 d in Directive 91/414/EEC [3], combined with other parameters if necessary, are alternatives that are more easily handled than flexible triggers.

A comparison of acute and reproduction test results was made to assess any relationships or possible predictability between tests. The available and suitable reproduction tests (n = 29) were compared with the results from acute tests (Table 1-6). For this comparison, only high-quality results were used, and an NOEC from the acute test had to be available. When only limit tests were available for acute tests, the reproduction tests were excluded from the evaluation. Only significant test results are listed here. The PEC estimation was done according to EPPO [4]. In the left column of Table 1-5, the ratio of LC50/PEC derived from the acute test is shown (n = 29), and in the right column, the percent of corresponding reproduction tests showing significant effects on the number of juveniles is listed. The evaluation shows that with a ratio of LC50/PEC < 20 in the acute test, the corresponding reproduction tests yielded a significant reduction in the number of juveniles. When the ratio of LC50/PEC was > 100, which is normally defined to be a safe margin for negative effects of a product to occur, 50% of the corresponding reproduction tests showed an effect on reproduction. Although the database of reproduction tests is small, these data show that even with a high LC50/PEC ratio, an effect on the number of juveniles in the reproduction test was observed.

Table 1-5   LC50/PEC ratios in acute tests (divided into 3 classes) and corresponding reproduction test results (n = 29)

Acute test	Reproduction test with 2 applications		
LC50/PEC	Effect in both rates	Effect in highest rate	No effect
< 10	100	0	0
10 to 20	100	0	0
> 100	33	17	50

In Table 1-6, the results of acute tests according to the ratio NOEC/PEC are compared with the reproduction test results. The NOEC is based primarily on bodyweight change in adults. The data show that when a ratio of NOEC/PEC was < 20, there was a significant reduction of juveniles in 33 to 53% of the tests. When an NOEC/PEC ratio was > 20 from the acute test, there were no corresponding effects in the reproduction test. The conclusion is that bodyweight change from the acute test could be a helpful endpoint and also

Table 1-6   NOEC/PEC ratios in acute tests (divided into 3 classes) and corresponding reproduction test results (n = 29)

Acute test	Reproduction test with 2 applications		
NOEC/PEC	Effect in both rates	Effects in highest rate	No effect
< 10	54	23	23
10 to 20	33	0	67
> 20	0	0	100

may help to select cases where a reproduction test is needed. Because the database of 29 tests is yet small, further data are needed to evaluate this relationship.

Conclusions

In consideration of the tiered system for testing effects of substances on earthworms, the existing guidelines appear suitable for most risk assessments needed for legislative purposes. The following actions should be taken to improve laboratory tests for risk assessment:

A range, and especially an upper limit, for the results of the reference compound in the acute test should be defined.

A validity criterion is needed for control bodyweight change in the acute test.

A biologically significant trigger should be defined for bodyweight decrease in treatments.

The role of bodyweight evaluation in the acute test should be defined.

A need for additional endpoints for the acute and reproduction tests should be clarified.

Criteria for risk assessment using limit tests should be developed.

Triggers that indicate the need for a sublethal test within risk assessment should be precise.

A need for testing of earthworm species other than *Eisenia fetida* and *Eisenia andrei* and other soil invertebrates, e.g., enchytraeids and collembola, must be defined.

Some of these actions need further research or further data to define general rules. In some cases, it is necessary to come to an agreement, e.g., about triggers, keeping in mind that there must be enough flexibility to deal with special questions in risk assessment procedures and to allow the inclusion of new research data.

Acknowledgment - I would like to thank Ms. H. Neugebauer for technical help and Dr. A. Wehling for valuable contributions to the manuscript.

References

1. [OECD] Organization for Economic Cooperation and Development. 1984. OECD guidelines for testing of chemicals: Earthworm acute toxicity test. OECD Guideline No. 207. Paris, France: OECD.

2. [ISO] International Standards Organization. 1996. Soil quality: effects of pollutants on earthworms (*Eisenia fetida fetida, Eisenia fetida andrei*). Part 2: Determination of effects on reproduction. Geneva, Switzerland: ISO. ISO DIS 11268-2.

3. Council Directive of 15 July 1991 concerning the placing of plant protection products on the market (91/414/EEC). Office Journal of the European Communities L 230, 19. August 1991.

4. [EPPO] European Plant Protection Organization. 1994. Decision-making scheme for the environmental risk assessment of plant protection products. EPPO Bulletins 23 and 24.

5. [ISO] International Standards Organization. 1993. Soil quality—Effects of pollutants on earthworms (*Eisenia fetida*). Part 1: Determination of acute toxicity using artificial soil substrate. Geneva, Switzerland: ISO. ISO DIS 11268-1.

6. [BBA] Biologische Bundesansstalt für Land-und Forstwirtschaft. 1994. Auswirkungen von Pflanzenschutzmitteln auf die Reproduktion und das Wachstum von *Eisenia fetida/Eisenia andrei*. Richtlinien für die Prüfung von Pflanzenschutzmitteln im Zulassungsverfahren, Teil VI, 2-2, Januar 1994.

7. Edwards CA. 1983. Development of a standardized laboratory method for assessing the toxicity of chemical substances to earthworms. Commission of the European Communities. Doc. EUR 8714 EN. 141 p.

8. Edwards CA. 1984. Report of the second stage in the development of a standardized laboratory method for assessing the toxicity of chemical substances to earthworms. Commission of the European Communities. Doc. EUR 9360 EN. 99 p.

9. Kula C. 1996. Development of a test method on sublethal effects of pesticides on the earthworm species *Eisenia fetida/Eisenia andrei*—comparison of 2 ring tests. *Mitt Biol Bundesanst* 320:50–81.

10. Hartenstein R, Neuhauser EF, Kaplan DL. 1979. Reproductive potential of the earthworm *Eisenia foetida*. *Oecologia* 43:329–340.

11. Van Gestel CAM, Dirven-Van Breemen EM, Baerselman R. 1992. Influence of environmental conditions on the growth and reproduction of the earthworm *Eisenia andrei* in an artificial soil substrate. *Pedobiologia* 36:109–120.

12. Stenersen J, Brekke E, Engelstad F. 1992. Earthworms for toxicity testing; Species differences in response towards cholinesterase-inhibiting insecticides. *Soil Biol Biochem* 24:1761–1764.

Chapter 2

Evaluation of factors influencing results from laboratory toxicity tests with earthworms

David J. Spurgeon and Jason M. Weeks

Laboratory toxicity tests, e.g., the Office of Economic Cooperation and Development (OECD) earthworm test, are conducted under fixed biotic and environmental conditions [1]. However, a conflict exists between increased standardization in the laboratory and the desire to maintain relevance to the field. To improve understanding of the factors that determine pollutant impact in the field, it is important to examine how variations in test conditions affect toxicity. Thus the effect of variations in biotic and abiotic conditions on metal (particularly Zn) toxicity for earthworms is assessed. Results indicate that soil properties, e.g., soil pH, organic matter (OM), clay type, and content, strongly modulate the toxicity of Zn, with the aging time of the contaminant in soil also being an important determinant. Biotic factors such as test species and life stage and environmental parameters such as temperature were of minor significance for determining effects. This study reinforces the concept that a standardized test, such as the OECD earthworm artificial soil test, serves as little more than a provisional benchmark exercise. The procedure designated by the OECD guideline is conducted under conditions that accentuate toxicity and must be supported by real or semi-field studies if realistic estimates of effects are to be provided.

In a review of the literature, Reinecke [2] concluded that test protocols, species, and parameters used for measuring the toxicity of chemicals to earthworms varied widely. However, of the suggested procedures, the test to measure the survival of *Eisenia fetida* in artificial soil developed by Edwards [3,4] and adopted by the OECD [1] and EEC [5] was by far the most commonly used.

The OECD artificial soil earthworm toxicity test is conducted under fixed biotic and environmental conditions that allow results from different laboratories to be compared [1,2,3]. However, for any test system, a conflict exists between increased standardization and the desire to maintain relevance to the field, where exposure applies to a range of species and life stages under diverse and variable environmental conditions [6]. To improve our understanding of the factors that determine pollutant impact in the field, it is important to examine how variations in test conditions affect toxicity. Such information can be used to improve the accuracy of laboratory-to-field extrapolation and to increase the opportunities for chemical- and site-specific risk assessments.

The effects of variations in standard conditions on the toxicity of selected pollutants are compared under different biotic and environmental regimes. This work has been conducted in both artificially contaminated OECD-type soil and polluted field soils that were collected from sites along a gradient of contamination from a smelting works (for details of the study site, see Hopkin [7] and Martin and Bullock [8]). The effects of variations in 7 standardized factors (species, life stage, test duration, temperature, soil OM content, soil pH, and clay type) have been considered. During the study, the potential harmful effects of Cd, Cu, Pb, and Zn were examined. This chapter, however, focuses primarily on the effects of Zn, since comparing toxicity values for the 4 metals in artificial soil with concentrations in soils found around the smelter indicates that Zn is most likely to cause deleterious effects on earthworm populations [9,10].

Test species

The choice of test species has been a major source of criticism of the OECD protocol. *Eisenia fetida* was selected primarily because it is robust and can be cultured at high densities, rather than being selected on the basis of sound taxonomy [11] or the representative nature of its ecology. (It is also not a natural soils species; it inhabits organic rich habitats such as compost and manure heaps.) To assess the relevance of toxicity values derived with *Eisenia fetida*, a series of OECD artificial soil tests were conducted to determine the comparative sensitivity to Zn of *Eisenia fetida* with those for the soil species *Lumbricus rubellus* and *Aporrectodea rosea* [10].

Results from these tests indicated that *Eisenia fetida* was the least sensitive of the 3 species, with a 14-d LC50 (95% confidence interval) of 1106 (1010 to 1212) µg Zn g^{-1}. *Lumbricus rubellus* showed intermediate sensitivity with an LC50 of 728 (627 to 845) µg Zn g^{-1}, while *Aporrectodea rosea* was most sensitive with an LC50 of 561 (499 to 633) µg Zn g^{-1}. Thus, the difference in the toxicity between the most- and least-sensitive species was approximately a factor of 2. Although these differences in sensitivity are relatively small and it is possible that they may result from within-test variability [4], it should be noted that even small differences in sensitivity can be important for populations in areas contaminated by a point source. In such areas, soil concentrations decrease exponentially with distance from the origin of pollution. Such a pattern of spatial distribution will act to magnify the ecological importance of any observed sensitivity differences between species [10]. Larger differences in sensitivity (up to a factor of 10) have been found between *Eisenia fetida* and *Lumbricus terrestris* for selected pesticides [12,13,14].

Relative sensitivity of different life stages

In the OECD earthworm toxicity test protocol, the use of adult worms is recommended. However, because of their lengthy life span and the continuous iteroparous nature of their reproduction, earthworm communities in nature always consist of a mixture of adults, large sub-adults, juveniles, newly hatched individuals, and cocoons [15]. If a pollutant has a particularly severe effect on any life stage, e.g., impairment of hatching, retardation of growth and sexual development, or limitation of reproduction, this could have adverse effects on populations. Differences in the sensitivity between life stages could therefore be an important source of variation in the responses of earthworms to pollutants under natural conditions.

To study the relative sensitivity of different earthworm stages, 2 experiments were conducted. For the first test, newly hatched worms were added to a control soil and to 7 contaminated field soils collected from sites along the gradient of contamination from the

Avonmouth smelter. Survival, growth (mean weight), and development were recorded at intervals over a 20-week exposure period [16]. For the second test, adult worms were exposed to similar soils. Survival and cocoon production were measured after 21 d of exposure [9]. Thus different exposure times were used in the 2 tests, although it is unlikely that this factor would greatly affect the toxicity, since Zn rapidly reaches equilibrium in the tissue of exposed *Eisenia fetida*. Results for the test conducted with juveniles indicated significant effects on survival in the 2 most-contaminated soils after 35 d, on sexual development after 56 d, and on growth after 35 d in all 7 contaminated soils. In the study with adults, no effect on survival occurred in any of the selected soils, while cocoon production was significantly reduced in the 4 most contaminated soils (Table 2-1). These results clearly suggest an increased sensitivity for juveniles, since their mortality was significant in the most-contaminated soils, in which no effects on adult survival were found. Furthermore, calculations of effect concentrations based on Zn levels in the collected soils indicate a lower EC50 for sexual development of juveniles (number of adults at 56 d) of 1860 (480 to 4190) μ Zn g^{-1} than the value of 3605 μg Zn g^{-1} (95% CI not available) for effects on cocoon production in adults. Thus, toxicity values for the most-sensitive measured parameter were lower in juveniles than for adults by a factor of 1.9.

Table 2-1   Effects of contaminated field soils on survival and cocoon production of adult *Eisenia fetida*, and survival, mean weight, and number of adults grown from juvenile *Eisenia fetida* exposed to contaminated field soil collected from the Avonmouth area

| | | Test with adults | | Test with juveniles | | |
		% adult survival (14 d)	Cocoons/ worm/week	Juvenile survival (35 d)	Mean weight at 35 d	% adults at 54 d
Increasing metal contamination ↑	Site 1	100	0.033***	80***	0.038***	0***
	Site 2	100	0.017***	47.5***	0.044***	0***
	Site 3	100	0.042***	97.5	0.07***	2.5***
	Site 4	100	0.133*	97.5	0.098**	18*
	Site 5	92.5	0.288	97.5	0.108**	15*
	Site 6	100	0.25	100	0.143*	17.5*
	Site 7	100	0.345	97.5	0.119**	18*
	Site 8	100	0.375	100	0.163	30

* = significantly different from controls at $P < 0.05$, ** = $P < 0.01$, *** = $P < 0.001$. Non-dose response for juvenile 35-d survival and adult cocoon production result from the high bioavailable Zn concentration found in Site 2 soil [9].

Length of the exposure period

For the OECD artificial soil test, a standard exposure period of 14 d is recommended. However, this period may be insufficient to allow internal concentration to reach equilibrium, and as a result, toxic effects may not be as severe as those found after long-term exposure in the field. To examine the relationship between length of exposure, internal concentration, and potential toxic effects, time-dependent uptake and elimination of Cu, Zn, Cd, and Pb were studied in *Eisenia fetida* exposed to a clean and a contaminated field soil. The polluted soil was collected approximately 500 m from the Avonmouth smelter and contained approximately 3000 μg Cu g^{-1}, 30,000 μg Zn g^{-1}, 300 μg Cd g^{-1}, and 20,000 μg Pb g^{-1}. Clean soil was collected at a distance > 100 km from the factory. Worms were exposed to the 2 soils for 42 d to allow metal accumulation. After this time, surviving worms were transferred to clean soil to measure their excretions.

No significant metal uptake was recorded for worms exposed to control soil (data not shown). However, all 4 metals accumulated in the polluted medium (Figure 2-1 a–d). Different accumulation and elimination profiles were found for essential and nonessential metals, primarily as a result of differences in elimination rates. For the essential metals Cu and Zn, elimination was high. Thus, these metals reached equilibrium during the accumulation study and returned quickly to normal essential levels in the excretion phase (Figures 2-1b and 2-1d). For the nonessential metals Cd and Pb, elimination was slow. Thus these metals did not reach equilibrium in the accumulation phase and declined only slowly after exposure ceased (Figures 2-1a and 2-1c).

The distinctive patterns of uptake for essential and nonessential elements indicate different time-dependent patterns for metal toxicity. For Cu and Zn, exposure over 14 d would allow internal concentrations to reach threshold level. Since toxic effects are primarily dependent on the internal concentration of a pollutant [17], this suggests that tests for Cu and Zn of the recommended 14-d duration should be suitable for predicting the effects of long-term exposure. For Cd and Pb, internal concentration does not reach equilibrium, even after 42 d. Thus for these metals, a 14-d test would be insufficient for predicting toxic effects on earthworms in the field, and an extended exposure period should be considered. That some chemicals require an extended exposure period to reach internal equilibrium should be an important consideration when hazard assessments of contaminated soil are conducted. Given that such soils usually encompass a mixture of contaminants, long exposure periods should be used to ensure that all chemicals reach an internal equilibrium during the test.

Temperature

Artificial soil tests are conducted at a standard temperature of 20 °C. However, exposure of earthworms in the field can take place at any number of soil temperatures. To examine how temperatures can affect the sensitivity of earthworms to pollutants, 3 toxicity tests with Zn were conducted to measure the survival of *Eisenia fetida* in artificial soil at 15, 20, and 25 °C [18]. Results of this work indicated a greater toxicity of Zn at higher temperatures. Thus, the lowest LC50 of 1131 μg Zn g^{-1} was found for the 25 °C test, the value at 20 °C of 1235 (811 to 2855) μg Zn g^{-1} was intermediate, and the highest value of 1598 (1460 to 1760) μg Zn g^{-1} was observed for the test at 15 °C. A comparison of the toxicity values calculated for these tests indicated that variations in temperature across a 10 °C range gave toxicity values that differed by a factor of 1.4 [18].

Soil organic matter content and pH

The soil OM content and pH recommended for the artificial soil used in OECD tests are 10% and 5.5 to 6.5, respectively. However, the OM content and pH of natural soils can

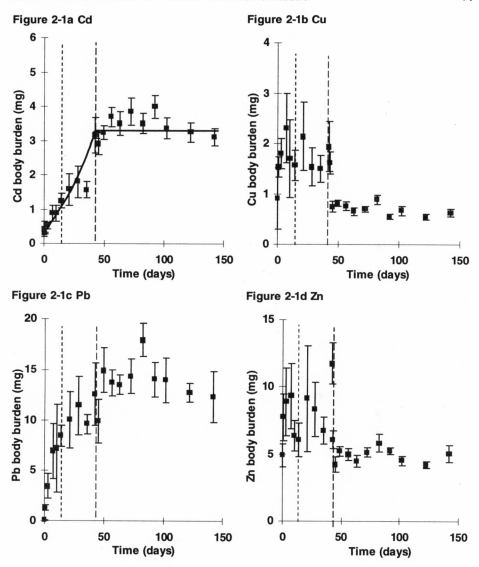

Figure 2-1a-d   Accumulation and elimination of Cd, Cu, Pb, and Zn by *Eisenia fetida* (± SE). Legend: Large-dash line indicates transfer to clean soil, and the small-dash line indicates the 14-d duration of the OECD artificial soil test for mortality. Original data are given as means with standard errors. Solid lines indicate the fit for a 1-compartment accumulation model.

vary widely. To investigate how differences in OM and pH can affect toxicity, experiments were undertaken in which *Eisenia fetida* was exposed to Zn in artificial soil with OM contents of 5, 10, 15% and pH of 4, 5, 6 respectively. These levels were used in factorial combination to give a total of 9 tests [16].

Earthworm survival decreased in soils with the highest Zn concentrations in all test soils, and it was possible to calculate LC50 values from these dose-response relationships. To permit the relative toxicity in the different tests to be compared, the LC50s for the 3 tests conducted at each OM content were been plotted in Figure 2-2 against soil pH (5% OM values are shown as triangles, 10% OM as squares, and 15% OM as circles), and a regression fit for these values was calculated (Figure 2-2). Regression lines have been fitted for each set of toxicity values: small dashes = 5% OM; large dashes = 10% organic solid line = 15% OM. Results from these plots indicated that the effects of Zn were most severe in soils in which OM content and pH were low. Furthermore, the LC50 value in each test was dependent both on organic content and acidity of the test soil. For example, increasing soil pH by 1 unit in the 3 soils resulted in a larger reduction in toxicity in the high OM soils than in the low OM soil. This is demonstrated by the steeper regression relationship found between LC50s and pH in the 15% OM soils (Figure 2-2).

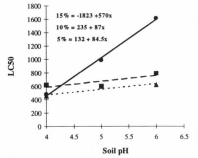

Figure 2-2   LC50 values for the earthworm, *Eisenia fetida*, exposed to a geometric series of Zn concentrations in artificial soils.

Interactions between soil OM levels and pH, which govern Zn toxicity, can be explained if the impact of changes in soil composition on the cation exchange properties of the medium are considered. The 2 most important fractions that bind metals within the soil matrix are the organic fraction and clay. Thus a reduction in the soil OM content will increase metal solubility, resulting in higher bioavailability and thus toxicity for earthworms. Variation in pH also increases toxicity by altering cation exchange properties. Metals are bound to negative charges on the surface of clay minerals and OM. In artificial soil, the charges present on the surface of kaolin and OM are temporary and can be broken down under acidic conditions. Thus, in low pH soils, the number of surface charges was reduced, resulting in a decrease in soil-bound Zn and an increase in the available fraction. It is well known from studies reported in the literature that soil pH influences metal bioavailability [19], and it is clearly both this factor and the soil OM content that affect metal toxicity under the varied conditions that exist in the field [16].

Clay type and aging of contaminant
Clay minerals are important soil components that bind pollutants. The clay type selected for use as the OECD guideline's artificial soil is kaolin. This mineral was selected prima-

rily because of its commercial availability rather than its environmental relevance, since it is rare in many European soils. As a result, the effects of metals may differ in natural soils, since these soils contain a mixture of clays with different sorption capacities. The influence of clay type on the toxicity of metals is evidenced by results of 2 experiments by Spurgeon and Hopkin [9], which compared the impact of Cd, Cu, Pb, and Zn in 2 soil types on the survival of *Eisenia fetida*. Although these experiments were not designed explicitly to measure how clay type alters toxicity, the results do suggest that mineral type may be an important determinant of effects in the field. For the first experiment, the effects of soils from a control site and 7 sites around the Avonmouth factory on the survival of *Eisenia fetida* were recorded. The second test measured the responses of worms to the same metal concentrations added as a mixture to artificial soil.

No significant mortality occurred in any of the smelter-polluted soils (Figure 2-3), despite measured Zn concentrations at the most-polluted site 30 × the 14-d LC50 for Zn in OECD tests. For worms exposed to the same concentrations in artificial soil, dramatic effects on survival were observed. All worms exposed to the 4 most-contaminated soils died, and significant mortality was also found in 2 later treatments (Figure 2-3). Although a mixture of pollutants was present in the test soil, and thus some additive effects anticipated, it was decided to compare relative toxicity in the 2 test soils by calculating LC50 values from levels of the predominant toxic metal Zn. Such calculations indicated that the 14-d LC50 of 1730 (1599 to 1830) µg Zn g^{-1} for the artificial soil test was lower than the zinc level in the most polluted-test soil, in which all worms survived, by a factor of 19.4. Thus, large differences exist for the impact of zinc in the 2 soil types.

The increased sensitivity to Zn of worms in artificial soil can be attributed to the physical and chemical characteristics of the respective soils. Although the pHs of the soils were similar (6.1 in the artificial soil, 5.5 to 7.4 in the field soils), the artificial soil had a lower

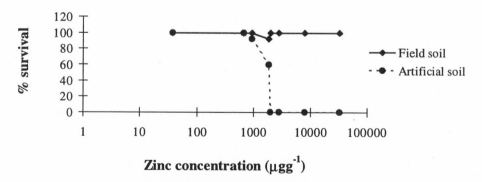

Figure 2-3   Survival of *Eisenia fetida* in soils collected from sites along a contamination gradient from a smelting works (see [9] for description of sites used) and artificial soil contaminated with similar metal levels. Legend: For both experiments, survival has been plotted against soil Zn concentration because comparative studies suggest this metal has the primary toxic effect.

OM content than did any of the field soils (10% in the artificial soil, 12.9% to 27.15% in the field soils). The results indicate the importance of OM in determining the toxicity of metals. However, it is unlikely that the increased organic content of the field soils can explain the large differences in effects found in this study. Differences probably result from a combination of factors, including the types of OM present in natural soils, the increased aging time of the contaminant in the field, and the lower binding capacity of the kaolin clay used in OECD tests compared to more common clay minerals (e.g., montmorillonite, illite, and vermiculite) [20]. In a comparison of the effects of these factors on the toxicity of Zn, Smit [21] concluded that the longer aging time of contaminant in the field is the key factor that explains the lower toxicity found under field conditions.

Summary of factors that hinder extrapolation of artificial soil test data to the field

Results indicate that variations in biotic, environmental, and soil conditions may all affect the toxicity of a pollutant. However, it is also clear that such variation has widely differing effects on toxicity. The potential effects of nonstandardized conditions on the toxic effects of metal pollutants in the field are illustrated in Table 2-2. The most important effects on the toxicity of Zn (and other metals) are determined by soil properties. Thus in natural soils, the longer aging time of contaminants and the presence of clays with a higher binding capacity than the kaolin used in the OECD test will lessen the impact of Zn, compared to the impact anticipated from a laboratory test. Soil organic content and pH are also key factors that determine the impact of metals. However, since natural soil can contain more or less OM and be more or less acidic than artificial soil, differences in the factors can both increase and decrease toxic effects.

Biotic factors also appear to determine the toxicity of metals that may occur in the field. For example, the presence of a mixture of species at polluted sites will result in the exposure of earthworms with different sensitivities, and this may cause effects on natural populations that differ from those predicted from the laboratory tests. The study of species sensitivity outlined in this chapter found that the natural soil species *Lumbricus rubellus* and *Aporrectodea rosea* were both more sensitive to Zn than *Eisenia fetida*. Thus, it is anticipated that the impact of Zn on natural populations may be greater than the impact predicted from the results of an OECD test. Studies of the relative sensitivities of juveniles and adults indicated greater sensitivity for young worms. Thus, in a natural population, effects on immature individuals will occur that would not be predicted from the standard laboratory test.

Abiotic environmental factors may also alter the toxicity of metals under field conditions. Results of the temperature study indicated a higher toxicity of Zn when the temperature was elevated [18]. The OECD test is conducted at 20 °C, a temperature rarely exceeded in northern temperate soils. Since in most circumstances the exposure of earthworms will occur at temperatures below the recommended value, the usual influence of temperature is to decrease the impact of metals below what would be predicted from the results of a laboratory test. The effects of temperature on toxicity gave the weakest effects

Table 2-2   Potential effect of nonstandardized conditions on the toxic effects of Zn for earthworms exposed in the field.

Standardized factor in OCED test	Potential effect of variation on sensitivity*	Comments
Test species	↑	*Eisenia fetida* appear relatively insensitive to Zn in comparison with soil species
Life stage	↑	Higher toxicity expected for juveniles
Duration of exposure	No effect or ↑	Long-term exposure will only increase toxicity if equilibrium concentrations are not reached after 14 d.
Temperature	↑ or ↓	Temperatures above 20 °C would occur only rarely in temperate regions. Under normal field conditions toxicity would be lower.
Soil pH	↑+ or ↓+	Toxicity would increase in acidic soils and decrease under alkaline conditions
Soil OM content	↑+ or ↓+	Zn toxicity will be lower in organic-rich soils, but higher when soil organic content is below 100%
Clay type	↓++	Clay type is an important factor altering toxicity. natural clays reduce effects due to their high binding capacity.
Contaminant aging time	↓+++	Smit [21] has demonstrated that the longer aging times for contaminants in the field can greatly decrease toxicity.

* Arrows pointed upward indicate that (on the basis of cited data) deviation in a given factor from the standardized test condition in the field will result in increased toxicity. Downward arrows indicated a reduction in toxicity. Pluses indicate relative importance of the potential effects.

of all the measured parameters in this study. Temperature is less important for determining effects in the field than is either soil or biotic variables.

The OECD standard earthworm toxicity test was not necessarily developed to enable the direct extrapolation of any toxicity results into a field situation. Furthermore, it is clear from the results outlined in this chapter that numerous factors, both biotic and abiotic, alter the toxicity of a chemical when species are exposed under field conditions. The fact that the range of factors affecting the toxicity of chemicals in the field is not represented in the OECD test reinforces the concept that the procedure serves as little more than a provisional benchmark exercise. Furthermore, with respect to the key factors influencing metal toxicity such as soil OM content, pH, clay type, and contaminant aging time, the test represents conditions that accentuate toxicity. Soil-binding capacity is low, and short aging times are used [16,21]. Therefore, if estimates of potential environmental effects of new or existing compounds are to be realistic, further real or semi-field testing is required. This is well understood within the pesticides registration community, although current procedures may need substantial modification [15]. However, the importance of semi-field and field studies for the risk assessment of contaminated land is being realized only now [22,23].

Acknowledgments - This work is supported by a research grant from the Leverhulme Trust.

References

1. [OECD] Organization for Economic Cooperation and Development. 1984. OECD guidelines for testing of chemicals: earthworm acute toxicity test. Organization for Economic Cooperation and Development Guideline No. 207. Paris, France: OECD.

2. Reinecke AJ. 1992. A review of ecotoxicological test methods using earthworms. In: Becker H, Greig-Smith PW, Edwards PJ, Heimbach F, editors. Ecotoxicology of earthworms. Andover UK: Intercept Ltd. p 7–19.

3. Edwards CA. 1983. Development of a standardised laboratory method for assessing the toxicity of chemical substances to earthworms. EUR 8714 EN. Report of Commission of European Communities.

4. Edwards CA. 1984. Report of the second stage in development of a standardized laboratory method for assessing the toxicity of chemical substances to earthworms. EUR 9360 EN. Report for Commission of European Communities.

5. [EEC] European Economic Community. 1985. EEC Directive 79/831. Annex V. Part C. Methods for the determination of ecotoxicity. Level I. C(II)4: Toxicity for earthworms. Artificial soil test. DG XI/128/82. Commission of European Communities.

6. Forbes VE, Depledge MH. 1992. Predicting population responses to pollutants: the significance of sex. *Funct Ecol* 6:376–381.

7. Hopkin SP. 1989. Ecophysiology of metals in terrestrial invertebrates. London UK: Elsevier.

8. Martin MH, Bullock RJ. 1994. The impact and fate of heavy metals in an oak woodland ecosystem. In: Ross SM, editors. Toxic metals in soil-plant systems. Chichester UK: Wiley. p 327–365.

9. Spurgeon DJ, Hopkin SP. 1995. Extrapolation of the laboratory-based OECD earthworm test to metal-contaminated field sites. *Ecotoxicology* 4:190–205.

10. Spurgeon DJ, Hopkin SP. 1996a. The effects of metal contamination on earthworm populations around a smelting works—quantifying species effects. *Appl Soil Ecol* 4:147–160.

11. Bouché MB. 1992. Earthworm species and ecotoxicological studies. In: Becker H, Greig-Smith PW, Edwards PJ, Heimbach F, editors. Ecotoxicology of earthworms. Andover UK: Intercept Ltd. p 20–35.

12. Heimbach F. 1985. Comparison of laboratory methods, using *Eisenia fetida* and *Lumbricus rubellus*, for the assessment of the hazard of chemicals. *Z Pflanzenkr Pflanzenschutz* 92:186–193.

13. Heimbach F. 1992. Correlation between data from laboratory and field tests for investigating the toxicity of pesticides to earthworms. *Soil Biol Biochem* 24:1749–1753.

14. Van Gestel CAM, Dirven-Van Breemen EM, Baerselman R, Emans HJB, Janssen JAM, Postuma JR, Van Vliet PJM. 1992. Comparison of sublethal and lethal criteria for nine different chemicals in standardized toxicity tests using the earthworm *Eisenia andrei*. *Ecotox Environ Saf* 23:206–220.

15. Edwards CA, Bohlen PJ. 1996. The biology and ecology of earthworms. 3rd ed. London UK: Chapman and Hall.

16. Spurgeon DJ, Hopkin SP. 1996b. Effects of variations in the organic matter content and pH of soils on the availability and toxicity of zinc to the earthworm *Eisenia fetida*. *Pedobiologia* 40:80–96.

17. Van Wensem J, Vegter JJ, Van Straalen NM. 1994. Soil quality criteria derived from critical body concentrations of metals in soil invertebrates. *Appl Soil Ecol* 1:185–191.

18. Spurgeon DJ, Hopkin SP, Tomlin MA. 1997. Influence of temperature on the toxicity of zinc to the earthworm *Eisenia fetida*. *Bull Environ Contam Toxicol* 58:283–290.

19. Van Gestel CAM, Rademaker MCJ, Van Straalen NM. 1995. Capacity controlling parameters and their impact on metal toxicity in soil invertebrates. In: Salomons W, Stigliani WM, editors. Biogeodynamics of pollutants in soils and sediments: risk analysis of delayed and non-linear responses. Berlin, Germany: Springer-Verlag. p 171–192.

20. Wild A. 1993. Soils and the environment: an introduction. Cambridge UK: Cambridge Univ.

21. Smit CE. 1997. Field relevance of the *Folsomia candida* soil toxicity test [Ph.D. thesis]. Amsterdam, The Netherlands: Vrije Universiteit.

22. Posthuma L. 1997. Effects of toxicants on population and community parameters in field conditions, and their potential use in the validation of risk assessment methods. In: Van Straalen NM, Løkke H, editors. Ecological risk assessment of contaminants in soil. London: Chapman and Hall. p 85–117.

23. Spurgeon DJ, Sandifer RD, Hopkin SP. 1996. The use of macro-invertebrates for population and community monitoring of metal contamination—indicator taxa, effect parameters and the need for a soil invertebrate prediction and classification scheme (SIVPACS). In: Van Straalen NM, Krivolutsky DA, editors. Bioindicator systems for soil pollution. Dordrecht, The Netherlands: Kluwer Academic. p 95–110.

Chapter 3

Improving the accuracy and precision of earthworm laboratory experiments through the use of pretreatment measurements

E.C. McIndoe, J.D. Bembridge, and P. Martin

Laboratory data on the effects of pesticides on earthworms are important in risk assessment. Although the acute study is mainly designed to assess acute toxicity, measurements on sublethal endpoints, e.g., earthworm body weight, are important as indicators of potential chronic effects. This chapter describes a method by which pretreatment measurements on potentially influential endpoints, e.g., body weight, can be used to design experiments so that any sublethal effects can be assessed with greater accuracy and precision. Data are presented from laboratory studies carried out with agricultural chemicals, showing the influence of this method on the precision of the experiment. This simple approach should be extended to the design of other laboratory studies such as sublethal effects studies. This approach requires knowledge of which factors, other than treatment, have the potential to influence key endpoints. Further research should be carried out to identify these factors.

Laboratory data on the effects of pesticides on earthworms are important in risk assessment. A tiered system of testing from the laboratory to the field has been generally adopted to provide this information [1,2,3]. The first level of testing is a laboratory acute test based upon the Organization for Economic Cooperation and Development (OECD) Guideline 207 [4] or the European Economic Community (EEC) Commission Directive 88/302/EEC [5]. The principal aim of this test is to produce an LC50 value for the test substance, though bodyweight changes along with behavioral and pathological symptoms must also be recorded, as these may indicate sublethal effects. Such data must be interpreted carefully, however, because the acute test is not specifically designed for this assessment [6,7]. A more specialized study has therefore been developed to investigate sublethal effects in the laboratory, and guidelines have been issued by the International Standards Organization (ISO) [8] and Biologische Bundesansstalt Für Land-Und Forstwirtschaft (BBA) [9]. Since the results from these laboratory studies are used to predict the likelihood of sublethal effects in the field, it is important that all such studies be designed and analyzed in a manner that allows the best interpretation of the data.

Biological data can exhibit high levels of natural variability. This variability has not been controlled by study design and can mask important treatment effects. Some of this variability, however, can be due to differences between experimental units prior to treatment application. When relevant pretreatment measurements are identified, simple experimental design and statistical analysis methods can be used to account for pretreatment differences between experimental units, resulting in more accurate and precise measurement of treatment effects.

Earthworm body weight in laboratory acute studies

Although not specifically designed to assess sublethal effects, the laboratory acute study requires the measurement of initial earthworm body weights (before treatment application) and final earthworm body weights (14 d after treatment application), and such data could indicate potential chronic effects. Figure 3-1 shows a plot of final body weight versus initial body weight from an acute study that used 2 rates of a test substance and a control. Each experimental unit consists of 10 randomly selected earthworms, with 4 replicate units assigned randomly to each of the treatments and the control.

The graph illustrates the strong correlation typically observed between initial and final weight. Closer inspection of the data, however, reveals that despite random assignment of experimental units to treatments, 3 of the 4 highest initial weights were assigned to the control, while there is no overlap between the initial weights of the experimental units assigned to the control and those assigned to the low rate. Because such an arrangement could create problems of data interpretation, a method must be employed that ensures each treatment is applied to earthworms with a similar range of initial weights.

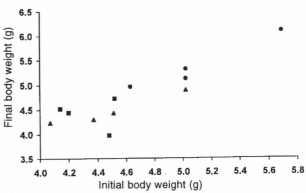

Figure 3-1   Plot of final body weight versus initial body weight from an acute study with 2 rates of a test substance and a control
● Control ■ 10 mg/kg ▲ 100 mg/kg

Blocking procedure

Figure 3-2 illustrates a procedure for ensuring that the replicate units for each treatment cover a similar range of initial weights for an experiment with 3 treatments and 4 replicates. This procedure is called "blocking" and can be automatically extended to any number of treatments and replicates.

The purpose of blocking is to make the experimental material within each block as similar as possible. This produces a "randomized block" design, and the resulting data can be analyzed by the analysis of variance (ANOVA) appropriate to such a design. The advantages of blocking are these:

·   Comparisons between treatments are more precise because within each block the treatments are applied to similar units.

a) Arrange the experimental units in order of increasing (or decreasing) initial weight.

experimental units

Increasing initial weight

b) Divide experimental units into 4 blocks of 3 units each. A, B, and C are the lightest 3 units while J, K, and L are the heaviest.

experimental units

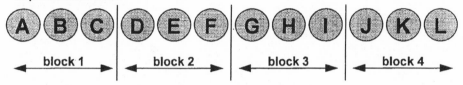

c) Randomly assign treatments to experimental units separately within each block.

experimental units

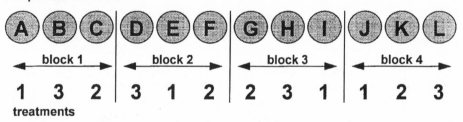

Figure 3-2   Randomization procedure for assigning treatments to experimental units using initial weight as a blocking factor. 3 treatments, 4 replicates.

· The risk of bias in the results is reduced. All treatments cover a similar range of initial weights, and the influence of initial weight is about the same for each.

Blocking by initial weight in acute studies

Figure 3-3 shows a plot of final body weight versus initial body weight from a study designed as a randomized block and carried out using 3 rates of a herbicide and a control and 4 replicates.

Final weight and the percentage weight change resulting from application of the treatment were analyzed by ANOVA, using a model appropriate to a randomized-block design. This analysis allows a comparison to be made between the mean final weight in each block by examining the F-ratio for blocks, $F = MS_{Block} / MS_{Residual}$. If this ratio is much larger than would be expected purely by chance, it implies that the blocking strategy has had a large effect and has reduced the effect of the pretreatment variability, thereby increasing the precision of the treatment comparisons. The results are summarized in Tables 3-1a and 3-1b.

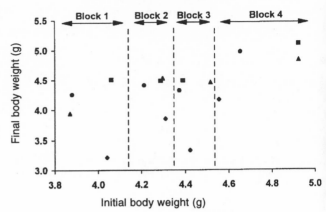

Figure 3-3   Plot of final body weight versus initial body weight from an acute study designed as a randomized block
● Control ■ 100 mg/kg ▲ 360mg/kg ◆ 640 mg/kg

Both tables show the effect of blocking to be highly significant, indicating that the blocking strategy has successfully controlled the variability caused by differences in initial weight. The result for percentage weight change is particularly interesting: this quantity is commonly analyzed in standard studies (where blocking is not used) in the belief that it accounts for the differences in initial weight. These results, however, suggest that simply analyzing percentage weight change, absent blocking, may not be sufficient to remove the effect of these differences. In any case, a percentage change can be calculated only if the pre- and posttreatment assessments are of the same type, e.g., weight. In sublethal experiments, where other endpoints such as reproduction may be of interest, this is not necessarily the case.

Table 3-1a   ANOVA on final body weights from the acute study of Figure 3-3

	Degrees of freedom	Sums of squares	Mean square	F-ratio	
Block	3	1.0601	0.3534	22.64	*
Treatment	3	1.5766	0.5255	33.68	*
Residual	9	0.1404	0.0156		
Total	15	2.7771			

* denotes statistically significant effect ($P = 0.05$).

Table 3-1b   ANOVA on percentage weight change from the acute study of Figure 3-3

	Degrees of freedom	Sums of squares	Mean square	F-ratio	
Block	3	102.57	34.19	4.27	*
Treatment	3	775.16	258.39	32.31	*
Residual	9	71.98	8.00		
Total	15	949.72			

* denotes statistically significant effect ($P = 0.05$).

Blocking by initial weight in sublethal studies

The sublethal study is designed to investigate effects on earthworm body weight and reproduction. Figures 3-4a and 3-4b show plots of final body weight and total juvenile production, respectively, compared to initial body weight from a study designed as a randomized block and using 5 rates of benomyl and a control and 4 replicates. Each experimental unit consisted of 10 randomly selected earthworms. Both final body weight and $\log_{10}$ (total number of juveniles) were analyzed by ANOVA, summaries of which are shown in Tables 3-2a and 3-2b.

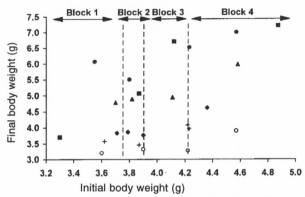

Figure 3-4   Plot of final body weight versus initial body weight from a sublethal study designed as a randomized block ● Control ■ 0.125 kg ai/ha ▲ 0.25 kg ai/ha ◆ 0.5 kg ai/ha + 1.0 kg ai/ha o 2.0 kg ai/ha

The analysis of final weight (Table 3-2a) shows the effect of blocking to be statistically significant, reinforcing the findings from the acute study. The analysis of the reproduction data in Table 3-2b does not demonstrate a statistically significant block effect. Nevertheless, blocking has helped reduce the residual variance by approximately 14%, indicating that it may prove useful in future studies. Van Gestel et al. [10] suggest that there may be a relationship between initial body weight, final body weight, and juvenile production.

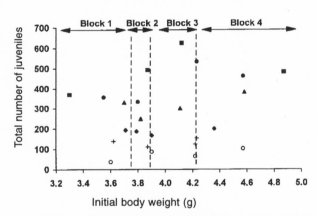

Figure 3-4b   Plot of total juvenile production versus initial body weight from a sublethal study designed as a randomized block ● Control ■ 0.125 kg ai/ha ▲ 0.25 kg ai/ha ◆ 0.5 kg ai/ha + 1.0 kg ai/ha o 2.0 kg ai/ha

Increasing precision further: analysis of covariance

In addition to blocking, it may be possible to further increase the accuracy and precision of treatment comparisons by including initial weight as a covariate in the ANOVA [11]. In its simplest form, this is equivalent to fitting a straight-line regression model to the relationship between the response and initial weight for each treatment with the regression lines having identical slopes. The fitted regression lines are then used to estimate adjusted-mean responses, which can be interpreted as estimates of the final means that would have been obtained had all the initial weights been the same. These adjusted means can be compared for evidence of treatment effects. Table 3-1c shows the effect of including initial weight as a covariate in addition to the analysis of final weight already presented in Table 3-1a.

Table 3-2a   ANOVA on final body weights from the sub-lethal study of Figure 3-4a

	Degrees of freedom	Sums of squares	Mean square	F-ratio	
Block	3	5.843	1.948	5.75	*
Treatment	5	26.222	5.244	15.49	*
Residual	15	5.077	0.338		
Total	23	37.142			

* denotes statistically significant effect ($P = 0.05$).

Table 3-2b   ANOVA on $\log_{10}$(number of juveniles) from the sub-lethal study of Figure 3-4b

	Degrees of freedom	Sums of squares	Mean square	F-ratio	
Block	3	0.0580	0.0193	2.07	
Treatment	5	2.1226	0.4245	45.31	*
Residual	15	0.1405	0.0094		
Total	23	2.3212			

* denotes statistically significant effect ($P = 0.05$).

Table 3-1c   ANOVA on final body weights from the acute study of Figure 3-3 with initial body weight included as a covariate

	Degrees of freedom	Sums of squares	Mean square	F-ratio	
Block	3	1.0601	0.3534	28.05	*
Treatment	3	1.5766	0.5255	41.72	*
Covariate	1	0.0397	0.0397	3.15	
Residual	8	0.1008	0.0126		
Total	15	2.7771			

* denotes statistically significant effect ($P = 0.05$).

Analysis shows that inclusion of initial weight as a covariate has reduced the residual SD by a further 10% compared to the analysis in Table 3-1a; this effect is not statistically significant. The covariate has already been used to form blocks that account for most of the variability, and there are only 8 degrees of freedom remaining for the estimation of residual variance. In this example, therefore, sufficient accuracy and precision may be achieved by blocking alone. However, there may be some occasions when analysis of covariance is complementary to blocking, so use of this technique should not be ruled out altogether.

Summary and recommendations

Pre- and posttreatment measurements in earthworm laboratory experiments can be correlated. Consequently, incorporating pretreatment information such as body weight into both the study design and the analysis of the subsequent data can substantially increase the accuracy and precision of the results.

Blocking of experimental units should be routinely adopted for acute and sublethal studies. This simple modification to current designs can successfully reduce the variability and increase the precision of treatment comparisons, even for percentage weight change. Results from a sublethal study support this and suggest that the approach can also improve the assessment of effects on earthworm reproduction.

Successful blocking of experimental material requires knowledge of those factors, other than treatment, that have the potential to influence the endpoints of interest. Because earthworm body weight is the only pretreatment factor currently recorded in laboratory studies, new research must identify other influential factors. There is no need for such factors to be quantitative; qualitative assessments may be sufficient so long as the experimental units can be blocked meaningfully prior to the application of treatments.

In addition to blocking, other methods of statistical analysis may be worthwhile when evaluating test results. The use of pretreatment measurements as covariates in ANOVA, for example, can increase the accuracy and precision of treatment comparisons and enhance data interpretation.

References

1. [EEC] European Economic Community. 1991. Council directive 91/414/EEC. Concerning the placing of plant protection products on the market. Official Journal of the EEC L133, Brussels, Belgium.
2. Kokta C, Rothert H. 1992. A hazard and risk assessment scheme for evaluating effects of pesticides on earthworms: The approach in the Federal Republic of Germany. In: Becker H, Edwards P, Greig-Smith PW, Heimbach F, editors. Ecotoxicology of earthworms. Andover, UK: Intercept. p 169–176.
3. [EPPO] European Plant Protection Organization. 1993. Decision making scheme for the environmental risk assessment of plant protection products. EPPO Bulletin 23. p 131–149.
4. [OECD] Organization for Economic Cooperation and Development. 1984. OECD guidelines for testing of chemicals: earthworm acute toxicity test. Organization for Economic Cooperation and Development Guideline No. 207. Paris, France: OECD.
5. [EEC] European Economic Community. 1988. Commission directive 88/302/EEC. Adapting to technical progress for the ninth time. Council directive 67/548/EEC on the approximation of laws, regulations and administrative provisions relating to the classification, packaging and labeling of dangerous substances, part C, Toxicity for earthworms: Artificial soil test. Official Journal of the EEC L133, Brussels, Belgium.
6. Kokta C. 1992. A laboratory test on sublethal effects of pesticides on *Eisenia fetida*. In: Becker H, Edwards P, Greig-Smith PW, Heimbach F, editors. Ecotoxicology of earthworms. Andover, UK: Intercept. p 213–216.

7. Kula C. 1996. Development of a test method on sublethal effects of pesticides on the earthworm species *Eisenia fetida/Eisenia andrei*—comparison of two ring tests. In: Riepert F, Kula C, editors. Development of laboratory methods for testing effects of chemicals and pesticides on Collembola and earthworms. Berlin, Germany: Parey.

8. [ISO] International Standards Organization. 1996. DIS11268-2: Soil Quality - Effects of pollutants on earthworms (*Eisenia fetida*). Part 2: Determination of effects on reproduction.

9. [BBA] Biologische Bundesansstalt Für Land-Und Forstwirtschaft. 1994: Guidelines for pesticide testing in the registration procedure. Part VI 2-2: Effects of pesticides on the reproduction and growth of *Eisenia fetida/Eisenia andrei*. Ribbesbuttel, Germany: Saphir-Verlag.

10. Van Gestel CAM, Dirven-Van Breemen EM, Baerselman R. 1992. Influence of environmental conditions on the growth and reproduction of the earthworm *Eisenia andrei* in an artificial soil substrate. *Pedobiologia* 36:109–120.

11. Steel RGD, Torrie JH. 1981. Principles and procedures of statistics: a biometrical approach. Singapore: McGraw-Hill.

Chapter 4

Effects of pesticide application methods on reproduction of *Eisenia fetida* in laboratory tests

Anke Litzel and Christine Kula

A laboratory reproduction test has been developed to assess sublethal effects of plant protection products on earthworms. According to this method, the application of plant protection products should resemble field conditions. The number of applications relevant for field use has limitations because the test duration might be shorter than the application period in the field. For this reason, the full dosage is normally applied once in the beginning of the test. This does not correspond to the field, where several applications and smaller application rates might occur. To investigate the consequences of these different application methods on the results of the reproduction test, studies with different modes of application were done: the full seasonal dose was applied once, and the results were compared to those of several single doses. The experience with the 3 fungicides tested reveals that the effects of both application types on reproduction of earthworms are comparable.

Risk assessment of plant protection products for legislative purposes is mainly based on standardized laboratory test guidelines, e.g., from the European Union [1] and the European Plant Protection Organization [2]. An international guideline is available for testing acute toxic effects of substances on earthworms [3]. Because the duration of this test is 14 d and the aim of the test is to evaluate LC50, effects on reproduction and other sublethal effects cannot be evaluated.

A new guideline was developed for testing sublethal effects of plant protection products on earthworm reproduction in the laboratory [4], with species and soil substrate identical to those in the acute test guideline. However, for pesticides, the application of the test substance in the sublethal test must closely mimic field conditions, whereas in the acute test the substance is mixed into the soil, which is not similar to field conditions. Because plant protection products usually are applied on the crop surface in the field, the substance must be applied on the soil surface in the reproduction test. This approach is part of the international harmonization procedure [5]. The test duration is 8 weeks, with an exposure of adults of 4 weeks, and the application of the test substance must be adjusted to this time frame; therefore, repeated applications are summed and applied once at test start. The question arises whether this approach leads to an appropriate risk assessment compared to the field situation. Comparative laboratory tests were done to evaluate the differences between one single application of the full dosage and several single applications.

Methods

Three fungicides were selected for testing, based on either their toxicity in the acute test, the recommended number of applications in the field, or the stability of the active substance in soil. These criteria are relevant triggers for requesting a reproduction test with earthworms in the German authorization procedure for plant protection products.

The plant protection product Brestan 60 is a combination product with the active substances fentin-acetat (540 g/kg) and maneb (180 g/kg). According to the registration guidelines in Germany, the product can be applied once per season, and it is labeled "harmful to earthworms" because of its effects on reproduction. The half-life in soil is higher than 47 d [6] for fentin-acetat. The organic part of the molecule is stable in soil. For maneb, the disappearance time of 50% of the substance (DT50) in a loamy sand is about 25 d [6]. The product Du Pont Benomyl contains the active substance benomyl (524 g/kg), which in a first step degrades to carbendazim; its half-life in soil is 6 to 12 months [6]. According to the registration guidelines in Germany, it can be applied up to 3 times per year, and it is labeled "harmful to earthworms" because of acute and chronic effects. The product Systhane 6W contains 60 g myclobutanil/kg. The half-life of this active substance is higher than the duration of the reproduction test. According to the registration guidelines in Germany, the product can be applied up to 6 times per year.

The guideline for testing reproduction of earthworms was modified in this study to allow a longer period of exposure for repeated applications. In Table 4-1, the main principles and the modifications to the guideline are summarized. To allow comparison of a single application of the full dosage with repeated applications, the exposure period of adults was extended to 8 weeks. In the case of repeated applications, these exposures were done weekly. There were at least 3 weeks between the last application and the end of exposure in adults. All studies were conducted with 2 application rates and were those recommended for field use. The single field rate for fentin-acetat and maneb was 0.3 L formulated product/ha; for benomyl, 0.45 L formulated product/ha; and for myclobutanil,

Table 4-1   Reproduction test with *Eisenia fetida/Eisenia andrei* and modifications to allow repeated applications

	Guideline specification (BBA 1994) (ISO DIS 11268-2)	Modification
Test organism	*E. fetida/E. andrei*	--
Test substrate	Artificial soil	—
Application of test substance	Dosage applied once on soil surface	dosage applied weekly on soil surface in up to 5 rates
Exposure of adults	4 weeks	8 weeks
Hatching period of juveniles	4 weeks	—
Duration of test	8 weeks	12 weeks
Food	Cattle manure	—
Endpoints	Number of juveniles	—
	Mortality of adults	
	Bodyweight of adults	

2.25 L formulated product/ha. The tests were repeated at least twice, with the exception of benomyl, for which only 1 test was done. The comparison of the application types was always done within the same experiment, so a direct comparison is possible.

The artificial soil had a moisture content of 36 to 38% dry weight substrate. The pH of the test soil was 6.5 at the beginning and about 7.3 at the end of the test. The age of the test animals was 12 months. The mean start weight of the earthworms was 690 mg.

The endpoints evaluated were the number of juveniles after 12 weeks, change in adult bodyweight after 8 weeks and adult mortality after 8 weeks of exposure. ANOVA and Dunnet-test were used to evaluate statistical significance.

Results

No significant differences were found between treatment types. One typical result of each product is presented in Figures 4-1 to 4-3. With the combination product of fentin-acetat and maneb, 4 applications of the single dose of 0.3 l product/ha led to a significant decrease to 30 juveniles in the treated boxes compared to 71 in the control boxes. This corresponds to a reduction to 43% of the control juvenile numbers (Figure 4-1, columns Ia and Ib). The effect of the single application of the full dosage is almost equal to the effect of several smaller applications. A double dose corresponding to 4 applications of the 2-fold rate and 1 application of the 8-fold rate resulted in a complete inhibition of reproduction in both types of application (Figure 4-1, columns IIa and IIb). Whereas reproduction was a sensitive parameter, there was no dose-related effect of the substance on bodyweight development of the adults. An increase in weight, which was not dose-related, was observed in the single-dose treatments. No mortality of adults occurred in this test. This test was conducted twice; another test (not presented here) conducted with the same product and the same application rates revealed nearly identical results.

With benomyl, 3 applications of the 0.5-fold field rate led to a significant reduction to 66% of control juvenile numbers, whereas 1 application of the 1.5-fold field rate led to a reduction of 49% of control juveniles numbers (Figure 4-2, columns Ia and Ib). Doubling these application rates resulted in total inhibition of reproduction (Figure 4-2, columns IIa and IIb). No effects on bodyweight and mortality of adults were observed.

With myclobutanil (Figure 4-3), the single field rate did not cause a significant reduction in juvenile numbers. In the field, the product can be applied up to 6 times. In the test, 3 times the double rate was applied (column IIa in Figure 4-3) and compared to 1 application of the 6-fold rate (column IIb). Both applications led to a significant reduction to 4.7% and 3.9% of the control juvenile numbers. No effects on mortality and bodyweight of adults were observed. Another test conducted with the same product and the same application rate showed comparable results.

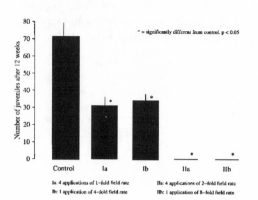

Figure 4-1   Effects of summed and repeated applications of the combination product of fentin-acetat and maneb on number of juveniles in the reproduction test (single field rate: 0.3 l formulated product/ha)

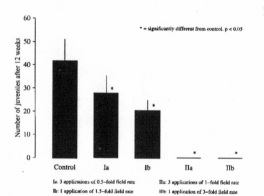

Figure 4-2   Effects of summed and repeated applications of benomyl on number of juveniles in the reproduction test (single field rate: 0.45 l formulated product/ha)

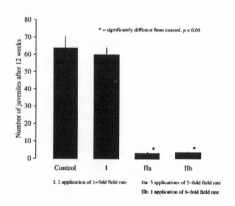

Figure 4-3   Effects of summed and repeated applications of myclobutanil on number of juveniles in the reproduction test (single field rate: 2.25 l formulated product/ha)

Discussion

The studies reported in this chapter show that different application methods produce comparable results: the reductions in number of juvenile earthworms were almost equal, whether the full dose was applied once or in several smaller amounts. In all tests, the number of juveniles was the most sensitive parameter, compared to bodyweight and mortality of adults.

The very similar results with fentin-acetat and maneb and myclobutanil might mean that the time between the applications was too short to establish any differences in reaction to the toxicant. In the tests, the application was made weekly with a maximum of 4 applications. In practice, the applications can be made across a broader time span, and the earthworms might not experience conditions consistent with those in the laboratory (e.g., equal food supply and constant soil moisture).

Only with benomyl was there a difference in treatment types: the summed applications showed a higher decrease in juvenile numbers than did to the single applications. Because benomyl has several modes of action concerning effects on earthworms (e.g., acute toxicity, reproduction toxicity, repellent effect), the summed application might have reached the level of acute effects, although no mortality occurred.

Because the substances tested were relatively stable, it is not yet clear whether the results of this experiment can be generalized, e.g., to substances with a short half-life. The available data allow the conclusion that the way of handling the type of application and the number of applications in the guideline will not yield results in very different orders of magnitude compared to agricultural practice. Whether this is also true for field populations, which may live in very different physiological conditions, remains to be clarified in the future.

Acknowledgments- We would like to thank Ms. C. Behrend for her excellent technical help and Dr. T. Kampmann for initiation of the project and valuable contributions.

References

1. Council Directive 91/414/EEC concerning the placing of plant protection products on the market. Official Journal of the European Communities, No. L 230, 18 August 1991.

2. [EPPO] European Plant Protection Organization. 1993. Decision-making scheme for the environmental risk assessment of plant protection products. EPPO Bulletins 23, 131–149.

3. [OECD] Organization for Economic Cooperation and Development. 1984. OECD guidelines for testing of chemicals: earthworm acute toxicity test. Organization for Economic Cooperation and Development Guideline No. 207. Paris, France: OECD.

4. [BBA] Biologische Bundesanstalt Für Land-Und Forstwirtschaft. 1994. Auswirkungen von Pflanzenschutzmitteln auf die Reproduktion und das Wachstum von *Eisenia fetida/Eisenia andrei*. Richtlinien für die Prüfung von Pflanzenschutzmitteln im Zulassungsverfahren, Teil VI, 2-2, January 1994.

5. [ISO] International Standards Organization. 1996. Soil quality–effects of pollutants on earthworms (*Eisenia fetida fetida*, *E. fetida andrei*); Part 2: Determination of effects on reproduction. ISO DIS 11268-2.2

6. Perkow W, Ploss H. 1994. Wirksubstanzen der Pflanzenschutz- und Schädlingsbekämpfungsmittel. Berlin, Germany: Parey.

Chapter 5

Application of body residues as a tool in the assessment of soil toxicity

Roman P. Lanno, Sherry C. LeBlanc, Bradley L. Knight, R. Tymowski, and Dean G. Fitzgerald

This chapter examines body residues of chemicals in soil organisms as an alternative to soil chemical concentrations in assessing bioavailability and expressing the toxicity of chemicals in soil systems. Earthworms were used as a model soil organism and pentachlorophenol (PCP) and Cd provided examples of organic chemicals and metals, respectively. Specifically, the relationship between critical body residues (CBRs) of PCP at lethality, soil PCP concentrations, and modifying factors of toxicity, e.g., body size, species, temperature, and soil type were investigated during 28 bioassays with earthworms. *Eisenia fetida* (0.2 to 0.4 g) were exposed to PCP at 15 and 24 °C in artificial soil and at 24 °C in Brookston clay. *Lumbricus terrestris* (3.3 to 4.6 g) and *Eudrilus eugeniae* (1.5 to 2.7 g) were exposed to PCP in artificial soil at 15 and 24 °C, respectively. Soil PCP concentrations ranged from 0 to 6.75 mmol/kg soil (dry weight), and body residues were determined by gas chromatography. Toxicity tests were also conducted with *E. fetida* exposed to Cd (as CdCl$_2$) at soil concentrations ranging from 0.90 to 44.5 mmol/kg (dry weight) in artificial soil at 24 °C for 32 d. Cadmium residues were determined by flame atomic absorption spectroscopy in pellet and supernatant fractions of worms exposed to 26.7 and 35.6 mmol Cd/kg. Comparisons of toxicity were made using incipient lethal levels (ILLs) and CBRs at lethality. ILLs for *L. terrestris* (0.72 mmol/kg soil) and *E. eugeniae* (0.63 mmol/kg soil) were similar and significantly higher than those for all *E. fetida* treatments. ILLs for *E. fetida* exposed to PCP in artificial soil at 15 and 24 °C were similar and lower than the ILL for *E. fetida* exposed to PCP in Brookston clay. Toxicity half-lives were similar for all *E. fetida* treatments but were higher than for the larger worms, which were similar. Mean CBRs at lethality for *E. fetida* tested in artificial soil at 24 °C ranged from 0.33 to 0.74 mmol/kg (wet weight) over the range of PCP exposure concentrations from 0.12 to 3.75 mmol/kg. A mean CBR of 2.65 mmol/kg was measured in *E. fetida* exposed to 6.75 mmol PCP/kg under the same conditions. Mean CBRs for *E. fetida* exposed to PCP in Brookston clay at 24 °C ranged from 0.51 to 1.59 mmol/kg. CBRs for *L. terrestris* and *E. eugeniae* ranged over 0.47 to 2.18 and 0.41 to 0.61 mmol/kg, respectively. Similar CBRs in 3 species of earthworms differing 10-fold in body mass and exposed to PCP at concentrations ranging over an order of magnitude demonstrate that body residues can act to more effectively integrate the relationship between test organism, exposure, and mortality than soil toxicant concentrations alone. *E. fetida* exposed to Cd exhibited a sharp threshold for mortality occurring between 26.5 and 35.6 mmol/kg, with an ILL of approximately 28 mmol Cd/kg. Over the course of an 8-d exposure, Cd residues in the supernatant fraction increased from 0.1 to 0.5 mmol/kg while residues in the pellet fraction remained relatively constant at approximately 0.1 mmol/kg. Mean total Cd CBRs at lethality ranged from 1.0 to 9.1 mmol/kg, dry weight. Critical body residues for Cd appear much more variable than for PCP but could still provide useful information in assessing the toxicity of metals in soils.

The toxicity of a chemical in soil is defined by the interaction of a chemical with an organism and time [1]. Except for unique classes of chemicals that act by different modes of action (e.g., membrane irritants) or extremely elevated chemical concentrations that would destroy epithelial tissues upon contact, most chemicals need to be taken up by an organism in order to be toxic. If there is no uptake, there is no toxicity, regardless of the concentration of the chemical in the soil. This interaction between a chemical and an organism has been termed "environmental bioavailability" [1]. Even if uptake of the chemical occurs, simple presence in the organism does not imply toxicity. The level in the organism must reach some threshold level at the target site for toxicity (e.g., cell membranes for narcotic agents) and is defined as "toxicological bioavailability" [1]. The most important determinant of toxicity is how many molecules of a chemical actually reach the sites of toxic action in an organism. The toxic dose could be expressed as the number of moles of chemical/the number of receptor sites for that chemical that result in a toxic effect. In reality, it is very difficult, if not impossible, to express the dose received by the organism in this manner. So what alternatives exist? In ecotoxicology, the most common way of expressing toxicological bioavailability is as a concentration in the environment (e.g., mg chemical/kg of soil), which can be called "exposure dose." The exposure dose is only a surrogate for toxicological bioavailability, but it is the primary measure used by regulators and ecotoxicologists in determining whether a soil is contaminated or toxic. This surrogate dose often fails to incorporate differences in bioavailability due to modifying factors of toxicity such as physical (e.g., soil type [2], weathering), chemical (Kow), and biological (e.g., species, body size) properties of the test system.

Most often, concentration-based estimates of soil toxicity are represented by the total chemical measured during a vigorous extraction method, not by the fraction of total chemical that is actually bioavailable [3]. Organism-based estimates of dose would directly address bioavailability issues and may provide an alternative means of expressing toxicity. The use of CBRs may reduce some of the variability in relating bioassay endpoints to measures of chemical exposure for test results obtained with different worm species in various soil types. A CBR can be defined as a whole-body concentration of a chemical that is associated with a specific adverse response, e.g., lethality [1]. This toxicokinetic approach assumes that the toxicant concentration at the sites of toxic action in an organism is relatively constant for an overt biological response such as lethality, regardless of the physicochemical properties of the chemical and the exposure matrix [4,5]. Additionally, uptake of the contaminant follows Fick's Law of Diffusion [6] and is first order or proportional to the external concentration. To apply this approach, the initial time–toxicity relationship for lethality is described using a 1-compartment first-order kinetics (1CFOK) model [2,7]. Comparisons of toxicity based on soil chemical concentrations can thus be made at steady state with respect to the time–toxicity curve. In practical terms, bioassays are conducted for a period of time sufficient for the concentration causing 50% mortality (LC50) to become time independent or reach an incipient lethal level (ILL) or acute lethality [7–9] (Figure 5-1). Body residues associated with lethality (CBRs) are then compared at various times, including steady state.

Critical body residues are often better estimates of the amount of chemical at the sites of toxic action in an earthworm than ambient concentrations because CBRs explicitly consider bioavailability. Whole-body residues of a contaminant are good estimates of the dose received by an earthworm, but chemicals often partition into specific phases of the organism. As an example, nonpolar narcotic compounds may partition into the neutral storage lipid fraction of an earthworm or into cell membranes where they elicit their toxic effect [5]. Similarly, metals could partition into sequestered (e.g., chloragogenous tissue) and available fractions, since many invertebrates have the ability to sequester metals in a biologically unavailable form in various tissues [10]. If the total body residue of metals could be separated into sequestered and bioavailable fractions, a better estimate of the true amount of metal at the sites of toxic action could be made [11].

This chapter discusses body residues of chemicals in soil organisms as an alternative to soil chemical concentrations in assessing bioavailability and expressing the toxicity of chemicals in soil systems. Earthworms were used as a model soil organism and pentachlorophenol (PCP) and Cd provided examples for organic chemicals and metals, respectively.

Materials and methods

The data for PCP tests in this chapter have been adapted from previous work by the authors [7], and readers are referred to that paper for a detailed description of the methods. In the PCP toxicity tests, *Eisenia fetida*, *Lumbricus terrestris*, and *Eudrilus eugeniae* were exposed to soil PCP concentrations ranging from 0 to 6.75 mmol/kg over 28 d in artificial soil (clay, 20%; organic matter (OM), 10%; sand, 69%; $CaCO_3$, 1%; pH, 7.5). A toxicity test with *E. fetida* in a natural clay soil (Brookston clay [2]) was also conducted. Tests with *E. fetida* were conducted at 15 and 24 °C, while those with *L. terrestris* and *E. eugeniae* were conducted at 15 and 24 °C, respectively. Toxicity tests were also conducted with *E. fetida* exposed to Cd (as $CdCl_2$) at soil concentrations ranging from 0.90 to 44.5 mmol/kg (dry weight) in artificial soil at 24 °C. The general test methods and conditions were the same as for PCP tests. Comparisons of toxicity were based on ILLs derived from nonlinear regressions (1CFOK models) of time–toxicity data using nominal soil concentrations or on whole-body PCP or Cd concentrations.

Cd was added to the soil as an aqueous solution and toxicity tests were conducted in duplicate with 10 worms per jar and lasted 32 d, with mortality readings after 1, 2, 4, 8, 16, and 32 d of exposure. The experimental design for this study differed from the PCP study in that independent replicates of all Cd concentrations were prepared for each sampling time. Worms were presumed dead when no contractile response was observed following a light prodding with a blunt probe. All mortalities were frozen at –80 °C for residue analysis. Soil chemical concentrations are reported as nominal values throughout.

Initial nominal chemical concentrations in soils were used to determine LC50s from mortality data by trimmed Spearman-Karber analysis [12] for each time in a geometric series when > 50% mortality occurred in at least 1 concentration. Incipient lethal levels were estimated by fitting a 1CFOK model to the LC50-time data, using nonlinear regression analysis and calculating the asymptotic value of the exponential curve of the inverse of the LC50 versus time data (Figure 5-1).

Whole-body PCP concentrations were determined for *E. fetida* at 24 °C in both soil types for all exposure concentrations. Residues in the other species were determined for worms from the distilled water control, from representative soil PCP concentrations associated with the estimated ILLs, and from 1 other concentration (2.10 mmol PCP/kg). PCP levels were determined by GC/ECD after cryoextraction [7].

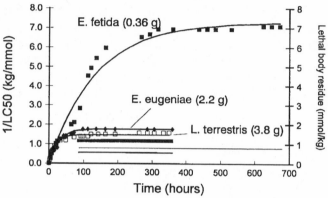

Figure 5-1   Toxicity curves for 3 species of earthworms exposed to PCP in artificial soil. Solid lines represent mean CBR values for different species of worms (narrow line - *E. fetida*, intermediate line - *E. eugeniae*, thick line - *L. terrestris*). Average body size of the earthworm species tested is provided in parentheses.

Cadmium residues were determined in pellet and supernatant fractions of worms exposed to 26.7 and 35.6 mmol Cd/kg in the following manner. Worms were weighed, placed in individual test tubes with 3.0 ml of distilled water, and homogenized. The suspensions were separated by centrifugation at 1500 g for 10 min. From each sample, 2.0 ml of the supernatant and the pellet were digested separately with trace metals grade nitric acid (5 ml) and hydrogen peroxide (0.2 ml, 30%). The digested samples were diluted to 10 ml, and Cd levels were determined by atomic absorption spectroscopy at a wavelength of 228.8 nm using a lean air-acetylene flame. Results were expressed in terms of mmol Cd/kg (wet mass of organism).

Results

Pentachlorophenol

Mortality and body residue data from Fitzgerald et al. [7] are summarized in Table 5-1. No *E. fetida* mortality occurred at either 15 or 24 °C in either soil type for treatment groups exposed to 0.038 or 0.068 mmol PCP/kg or in the artificial soil controls. *E. fetida* exposed in Brookston clay showed no mortality at 0.12 mmol PCP/kg. For all other *E. fetida* treatments, as well as for all tests with *E. eugeniae* and *L. terrestris*, all exposure con-

Table 5-1   Summary of test conditions, toxicological endpoints, and CBRs for earthworms exposed to PCP in different experimental regimes and for *E. fetida* exposed to Cd in artificial soil. Data are presented as mean with SDs in parentheses. Values in columns with a common letter are not significantly different ($P \geq 0.05$).

Species	Size (g)	Soil type	Temp (°C)	ILL (mmol/kg soil)	$t_{1/2}$ (h)	CBR range (mmol/kg, wet weight)
Pentachlorophenol (PCP) (from Fitzgerald et al. 1996)						
E. fetida	0.23 (0.023)	Brookston Clay	24	0.27^A (0.08)	72^A (21.5)	0.84-1.39
E. fetida	0.36 (0.06)	Artificial	24	0.14^B (0.001)	138^A (31)	0.33-2.65
E. fetida	0.36 (0.06)	Artificial	15	0.10^B (0.002)	129^A (5.7)	0.80-1.29
E. eugeniae	2.25 (0.35)	Artificial	24	0.63^C (0.12)	23^B (5.7)	0.41-0.61
L. terrestris	3.81 (0.35)	Artificial	15	0.72^C (0.12)	17^B (6.4)	0.47-2.18
Cadmium (Cd)						
E. fetida	0.3	Artificial	24	28.5 (2.38)	-	1.0-9.1

centrations showed some mortality over the 28-d duration of the tests. For *E. eugeniae* and *L. terrestris*, mortalities at exposure concentrations below 0.12 mmol PCP/kg and in controls, however, consistently occurred after 14 d of exposure, by which time thresholds of toxicity (ILLs) were clearly established. As such, only the data up to 14 d were used to model toxicity. A significant ($P < 0.0001$) treatment effect on ILLs and toxicity half-lives ($t_{1/2}$, the time required to reach half the ILL) was evident for PCP exposures (Table 5-1). The $t_{1/2}$ for all *E. fetida* treatments were similar ($P = 0.13$), as were half-lives for the 2 larger species of worms, *L. terrestris* and *E. eugeniae*. However, $t_{1/2}$ values for the larger species were significantly lower ($P = 0.010$) than for *E. fetida*. The ILL for *E. fetida* exposed to PCP in Brookston clay was significantly higher ($P = 0.026$) than that for the other *E. fetida* treatments. ILLs for *L. terrestris* and *E. eugeniae* were not statistically different ($P = 0.93$), but were significantly higher than those for all *E. fetida* treatments ($P = 0.0036$).

Critical body residues at lethality for *E. fetida* tested in artificial soil at 24 °C ranged from 0.33 to 2.65 mmol PCP/kg (wet weight) over the entire range of soil PCP exposure concentrations from 0.12 to 6.75 mmol PCP/kg (dry weight). Only CBRs in worms exposed to 6.75 mmol PCP/kg were significantly higher than those shown by the other PCP treatments ($P < 0.0001$). PCP residues in surviving worms were either similar to or lower than residues in dead worms. Body residues of worms from the 2 control treatments did not differ and were significantly lower than residues in all PCP-exposed worms ($P < 0.0001$). Mean CBRs for *E. fetida* exposed to PCP at 24 °C in Brookston clay ranged from 0.51 to 1.59 mmol PCP/kg (wet weight) over the same range of soil PCP concentrations and were not significantly different. Critical body residues for *E. fetida* (15 °C), *L. terrestris*, and *E. eugeniae* tested in artificial soil at PCP concentrations near their respective ILLs and 2.10 mmol PCP/kg were not significantly different (Table 5-1).

Cadmium

E. fetida exposed to Cd exhibited a sharp threshold for mortality occurring between 26.5 and 35.6 mmol/kg (3000 and 4000 mg Cd/kg). The ILL for *E. fetida* exposed to Cd was approximately 28 mmol Cd/kg (3150 mg/kg). Tests conducted with *E. fetida* exposed to

Cd showed that body residues of dead worms were much higher than those of live worms exposed to the same Cd concentration in soil. An interesting relationship was noted in live worms exposed to 26.5 mmol Cd/kg (Figure 5-2). Over the course of an 8-d exposure, Cd residues in the supernatant fraction increased from 0.1 to 0.5 mmol/kg, while residues in the pellet fraction remained relatively constant at approximately 0.1 mmol/kg. Mean total Cd CBRs at lethality ranged from 1.0 to 9.1 mmol/kg, dry weight (Table 5-1).

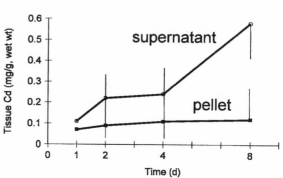

Figure 5-2   Cd concentrations in earthworm supernatant and pellet fractions (centrifugation at 1500 g for 10 min.) of *E. fetida* exposed to 3000 mg Cd/kg. (Error bars are 95% CI; n = 8 for each d

Discussion

Pentachlorophenol

Earthworms exposed to PCP had similar CBRs despite variable toxicity test conditions such as earthworm test species, soil type, and a wide range of PCP concentrations spanning approximately 3 orders of magnitude (Table 5-1; Figure 5-1). These results suggest that despite varying modifying factors, the internal dose of chemical that is lethal to the earthworm remains relatively constant. However, a difference of 9 °C in test temperature affected neither ILLs nor CBRs for *E. fetida*.

The earthworms tested in this study represented 3 genera from 2 separate families (Lumbricidae and Eudrilidae) and differed 10-fold in body mass (Table 5-1). Body size has long been known to affect metabolic processes and alter toxicant kinetics [13]. At first glance, it appears that the worm with the smallest body mass, *E. fetida*, is the most sensitive to artificial soil spiked with PCP. When toxicity is expressed in terms of soil concentration (ILL, Table 5-1), *E. fetida* appears to be approximately 6 to 7 times more sensitive to PCP than are the 2 larger worms. Upon closer examination, when comparisons of toxicity are based upon CBRs, *E. fetida* does not appear to be the most sensitive, with the ranges of CBRs for all 3 species overlapping. This would suggest that when ILLs are compared, differences in sensitivity to PCP merely are due to expressing toxicity as a soil concentration rather than as a toxicologically bioavailable concentration represented by the internal dose in the organism.

Other studies have compared the toxicity of various chemicals to different species of earthworms. Neuhauser et al. [14] observed that the 4 earthworm species tested (*E. fetida, E. eugeniae, Allolobophora tuberculata*, and *Perionyx excavatus*) varied in their sensitivity to a given contaminant, based upon soil concentrations of the chemicals and time-dependent LC50s. Similarly, van Gestel and Ma [15] found up to a 10-fold difference in the toxicity of various chlorophenols between 2 species of earthworms (*E. fetida andrei* and *Lumbricus rubellus*). Additional research comparing CBRs for different species of earthworms may provide information as to whether 1 species is truly more sensitive than the other to a given chemical. It may be that observed differences in sensitivity are only a function of expressing toxicity as an external concentration, such as time-dependent LC50, or a time-independent measure, such as an ILL.

Because physical modifying factors associated with the test matrix and chemical modifying factors associated with the test chemical (PCP) were controlled, observed differences in ILLs may be attributed to biological modifying factors associated with the test organisms. Body size, behavior, and physiological differences in toxicant metabolism can alter toxicant kinetics, resulting in differing ILLs. Body size has been observed to alter toxicant kinetics in poikilotherms [16], and in the current study, distinct differences in toxicant kinetics based upon the $t_{1/2}$, or time taken to reach half the ILL, were observed. The 2 larger worms (*L. terrestris* and *E. eugeniae*) reached an ILL about 4 to 5 times faster than *E. fetida*, even though the ILL was significantly lower for *E. fetida*. This could not be explained directly by conventional allometry, which suggests smaller invertebrates have a higher mass-specific metabolic rate and larger surface-area-to-volume ratios than larger invertebrates [17]. These results suggest a more rapid uptake of PCP by *E. fetida* than by the larger worms, allowing *E. fetida* to reach an ILL faster than larger worms. In contrast to this study, van Gestel and Ma [15] found no statistical differences in the toxicity of 2,4,5-trichlorophenol to *L. rubellus* despite a 3-fold difference in body mass of test worms, which may be too small a difference in body mass to reveal an effect of body size on toxicity.

Organism behavior during a toxicity test may also act as a modifying factor of toxicity. *E. fetida* was often found clumped together near the bottom of the test chamber during mortality checks, while the 2 larger species were usually found throughout the test chamber, perhaps indicating greater activity. The larger, more active worms may be continuously encountering renewed toxicant gradients, promoting more rapid diffusion of PCP into their tissues, resulting in the accumulation of large amounts of PCP in a short time. Although the nature of the testing medium and concentrations of PCP were held constant between tests, the worm biomass-to-soil ratios differed (*E. fetida*, 83; *E. eugeniae*, 13; *L. terrestris*, 10), which may also influence behavior.

Additionally, the larger worms may have found the organic matter in the test soil (sphagnum moss) more suitable for consumption, thus enhancing the rate of PCP uptake through the dietary route of exposure. The importance of the dietary uptake of contaminants on toxicity in earthworms is unknown, but it has been hypothesized that cutaneous uptake is the more important of the 2 routes of exposure [15,18]. Van Gestel and Ma [15] inferred the cutaneous route of uptake to be dominant when they observed that differences in the toxicity of chlorophenols in different natural soils, expressed on a mg/kg soil basis, disappeared once the exposure dose was calculated in mg/L pore water using sorption coefficients. In the current study, differences in PCP sorption would be minimized, since the same artificial soil was used in the tests comparing the 3 species.

Physiological differences may exist between species in their ability to regulate contaminant uptake through mechanisms such as mucous production or PCP metabolism. *E. fetida* may have a greater capacity for metabolizing and excreting PCP, hence increasing the ILL and the time required to attain a CBR. Earthworms appear able to bind xenobiotics or their metabolites in unextractable forms [19]. Specifically, Haque and Ebing [20] documented the formation of unextractable compounds in worms exposed to ^{14}C-PCP. Differences between species in the ability to bind PCP may also contribute to the observed differences in ILLs despite similarities in body residues.

In addition to biological modifying factors, physical modifying factors were also varied during our study. Based upon soil concentrations, differences in toxicity of an order of magnitude have been observed for earthworms in different soils [2]. Significant effects of soil type on the ILL were also observed in the current study. The ILL for *E. fetida* exposed to PCP in Brookston clay at 24 °C (0.25 mmol PCP/kg) was approximately 2-fold higher ($P = 0.0019$) than the ILL for those exposed at the same temperature in artificial soil (0.14 mmol PCP/kg; Table 5-1). A difference in ILLs was not surprising because the 2 test soils differed in physical and chemical properties [2]. The pH of the clay ranged between 5.7 and 6.5, while the pH of the artificial soil ranged between 6.9 and 8.1. The amount of OM and its composition, as well as the particle size distribution, also varied between the 2 soil types. The toxicity of PCP to *E. fetida andrei* has been reported to be influenced by soil pH and OM content [21]. The 14-d LC50 for worms exposed in a natural soil (Gilze; pH, 4.1; OM, 1.7%) was 52 mg/kg, 2-fold higher than the LC50 of 28.5 mg/kg shown by worms exposed in an artificial soil (pH, 7.0; OM, 7.7%). The interaction between modifying factors will determine the environmental bioavailability of PCP and the uptake kinetics from the soils, resulting in different ILLs or LC50 values. Although differences in ILLs were observed in the current study, CBR ranges were similar (Table 5-1).

Another modifying factor that can influence toxicity is temperature. As ambient temperature increases, the toxicity of chemicals to ectothermic organisms often increases due to increased activity, both metabolic and behavioral, as well as to enhanced diffusion across epithelial surfaces. Contrary to the effect of other modifying factors on ILLs, no effect of exposure temperature on the toxicity of PCP to *E. fetida* at 24 °C or 15 °C was observed. CBRs were similar in worms exposed at both temperatures (Table 5-1). Either temperature is not a major modifying factor of toxicity in *E. fetida* or the temperature differential in this experiment was too small to have an impact.

Whole-body PCP residues of *E. fetida* that survived the 28-d exposure in either artificial soil or Brookston clay were similar to those of worms that perished during the test. Similar results were observed by Belfroid et al. [22] with *E. fetida* exposed to chlorobenzenes in artificial soil. Since these worms are exposed to lower soil PCP concentrations (i.e., below the ILL), the amount of PCP available for uptake may be low enough that the animals established some physiological equilibrium between uptake, metabolism, and depuration that would allow nonlethal adaptation to residues similar to those residues measured in worms exposed to higher concentrations of PCP.

Cadmium

Heavy metals in soil can also bioaccumulate in earthworms. However, metal residues may not provide information similar to CBRs of organic chemical residues. Many organic chemicals have a nonspecific or baseline toxicity associated with a general residue level or a tissue-specific residue level [23]. Metal metabolism can be quite different, and an excellent model of metal toxicity based upon the organismal content of metals is given by Rainbow [10]. Once absorbed, metals may be partitioned into biologically available fractions or biologically unavailable or storage fractions. Biologically available metals can function in essential metabolism, or in the case of nonessential metals or excess essential metals, contribute to toxicity. Unavailable metals may be sequestered as intracellular inorganic complexes or bound to metallothionein. Just as total environmental concentrations may not correlate well with environmental bioavailability, total metal levels in an organism may not correlate well with toxicological bioavailability. For this reason, it may be advantageous to separate the bound or sequestered metals from the bioavailable fraction when determining CBRs. Jenkins and Mason [11] examined the relationship between waterborne Cd^{2+} and Cd accumulation in homogenates of the marine polychaete *Nereis areaceodentata*, separated by differential centrifugation. Cd levels were determined in homogenate fractions representing the cytosol, cellular debris, nuclear region, mitochondria-lysosomes, and microsomes. The cytosolic and cellular debris fractions accounted for about 90% of absorbed Cd. The cytosolic fraction contained 40 to 65% of the total accumulated Cd dose, and the majority of this Cd was associated with metallothionein. The Cd accumulation ratio (mol Cd/g tissue per mol Cd/ml seawater) increased at higher concentrations of waterborne Cd because of increased levels of Cd in the cytosolic fraction. Concomitant with increased cytosolic Cd was a cessation in reproduction by the polychaetes over the 11-week exposure period. Part of the investigation into the applications of metal residues was to conduct a simple fractionation of homogenized worms exposed to Cd to determine if changes in Cd levels could be observed in the supernatant and pellet fractions.

Preliminary results with *E. fetida* exposed to Cd in artificial soil suggest that supernatant Cd levels increase with time in response to exposure to 26.7 mmol Cd/kg (Figure 5-2). Cadmium levels in the supernatant increased from 0.2 mmol/kg at 2 d to 0.6 mmol/kg at 8 d. No difference was observed in Cd levels in the pellet fraction of the earthworm homogenates over the 8-d exposure period. Total Cd levels in dead worms ranged from 1.0 to 9.1 mmol/kg. Mean total Cd levels in live worms at 8 d were approximately 0.8 mmol/kg, which approaches the lower bound for CBRs for lethality observed in this study. It appears that the amount of Cd in the supernatant could be linked with mortality in a dose-dependent manner, while pellet–Cd concentrations could not. However, no datapoints were available for Cd levels in different fractions at times between 8 and 16 d, when all the worms were dead. It may be possible to establish a CBR for lethality from Cd exposure by employing supernatant Cd levels as a measure of toxicological bioavailability. More research needs to be conducted on the associations between the Cd levels in fractions achieved by differential centrifugation and specific biological endpoints before

CBRs can be established. These fractionation techniques may have greater applicability to essential metals that are relatively well regulated and may be deposited in other cellular fractions than the cytosol.

Nonessential metal residues (e.g., Cd) in an organism may provide an indication of its bioavailability from soil [24]. Earthworm body residue data reported in the literature along with data from the this test were examined to determine the relationship between total soil Cd concentration and whole-body Cd residues in earthworms. Data from many biomonitoring studies that reported Cd soil concentrations and body residues were available [25–29]. However, few of these studies reported relationships between Cd concentrations in the soil and the earthworm and related toxicity endpoints, such as mortality [30-33].

In total, data reported in 9 studies were summarized, listing species, soil properties, soil concentrations, earthworm concentrations, and critical effects. Seventeen different species of earthworms were found in soil containing Cd concentrations ranging from 0.003 to 0.285 mmol Cd/kg dry weight (0.322 to 32 mg/kg). Body burdens in these earthworms ranged from 0.003 to 0.81 mmol Cd/kg. Four studies in which *E. fetida* was exposed to Cd in either artificial soil or amended soil reported exposure concentrations from 0.044 to 44.5 mmol Cd/kg soil dry weight (5 to 5000 mg/kg). Body burdens ranged from 0.311 to 8.9 mmol Cd/kg in live earthworms and up to 387 mmol Cd/kg in dead earthworms. Earthworms collected from contaminated sites were assumed to be representative of the population at the sites, and only studies of > 10 d were used to determine body burdens.

The relationship between Cd residues in earthworms and total Cd soil concentrations was examined using linear regression. Log transformations were applied to normalize the data, since raw data may be positively skewed [34]. A significant regression ($P <$ 0.0001) with a positive slope was evident (Figure 5-3). Although the linear regression model was a good fit for this data set ($R^2 = 0.902$), it should be viewed with caution because of the limited number of datapoints that actually relate total body residues of Cd to a biological response. Most of the datapoints are derived from monitoring studies in which residues were not related to any biological effect. This general relationship is interesting, suggesting that for Cd, total soil levels may be good predictors of internal Cd residues in earthworms. However, more data are needed on CBRs for Cd at elevated soil concentrations and under different test and field conditions before the utility of this relationship is understood.

Summary

The concentration of a chemical in soil, external to the organism, often does not provide a good estimate of the hazard presented by the chemical. This is due to biological, chemical, and physical modifying factors that determine the chemical's bioavailability. Determining body residues of soil organisms provides a method that inherently accounts for the bioavailability of a chemical from soil. Body residues that are linked to ecologically relevant endpoints such as survival, growth, and reproduction are termed "critical body residues" [7,22] and may provide even more powerful tools to assist in the investigations of soil toxicity. The CBR approach provides an estimate of exposure dose for certain chemicals that is less variable than concentration-based toxicity estimates, such as ILLs. Because CBRs explicitly account for the bioavailability of a chemical, this expression of the internal dose of contaminant received by an organism has distinct advantages over the traditional expression of dose as a concentration of contaminant in the test medium. Expression of toxicity in this manner accommodates the classical problem of modifying factors in toxicity testing and provides a mechanism for comparing toxicity between different test media, regardless of their physical and chemical characteristics. For these reasons, CBRs provide unique tools that can assist in understanding toxicological relationships in soil.

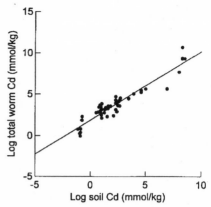

Figure 5-3 · Linear regression of whole-body Cd residues (mmol/kg) and total soil Cd concentration (mmol/kg, dry weight). Regression equation is Y = 1.89 + 0.83x R^2 = 0.902.

Acknowledgment - We thank L.S. McCarty for discussion and advice on the concepts and experimental design. We thank the following people for their technical assistance: K.E. Warner, A. Farwell, M. Rumpel, and D.G. Dixon, Department of Biology, University of Waterloo, Waterloo, Ontario, Canada.

References

1. Rand GM. 1995. Fundamentals of aquatic toxicology. Washington DC: Taylor & Francis.

2. Lanno RP, Wren CD, Stephenson GL. 1997. The use of toxicity curves in assessing the toxicity of soil contaminants to *Lumbricus terrestris*. *Soil Biol Biochem* 29:689–692.

3. Kelsey JW, Bennet DK, Alexander M. 1996. Selective chemical extractants to predict bioavailability of soil-aged organic chemicals. *Environ Sci Technol* 31:214–17.

4. Ferguson J. 1939. The use of chemical potentials as indices of toxicity. *Proc R Soc Lond Ser B* 127:387–404.

5. van Wezel AP. 1995. Residue-based effects of narcotic chemicals in fish and in lipid bilayers. Utrecht, Netherlands: University Utrecht. [Ph.D. thesis].

6. McCarty LS, Mackay D. 1993. Enhancing ecotoxicological modeling and assessment. *Environ Sci Tech* 27:1719–1728.

7. Fitzgerald DG, Warner KA, Lanno RP, Dixon DG. 1996. Assessing the effects of modifying factors on pentachlorophenol toxicity to earthworms: applications of body residues. *Environ Toxicol Chem* 15:2299–2304.

8. Sprague JB. 1970. Measurement of pollutant toxicity to fish. II. Utilizing and applying bioassay results. *Water Res* 4:3–32.

9. Lanno RP, McCarty LS. 1997. Worm bioassays: What knowledge can be applied from aquatic toxicity testing? *Soil Biol Biochem* 29:693–697.

10. Rainbow PS. 1996. Heavy metals in aquatic invertebrates. In: Beyer WN, Heinz GH, Redmon-Norwood AW, editors. Environmental contaminants in wildlife. Boca Raton FL: Lewis. p 405–425.

11. Jenkins KD, Mason AZ. 1988. Relationships between subcellular distribution of cadmium and perturbations in reproduction in the polychaete *Neanthes arenaceodentata*. *Aquat Toxicol* 12:229–244.

12. Hamilton MA, Russo RC, Thurston RV. 1977. Trimmed Spearman-Karber method for estimating median lethal concentrations in toxicity bioassays. *Environ Sci Technol* 11:714–719.

13. Peters RH. 1983. The ecological implications of body size. Cambridge England: Cambridge Univ. 329 p.

14. Neuhauser EF, Durkin PR, Malecki MR, Antra M. 1986. Comparative toxicity of ten organic chemicals to 4 earthworm species. *Comp Biochem Physiol* 83C:197.

15. van Gestel CAM, Ma W. 1988. Toxicity and bioaccumulation of chlorophenols in earthworms, in relation to bioavailability in soil. *Ecotox Environ Safety* 15:289–297.

16. McCarty LS, Mackay D, Smith AD, Ozburn GW, Dixon DG. 1993. Residue-based interpretation of toxicity and bioconcentration QSARs from aquatic bioassays: Polar narcotic organics. *Ecotox Environ Safety* 25:253–270.

17. Schmidt-Nielsen K. 1984. Scaling: Why is animal size so important? Cambridge UK: Cambridge Univ.

18. Belfroid AC, Seinen W, van Gestel CAM, Hermens JLM, van Leeuwen KJ. 1995. Modelling accumulation of hydrophobic organic chemicals in earthworms. *Environ Sci Pollut Res* 2:5–15.

19. Stenerson J. 1992. Uptake of metabolism of xenobiotics by earthworms. In: Greig-Smith PW, Becker H, Edwards PJ, Heimbach F, editors. Ecotoxicology of earthworms. Andover UK: Intercept Ltd. p 129–138.

20. Haque A, Ebing W. 1988. Uptake and accumulation of pentachlorophenol and sodium pentachlorophenate by earthworms from water and soil. *Sci Total Environ* 68:113–125.

21. van Gestel CAM, van Dis WA. 1988. The influence of soil characteristics on the toxicity of 4 chemicals to the earthworm *Eisenia fetida andrei* (Oligochaeta). *Biol Fert Soils* 6:262–265.

22. Belfroid A, Seinen W, van Gestel K, Hermens J. 1993. Acute toxicity of chlorobenzenes for earthworms (*Eisenia andrei*) in different exposure systems. *Chemosphere* 26:2265–2277.

23. Verhaar HJM, Busser FJM, Hermens JLM. 1994. A surrogate parameter for the baseline toxicity content of contaminated water: simulating bioconcentration and counting molecules. *Environ Sci Technol* 29:726–734.

24. van Straalen NM. 1996. Critical body concentrations: their use in bioindication. In: van Straalen NM, Krivolutsky DA, editors. Bioindicator systems for soil pollution. Dordrecht, The Netherlands: Kluwer Academic. p 5–16.

25. Rida AMMA, Bouche MB. 1995. The eradication of an earthworm genus by heavy metals in southern France. *Applied Soil Ecology* 2:45–52.

26. Beyer WN, Stafford C. 1993. Survey and evaluation of contaminants in earthworms and in soils derived from dredged material at confined disposal facilities in the Great Lakes region. *Environ Monitor Assess* 24:151–165.

27. Marino F, Ligero A, Cosin DJD. 1992. Heavy metals and earthworms on the border of a road next to Santiago (Galicia, northwest of Spain). Initial results. *Soil Biol Biochem* 24:1705–1709.

28. Beyer WN, Cromartie EJ. 1987. A survey of Pb, Cu, Zn, Cd, Cr, As, and Se in earthworms and soil from diverse sites. *Environ Monitor Assess* 8:27–36.

29. Carter A, Kenney EA, Guthrie TF, Timmenga H. 1983. Heavy metals in earthworms in non-contaminated and contaminated agricultural soil from near Vancouver, Canada. In: Satchell JE, editor. Earthworm ecology from Darwin to vermiculture. New York: Chapman and Hall. p 267–274.

30. Svendsen S, Meharg AA, Freestone P, Weeks JM. 1996. Use of an earthworm lysosomal biomarker for the ecological assessment of pollution from an industrial plastics fire. *Applied Soil Ecology* 3:99–107.

31. Honeycutt ME, Roberts BL, Roane DS. 1995. Cadmium disposition in the earthworm *Eisenia fetida*. *Ecotoxicol Environ Safety* 30:143–150.

32. Bengtsson G, Gunnarsson T, Rundgren S. 1986. Effects of metal pollution on the earthworm *Dendrobaena rubida* (Sav.) in acidified soils. *Water Air Soil Pollut* 28:361–383.

33. Tymowski RG. Unpublished data. Use of body residue analysis to predict the toxic responses of the earthworm *Eisenia foetida* to cadmium contaminated soils.

34. Morgan JE, Morgan AJ. 1988. Calcium-lead interactions involving earthworms. Part 1: The effect of exogenous calcium on lead accumulation by earthworms under field and laboratory conditions. *Environ Pollut* 54:41–53.

Chapter 6

Toxicity endpoints and accumulation of Cd and Pb in *Eisenia fetida* (Oligochaeta)

A.J. Reinecke and S.A. Reinecke

The literature is not consistent about the effects of Cd and Pb at low concentrations on various toxicity endpoints for earthworms. Two experiments were undertaken, one a short-term study in which growth and reproduction of *Eisenia fetida* were examined, and the other a 120-week study on the accumulation of Cd and Pb in *E. fetida*. It is concluded that Pb and Cd are accumulated differently in *E. fetida*. In the short-term study, Pb seems to have a stimulating effect on growth and cocoon production at certain concentrations. In the long-term study, both Pb and Cd are concentrated to a larger extent in the posterior section of the body. The digestive tract and surrounding tissue tend to contain the major part of the total body burdens of both Pb and Cd.

The availability of reliable ecotoxicological test methods and risk-assessment procedures is paramount for the successful implementation of regulatory guidelines for the classification of polluted soils and chemicals [1]. The risk assessment of chemicals for ecosystems is in practice based on extrapolations from the results of laboratory tests [2]. The sensitivity of the organisms tested in the laboratory is an indicator of the sensitivities to be expected in the field. Various sensitive endpoints have been identified in earthworm ecotoxicology, such as bodyweight change and reproduction success [3].

More and more heavy metals are reaching the environment because of human activities. The toxic effects of these metals and their bioavailability are very complex [4,5]. Earthworms are affected in different ways, ranging from effects on growth and cocoon production to effects on sperm counts and sperm ultrastructure [6,7,8,9]. Because earthworms form an important component of the soil biota and are sensitive to contaminants, their protection may provide a margin of safety for other fauna [10]. This needs to be confirmed once comparative toxicological data are available.

The greatest threat from Pb in the environment is due to pollution from manmade organolead compounds, especially the use of tetra alkylleads in petrol [11]. Soils around smelting works may also contain high levels of Pb [7]. Earthworms are efficient accumulators of Pb and other metals in soil [12,4], depending on various edaphic factors [13].

Spurgeon and Hopkin [7] concluded that Pb in artificial soils increased mortality and reduced growth and cocoon production in *E. fetida*, with cocoon production being the most sensitive parameter. They have calculated EC50 values for Pb (effect concentrations predicted to cause 50% increase in mortality or 50% reduction in cocoon production and growth rate). They also estimated no-observed-effect concentration (NOEC) values for mortality, growth rate, and cocoon production. No actual body burdens of Pb were determined.

These authors [7] concluded from laboratory exposures of worms in artificial soil that cocoon production was the most sensitive parameter of toxicity for heavy metals, including Pb. Reinecke and Reinecke [8] exposed *E. fetida* to 0.1% lead nitrate in food (wet-weight basis) and found that the exposed worms grew as well as the controls. They also matured at a comparable rate to the control worms and produced just as many cocoons, causing the authors to conclude that these parameters may not be such sensitive endpoints for Pb toxicity as is generally expected [3]. However, Reinecke and Reinecke [8] did find a decrease in cocoon viability.

Both Cd and Pb are nonessential metals with wide distributions. They are micropollutants with a range of ecological and physiological effects [14]. Although Cd has no metabolic role, it is accumulated by many organisms. High concentrations of Cd and Pb can affect the population density, cocoon production and viability, and growth and sexual development of earthworms [2,4,7,15,16]. Earthworms also have been considered as waste decomposers and are used to this effect. They have also been considered as a protein source for animal feeds [17,18,19]. However, the presence of heavy metals in municipal waste and the fact that earthworms can accumulate heavy metals in their tissues [20,21] have reflected negatively on their possible use as a protein source in animal feeds [22].

Although *E. fetida* does not occur naturally in many soils, it is widely considered a suitable model species [14] and is also prescribed by the Organization for Economic Cooperation and Development (OECD) [23] for toxicity testing. It differs from other species in sensitivity to metals to a limited extent only and is considered an ideal laboratory animal because of the ease with which it can be reared and handled [24]. Species differences in accumulating heavy metals have been documented [25]. Ireland [12], Lee [26], and Van Gestel [27] have also reviewed aspects of heavy metal uptake by earthworms. Although bioaccumulation of heavy metals in earthworms is well documented, very little information is available on the relative locality of accumulation in the earthworm body, apart from the contribution by Morgan et al. [28].

Morgan and Morris [29] have studied the intracellular compartmentalization of Cd and Pb in earthworms. Cadmium is mainly accumulated in granules in the chloragogenous tissue surrounding the digestive tract. Andersen and Laursen [30] found that Pb is eventually accumulated in waste nodules (brown bodies) in the posterior segments of the earthworm body. By amputating the hindmost segments, the earthworm is thought to rid itself of the accumulated contaminant, although there is no clear evidence that this is true for Pb.

The aims of this study were first, to study the effects of a sublethal concentration of Pb on growth, saturation, and cocoon production and second, to confirm a previous finding that low concentrations of Pb affect cocoon viability without necessarily affecting cocoon production. Another aim was to determine the level of accumulation of Pb and Cd in laboratory cultures that have been exposed to these metals for an extended period of more than 2 y. This included the opportunity to obtain a picture of metal distribution in the earthworm body after such an exposure.

Material and methods

The following section discusses issues and methods.

Short-term evaluation of growth and reproduction after exposure to Pb

The following section discusses growth and reproduction of earthworms after exposure to Pb.

Exposure conditions

Cocoons of *E. fetida* were harvested from laboratory cultures, placed in distilled water in Petri dishes, and incubated in a climate chamber at 25 °C. The water was replaced every second day, and emerging hatchlings were placed in a manure substrate until they were 20 d old.

The substrate consisted of urine-free cattle manure that was previously sundried, ground, and sieved (particle size 500 to 1000 μm). The substrate was wetted with distilled water to a moisture content of 75%. Plastic containers (dimensions 180 H 110 H 85 mm) with a surface area of 140 cm², a perforated lid, and a fine gauze covering were used for all exposures. Each container received 400 g of the wetted substrate.

The worms were weighed on an analytical balance at the age of 20 d, and 20 worms were placed in each of 5 containers. A sixth container with 20 worms served as control. The worms in each experimental container were fed weekly with 60 g of fresh cattle manure. The 5 experimental containers received manure into which 0.12 g lead nitrate was mixed to provide a concentration of 0.2% or 2000 mg/kg.

Every 14 d over a period of 56 d until the worms were 76 d old, they were handsorted and weighed individually before being replaced in the containers. The development of clitellums was noted, and cocoons were collected. The fresh food (into which the Pb salt was mixed) was spread over the surface of the substrate, and the worms were allowed to feed on demand. The substrate pH was measured every 2 weeks with a Crison micro pH 2001 (KCl electrode) by shaking up the substrate sample with distilled water of known pH (1:2) and measuring the pH directly; it remained between pH 6.8 and 7.3 throughout. The collected cocoons were incubated at 25 °C in repli-dishes containing distilled water [31].

Analysis of earthworms

Every 14 d from day 34 onward, three worms were removed from the control group and 1 worm was removed from each of 3 experimental containers and placed for 24 h at 25 °C to allow the depuration of their gut contents. Afterwards they were washed in distilled water, dried on paper towels, weighed, and frozen in sample vials. The samples from the 3 different experimental containers were pooled, defrosted, and digested for 2 h in 10 ml of 55% HNO_3 that was heated to 50 °C. The solution was cooled and 1 ml of 70% perchloric acid was added, and the mixture was kept at 150 °C for 1 h to complete the digestion process.

Cooled samples received 5 ml of milli-Q water filtered using 0.45 μm Millipore filters and were topped up with milli-Q water to a volume of 20 ml. Lead content was expressed as μg/g of worm material (wet mass). Freshly prepared 1, 5, and 10 mg/kg standards were used for calibration during analysis.

Statistical analysis

Statistical analysis was performed by applying t-tests [32] to evaluate differences in biomass at specific times between the control and experimental groups and also to evaluate differences in cocoon production and hatching success. Vertical bars in Figures 6-1 and 6-2 indicate SD of mean values for the total number of worms in the control and experimental groups. In order to meet the minimum weight criterion for digestion, the pooling of samples inevitably caused a loss of variance.

Longer-term exposure to Pb and Cd

Exposure conditions

Specimens of *E. fetida* were obtained from the laboratory stock that came from garden compost heaps originally. The worms were first placed in artificial soil substrate consisting of 10% peat moss, 20% kaolin, 69% silica sand, and 1% lime and wetted to 40% of the dry mass. Each of 3 plastic containers (30 cm × 30 cm × 30 cm) received 4000 cm^3 of this substrate and 200 pre-clitellate worms. Black plastic was used to cover the substrate surface to prevent moisture loss. The containers had perforated lids and were kept at 25 °C in an environmental chamber. Each container received fresh, urine-free cattle manure on a weekly basis as a food source. After 12 weeks, Pb was administered as lead nitrate to 1 container by thoroughly mixing it with the fresh manure to give a 0.1% (wet weight) concentration. The second container received cadmium sulfate mixed into the manure to give a 0.01% concentration. The third container served as control and received only the fresh manure. In all cases, the manure was placed on a weekly basis in a thin layer on the surface of each container.

After 90 weeks, 4 pre-clitellate worms were removed from each of the Cd and Pb exposure groups and 8 worms from the control group. The worms were placed in Petri dishes on moist filter paper at 25 °C for 24 h to allow for depuration of the gut contents. The worms were killed by immersion in heated water (60 °C), rinsed in distilled water, dried on a paper towel, and frozen individually in marked vials.

After 100 and 108 weeks, 8 more specimens were removed from the Pb and Cd exposure groups respectively. After allowing for the depuration of the gut contents, the worms were killed. The digestive tracts with the chloragogenous tissue were dissected out, weighed, and frozen. The remaining body tissue of each worm was also weighed and frozen. Samples had to be pooled to attain the minimum weight required for analysis.

To determine the longitudinal distribution of Pb and Cd in the body of the earthworms, 8 clitellate specimens were removed from the Pb and Cd exposure groups after 110 and 115 weeks respectively. After allowing gut depuration, the worms were killed and divided into 3 longitudinal sections consisting of an anterior section from the prostomium to the last clitellum segment, a medial section, and a posterior section comprising the last 20 segments. Similar sections of different worms had to be pooled to attain the minimum weight required for analysis.

After 120 weeks, 4 more worms were removed from each of the Pb and Cd groups and weighed individually along with substrate samples that were taken for digestion and analysis. Substrate samples were dried at 106 °C for 48 h and ground to a homogenous particle size before being weighed.

Analysis of earthworms

Worm samples weighing between 0.25 and 0.5 g were thawed and left overnight in 10 ml (55%) nitric acid. The samples were further treated and analyzed as mentioned above for the short-term exposure study.

Substrate samples were digested similarly, except 5 ml perchloric acid was used instead of 1 ml. Substrate samples were centrifuged at 3000 rpm for 5 minutes, and the supernatant were filtered. Substrate samples were made up to a volume of 25 ml.

Extractions were also performed to determine the concentrations of heavy metals in extractable forms considered to be bioavailable [33]. Soil samples of 1 g each were extracted with 30 ml 0.1 N hydrochloric acid and centrifuged at 110 rpm for 1 h. After the samples were filtered, they were analyzed as mentioned above.

All samples were stored in plastic containers to prevent adsorption of the metals to the container surface. A Varian AA-1275 Atomic Absorption Spectrophotometer was used to analyze all samples. Fresh Pb standards of 2, 10, and 15 mg/kg were used for calibration for worm samples. Lead standards of 10, 50, and 100 mg/kg were used to calibrate for the higher concentrations of the substrate samples.

Results

Short-term exposure to Pb

The following section discusses the results of short-term exposure of earthworms to Pb.

Growth and maturation

The biomass of the exposed worms differed significantly ($P < 0.001$) from the control group after approximately 3 weeks (Figure 6-1). The control group attained a mean mass of 0.45 ± 0.07 g per worm at the age of 76 d, while the experimental group attained a mean mass of 0.76 ± 0.12 g per worm. The worms that were exposed to the low concentrations of lead nitrate in their food grew noticeably better, but both groups showed a slight decline in mass at the age of 62 d when 100% maturation was attained. The maturation times of both groups showed no marked differences (Figure 6-2).

Cocoon production, hatchlings, and hatching success

Total cocoon production per worm was significantly ($P < 0.001$) higher in the exposed worms than in the control worms. The control group produced a mean of 3.75 cocoons per worm over the observation period, while the experimental group produced a mean of 8.81 cocoons per worm.

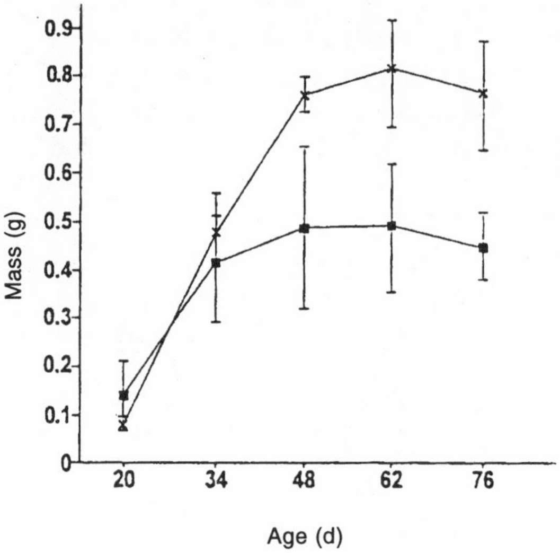

The hatching success of the cocoons produced by the worms exposed to low concentrations of lead nitrate in their food was significantly ($P < 0.001$) lower than that of the control worms. Cocoons produced by the control worms showed a hatching success of 71.4%, while the cocoons produced by the exposed worms showed only 52.6%.

Owing to the lower mean number of cocoons produced per worm, the mean number of hatchlings produced per worm over the production period was much lower in the control worms, in spite of the fact that the cocoons produced by these worms had a significantly higher hatching success than cocoons produced by the exposed worms.

Body burdens of Pb

The analysis of the Pb content of the exposed worms at different stages of the observation period showed that it reached a maximum of 20.28 ± 0.03 mg/kg when the worms were 62 d old. It decreased to 15.84 ± 0.11 mg/kg at the age of 76 d.

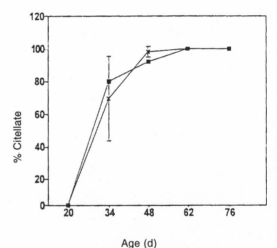

Figure 6-1   Mass of individual *E. fetida* exposed to control soil (■) and to soil with 2000 mg Pb/kg (x) for 76 days

Figure 6-2   True maturation of *E. fetida* exposed to control soil (■) and to soil with 2000 mg Pb/kg (x) for 76 days

Long-term exposure to Pb and Cd

The following section discusses the long-term effects of Pb and Cd on earthworms.

Body burdens of Pb and Cd

The worms collected after 90 weeks from the container in which they were exposed to food contaminated with 0.1% lead nitrate had a mean body burden of 35 ± 17 mg/kg. This was significantly higher ($P < 0.05$) than the Pb concentration in the control groups, which was 7.7 ± 7.3 mg/kg. The exposed worms had an even higher mean concentration of Pb of 108 ± 35 mg/kg after 120 weeks, which differed significantly ($P < 0.05$) from both the control and the 90-week sample.

Worms sampled after 90 weeks from the container in which they were exposed to Cd at 0.01% had a mean concentration of 106 ± 8.2 mg/kg in their bodies. This concentration was significantly higher ($P < 0.001$) than the concentration of Cd in the control worms, which was 4.0 ± 2.6 mg/kg. However the whole body concentrations for Cd did not differ significantly between samples taken at 90 and 120 weeks (Figure 6-3). After 120 weeks, the mean concentration of Cd was 165 ± 40.3 mg/kg.

Despite the fact that the Cd group was exposed to a much lower concentration than the Pb group, these worms had a significantly higher concentration of Cd after 90 weeks but the difference was not significant ($P < 0.1$) from worms sampled after 120 weeks.

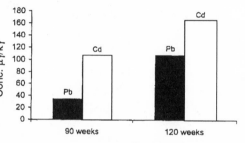

Figure 6-3   Concentration of PB and Cd in whole body of *E. fetida* after 90 and 120 weeks exposure to 1000 mg Pb/kg or 100 mg Cd/kg

Distribution of heavy metals in worms

The concentrations of Pb (wet-weight basis) in the digestive tract and the rest of the body tissues were 100 mg/kg and 58 mg/kg, respectively, based on the pooled samples (Figure 6-4). This indicated a concentration in the digestive tract that was about 2 times that of the rest of the body. In the case of Cd, also based on pooled samples, the digestive tract had a concentration of 95 mg/kg, and the rest of the body had 82 mg/kg.

The longitudinal distribution of the metals in the earthworm body differed between the anterior and posterior sections (Figure 6-5). Based on pooled samples from different worms, thereby masking possible variation between worms, the concentration of Cd in the posterior section was 211 mg/kg. This was much higher than the concentra-

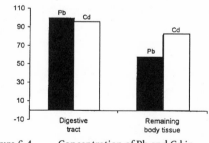

Figure 6-4   Concentration of Pb and Cd in body parts of *E. fetida* after 100

tions of 109 and 117 mg/kg (Figure 6-5) for the medial and anterior sections respectively.

The concentration in the posterior section was 1.8 times greater than that obtained for the anterior section.

In the case of Pb, a very similar increased concentration was found toward the posterior section of the worm (Figure 6-5). The anterior section had a concentration of 17.8, the medial section 46.4, and the posterior section 73.1 mg/kg.

Heavy metals in the substrate Substrate samples taken after 120 weeks showed mean concentrations (wet-weight basis) of 5200 ± 450 mg/kg and 387 ± 28 mg/kg for Pb and Cd exposures, respectively. Soil extractions to determine metal bioavailability yielded concentrations of 3880 ± 140 mg/kg for Pb and 349 ± 13.7 mg/kg for Cd.

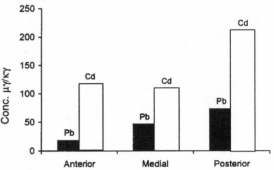

Figure 6-5   Concentration of Pb and Cd in body segments of *E. fetida* after 100 to 115 weeks exposure.

Concentration ratios/factors were calculated for Pb (0.02) and Cd (0.42) based on the method of acid digestion. This compared favorably with previous findings [8] of 0.05 and 0.34 respectively. Also, field samples of earthworms have shown very similar ratios for Cd [34]. The concentration ratio for Cd was almost 20 × that for Pb.

Discussion

Short-term exposure to Pb

The growth obtained after exposing *E. fetida* to low levels of Pb suggested that the exposed worms fed actively on the contaminated substrate. The palatability of the food did not seem to be affected by the presence of Pb at this low concentration of 0.2%. Previous observations [8] found no significant difference between the growth of *E. fetida* in controls and in experimental conditions of exposure to food with a concentration of 0.01%. Reinecke et al. [16], however, concluded that Pb at low concentrations could affect initial worm growth positively. Since the concentrations of metals in organisms can vary because of changes in the rates of uptake and depuration, which depend on factors such as pH and other soil characteristics, different members of a population in a heterogeneous substrate may exhibit different growth responses. The actual variation in body burdens of individual worms was masked by the pooling of samples, which was necessary to meet the weight criteria for analysis.

Growth is normally considered to be a sensitive endpoint of toxicity. Based on the present findings, the suitability of growth as a toxicity endpoint for low concentrations of Pb can be questioned. It is known that some toxicants inhibit growth at high exposure levels but stimulate growth at low levels [35]. Lead seems to be an example of a substance causing this hormetic effect. However, it is not clear what role the anion component of the metal salt had. The hormetic response may be an indirect response to nitrate.

The exposed worms also showed a higher cocoon production than the control worms. This is in agreement with the finding of Bengtsson et al. [36], who have shown that low metal concentrations can stimulate cocoon production. Spurgeon and Hopkin [7] on the other hand concluded from laboratory exposures of worms in artificial soil that cocoon production was the most sensitive parameter of toxicity for various metals, including Pb. Their EC50 value for cocoons of 1630 mg/kg and the estimated NOEC value of 608 mg/kg in soil are much lower than the initial exposure concentrations of 2000 mg/kg in food used in the present study. Differences in experimental design may explain the differences in findings. However, even cocoon production may not be a reliable toxicity endpoint for Pb at certain low, sublethal concentrations. The hatching success of cocoons produced by exposed worms was much lower than those for control worms in the present study; this finding is supported by 2 other studies [8,16]. In contradiction to the finding of Kokta [3] for other chemicals, the indications are that cocoon viability may be an even more sensitive endpoint of Pb toxicity, at least at low exposure concentrations.

Long-term exposure to Pb and Cd
The mean Pb content of worms from a culture that had been exposed for 90 weeks was 35 ± 17 mg/kg. This corresponds well with the concentration of 41.8 mg/kg obtained by Reinecke and Reinecke [16] who exposed E. fetida to a similar concentration of Pb for only 8 weeks. The significantly higher mean concentration of 108 ± 35 mg/kg in worms collected after 120 weeks may be indicative of continued long-term accumulation. It should be kept in mind that the sampled worms at both weeks 90 and 120 consisted of adult worms of varying age groups. Some of these worms could have hatched 7 or 8 weeks previously, while others could have been exposed to Pb for much longer. Metal accumulation can be influenced by age, diet, and possible genetic adaptations, and it is variable [37].

The results appear to indicate an increased concentration of Pb in the tissue of the digestive tract and the chloragogenous tissue surrounding it. Dendrobaena octaedra is known to accumulate heavy metals in the glycogen-rich chloragogenous tissue [38].

Both Cd and Pb appear to be associated with the posterior section of the worm, and this is probably associated with the brown bodies that form posteriorly in the coelomic fluid [39]. According to Morgan and Morris [29], Cd is stored in designated vesicular components, cadmosomes, where it is bound to a sulfur-rich component. Cadmium-binding metallothioneins are suspected to be responsible for increased tolerance [38,40].

Pb and Cd are accumulated differently in E. fetida. Pb seems to have a stimulating effect on growth and cocoon production at certain concentrations, and both Pb and Cd are concentrated to a larger extent in the posterior section of the body. The digestive tract and surrounding tissue tend to contain the major part of the total body burdens of both Pb and Cd.

Acknowledgments - The authors are indebted to the Foundation for Research Development for financial support. The authors also thank M.S. Maboeta and G. Gouws (research assistance) and P. Walters (A.A.-analysis)

References

1. Van Straalen NM, Løkke H. 1997. Ecological approaches to soil ecotoxicology. In: Van Straalen NM, Løkke H, editors. Ecological risk assessment of contaminants in soil. London: Chapman & Hall. p 3–21.

2. Van Gestel CAM. 1997. Scientific basis for extrapolating results from soil ecotoxicity tests to field conditions and the use of bioassays. In: Van Straalen NM, Løkke H, editors. Ecological risk assessment of contaminants in soil. London: Chapman & Hall. p 25.

3. Kokta C. 1992. Measuring effects of chemicals in the laboratory: Effect criteria and endpoints. In: Greig-Smith PW, Becker H, Edwards PJ, Heimbach F, editors. Ecotoxicology of earthworms. Hants UK: Intercept.

4. Morgan JE, Morgan AJ, Corp N. 1992. Assessing soil metal pollution with earthworms: indices derived from regression analyses. In: Greig-Smith PW, Becker H, Edwards PJ, Heimbach F, editors. Ecotoxicology of earthworms. Hants UK: Intercept. p 233–237.

5. Hopkin SP. 1993. Ecological implications of "95% protection levels" for metals in soil. *Oikos* 66:37–141.

6. Neuhauser EF, Malecki MR, Loehr RC. 1984. Growth and reproduction of the earthworm *Eisenia fetida* after exposure to sublethal concentrations of metals. *Pedobiolobia* 27:89–97.

7. Spurgeon DJ, Hopkin SP. 1995. Extrapolation of the laboratory-based OECD earthworm toxicity test to metal contaminated field sites. *Ecotoxicol* 4:190–205.

8. Reinecke AJ, Reinecke SA. 1996. The influence of heavy metals on the growth and reproduction of the compost worm *Eisenia fetida* (Oligochaeta). *Pedobiologia* 40:439–448.

9. Cikutovic MA, Fitzpatrick LG, Venablej BJ, Goven AJ. 1993. Sperm count in earthworms (*Lumbricus terrestris*) as a biomarker for environmental toxicology: Effects of cadmium and chlordane. *Environ Pollut* 81:123–125.

10. Greig-Smith PW. 1992. Recommendations of an international workshop on ecotoxicology of earthworms, Sheffield, UK (April 1991). In: Greig-Smith PW, Becker H, Edwards PJ, Heimbach F, editors. Ecotoxicology of earthworms. Hants UK: Intercept.

11. Hill SJ. 1992. Lead. In: Stoeppler M, editor. Hazardous metals in the environment, Vol. 12. Amsterdam, the Netherlands: Elsevier.

12. Ireland MP. 1983. Heavy metal uptake and tissue distribution in earthworms. In: Satchell JE, editor. Earthworm ecology from Darwin to vermiculture. London: Chapman & Hall. p 247–265.

13. Kiewiet AT, Ma W-C. 1991. Effect of pH and calcium on lead and cadmium uptake by earthworms in water. *Ecotoxicol Environ Safety* 21:32–37.

14. Spurgeon DJ, Hopkin SP. 1996. Effects of metal-contaminated soils on the growth, sexual development, and early cocoon production of the earthworm *Eisenia fetida*, with particular reference to zinc. *Ecotoxicol Environ Safety* 35:86–95.

15. Bengtsson G, Tranvik L. 1989. Critical metal concentrations for forest soil invertebrates. *Water Air Soil Pollut* 47:381–417.

16. Reinecke AJ, Maboeta MS, Reinecke SA. 1997. Stimulating effect of low lead concentrations on growth and cocoon production of *Eisenia fetida* (Oligochaeta). *S Afr J Zool* 32:72–75.

17. Schulz JR, Graff O. 1977. Bewerking von Regenwürmmehl aus *Eisenia foetida* (Savigny 1826) als Eiweissmittel - Landbou forschung. *Völkenrode* 27:216–218.

18. Sabine JR. 1978. The nutritive value of earthworm meal. In: Hartenstein R, editor. Utilization of soil organisms in sludge management. Syracuse: State University of New York. p 122–130.

19. Reinecke AJ, Alberts JN. 1994. Earthworm research in southern Africa since W. Michaelsen, with emphasis on the utilization of the earthworm (*Eisenia fetida*) as a protein source. *Mitt Hamb Zool Mus Inst* 89:23–36.

20. Ireland MP. 1975. Metal content of *Dendrobaena rubida* (Oligochaeta) in a base-metal mining area. *Oikos* 26:74–79.

21. Gish CO, Christensen RE. 1973. Cadmium, nickel, lead and zinc in earthworms from roadside soil. *Environ Sci Technol* 7:1060–1062.

22. Fleckenstein J, Graf O. 1983. Heavy metal uptake from the earthworm *Eisenia fetida* (Savigny 1826). *Animal Res Developm* 18:62–69.

23. [OECD] Organization for Economic Cooperation and Development. 1984. OECD guidelines for testing of chemicals: earthworm acute toxicity test. OECD Guideline No. 207. Paris, France: OECD.

24. Edwards PJ, Coulsen JM. 1992. Choice of earthworm species for laboratory tests. In: Greig-Smith PW, Becker H, Edwards PJ, Heimbach F, editors. Ecotoxicology of earthworms. Hants UK: Intercept. p 36–43.

25. Morgan JE, Morgan AJ. 1991. Differences in the accumulated metal concentrations in two epigeic earthworm species (*C. rubellus* and *D. rubidus*) living in contaminated soils. *Bull Environ Contam Toxicol* 47:296–301.

26. Lee KE. 1985. Earthworms: the ecology and relationships with soils and land use. Sydney, Australia: Academic.

27. Van Gestel CAM. 1992. The influence of soil characteristics on the toxicity of chemicals for earthworms: a review. In: Greig-Smith PW, Becker H, Edwards PJ, Heimbach F, editors. Ecotoxicology of earthworms. Hants UK: Intercept. p 44–54.

28. Morgan AJ, Morgan JE, Turner M, Winters C, Yarwood A. 1993. In: Dallinger R, Rainbow PS, editors. Metal relationships in earthworms. In: Ecotoxicology of metals in invertebrates. London UK: Lewis.

29. Morgan AJ, Morris B. 1982. The accumulation and intra-cellular compartementation of cadmium, lead, zinc and calcium in two earthworm species (*Dendrobaena rubida* and *Lumbricus rubellus*) living in highly contaminated soil. *Histochemistry* 75:269–285.

30. Andersen C, Laursen JJ. 1982. Distribution of heavy metals in *Lumbricus terrestris*, *Apporectodea longa* and *A. rosea* measured by atomic absorption and X-ray fluorescence spectrometry. *Pedobiologia* 24:347–56.

31. Reinecke AJ, Venter JM. 1985. Influence of dieldrin on the reproduction of the earthworm *Eisenia fetida* (Oligochaeta). *Biol Fertil Soils* 1:39–44.

32. Spatz C. 1993. Basic statistics. Pacific Grove CA: Brooks/Cole.

33. Hartenstein R, Neuhauser EF, Collier J. 1980. Accumulation of heavy metals in the earthworm *Eisenia fetida*. *J Environ Qual* 9(1):23–26.

34. Pascoe GA, Blanchet RJ, Linder G. 1996. Food chain analysis of exposures and risks to wildlife at a metals contaminated wetland. *Arch Environ Contam Toxicol* 30:306–318.

35. Stebbing ARD. 1982. Hormesis - the stimulating of growth by low levels of inhibitors. *Sci Tot Environ* 22:213–234.

36. Bengtsson G, Gunnarson T, Rundgren S. 1986. Effects of metal pollution on the earthworm *Dendrobaena rubida* (Sav) in acidified soil. *Water Air Soil Pollut* 28:361–383.

37. Pokarzhevskii AD, Van Straalen NM. A multi-element view of heavy metal biomagnification. *Appl Soil Ecol* 3:95–98.

38. Bengtsson G, Ek H, Rundgren S. 1992. Evolutionary response of earthworms to long term metal exposure. *Oikos* 63(2):289–297.

39. Morgan JE, Norey CG, Morgan AJ, Kay J. 1989. A comparison of the cadmium-binding proteins isolated from the posterior alimentary canal of the earthworms *Dendrodrilus rubidus* and *Lumbricus rubellus*. *Comp Biochem Physiol* 92C:15–21.

40. Spurgeon DJ, Hopkin SP, Jones DT. 1994. Effects of cadmium, copper, lead and zinc on growth reproduction and survival of the earthworm *Eisenia fetida* (Savigny): Assessing the environmental impact of point-source metal contamination in terrestrial systems. *Environ Pollut* 84:123–130.

Chapter 7

Use of an avoidance-response test to assess the toxicity of contaminated soils to earthworms

Gladys L. Stephenson, Aneel Kaushik, Narinder K. Kaushik, Keith R. Solomon, Tracey Steele, and Richard P. Scroggins

The presence of chemoreceptors in the prostomium and anterior segments of the earthworm and the distribution of tubercles within and among body segments render the earthworm highly sensitive to chemicals in its environment. This acute chemical sensitivity, coupled with its locomotory abilities, enables the earthworm to avoid unfavorable environments (i.e., soils containing substances that might be toxic and to which earthworms will respond). This chapter presents a novel approach to assessing the avoidance response of earthworms and compares the results of the avoidance test with those of the more traditional acute and chronic (reproduction) earthworm toxicity tests. The duration of the sublethal test (e.g., avoidance response) ranged from 7 to 72 h. The durations of the acute and chronic toxicity tests are substantially longer. If the results from a 24-h avoidance-response test could be used to predict the chronic toxicity of contaminated soils to earthworms, a very powerful tool would be added to the battery of tests currently available to risk assessors. With this in mind, a method was developed for assessing the avoidance response of earthworms to soils with contaminants.

A Plexiglas testing apparatus was designed and constructed that would allow earthworms to be presented with either 6 different dilutions of a contaminated soil, including a clean or uncontaminated control soil, or 6 different exposure concentrations of a chemical that had been added to the soil. Worms were placed into the central, soil-free chamber and, because of their phototactic response to light, they would move relatively quickly into the chambers with the soils. At the end of the exposure period, the location of the worms was recorded.

Preliminary test results suggest that this short-term, avoidance-response test has the potential to predict the responses of earthworms in more traditional tests used to assess the sublethal toxicity of contaminated soils (e.g., reproduction tests) and, as such, could prove useful as an initial screening test in ecological risk assessments.

Acute and chronic, single-species toxicity tests have been used traditionally to assess the toxicity of contaminants in soils to terrestrial invertebrates. Acute toxicity tests with earthworms can range in duration from 7 to 14 d [1,2], whereas the duration of chronic toxicity tests to assess effects of contaminants on reproduction and growth of earthworms might range from 28 to 55 d [3,4]. These test durations are sufficiently long, and they delay the decisions of risk managers or environmental engineers responsible for designing and implementing corrective action plans to address an environmental perturbation. Therefore, there is a need for rapid toxicity assessment of contaminants in soils.

With this in mind, we developed a relatively rapid, sublethal earthworm-avoidance test that has the potential to generate results predictive of the chronic responses of earthworms to contaminants in soils. If the results of a 24-h sublethal avoidance test could be used to predict these responses, a very powerful and useful tool would be added to the battery of tests currently available to risk assessors.

In 1995, the research phase of a 5-y program to develop a battery of terrestrial toxicity tests for use in the evaluation of contaminated sites and for site-specific risk assessment began at the University of Guelph. The research completed to date on the development of the avoidance-response test with earthworms is just 1 component of this research program. The objectives of this chapter are to:

1) describe the design of the test unit and the procedures for conducting an earthworm avoidance test,

2) briefly discuss a number of factors that might influence the outcome of a test,

3) discuss the use of 2 control soils that are used as diluents of the contaminated site soils, and

4) discuss the results of the behavior tests relative to results from the traditional acute toxicity tests and reproduction tests with earthworms.

Materials and methods

Test unit design and construction

The basic design of the test unit in this study was modified from a test chamber that was used to evaluate the feeding preferences of Tubificid worms [5] and from a similarly designed test unit used several years later to investigate substrate preferences of benthic invertebrates in streams (Kaushik, unpublished data). Three circular Plexiglas (6-mm gauge) test units (outer diameter 25.5 cm) were constructed by engineers in the Department of Engineering, University of Guelph, and each test unit was comprised of 6 truncated pie-shaped chambers (Figure 7-1B), which surrounded a central circular chamber (inner diameter 6.5 cm). Each of the 6 pie-shaped chambers was connected to the central circular chamber and to adjacent chambers by 3 arches, each 1.0 cm wide and 0.5 cm high, in series (Figure 7-1A). Each of the 6 chambers could hold approximately 300 g of soil dry weight or approximately 400 ml of wet soil. Test soils were placed into the test unit such that the worms could be presented with up to 6 different dilutions of a contaminated soil, including the appropriate control. Worms were added to the central, soil-free, circular chamber, 1 at a time or in a group of 5 or 10 individuals and, because of their phototactic response to light, they would move relatively quickly into the chambers with the soils. At the end of the exposure period, the chambers were isolated from each other by sliding thin (e.g., < 1-mm gauge) metal plates along the walls that partitioned the chambers (Figure 7-1B-5). With the metal plates in place, further movement of the worms into or out of the chambers was prevented, and the location of the worms within the test unit could be determined.

Figure 7-1   Test chamber without test soils (A) and with test soils (B) for assessing the avoidance response of earthworms. 1 Test unit lid; 2 Test unit wall; 3 Central chamber; 4 Holes for movement among compartments; 5 Metal plate for blocking movement of worms among compartments; 6 Test soil in 1 of 6 compartments.

Test organisms

Two test species were used to evaluate the avoidance behavior of earthworms exposed to contaminants in soils. The test species were the Red Wiggler or Compost Worm, *Eisenia fetida*, and the Common Dew Worm or Canadian Night Crawler, *Lumbricus terrestris*. *E. fetida* were purchased from either the Salmon River Worm Farm (Shannonville, Ontario) or the Worm Factory (Perth, Ontario). *L. terrestris* were purchased from Kingsway Sport (Guelph, Ontario). Both species were successfully maintained in the laboratory for relatively long periods of time in a mixture of bedding material consisting of either peat, garden soil, and the artificial control soil or peat, garden soil, and the field-collected reference soil. The compost worms were held in soil-filled, plastic containers with perforated, screened lids in a growth room with a constant temperature of 20 °C; they were fed cooked oatmeal on a regular basis (e.g., once or twice weekly, depending on the number of worms in the container). The night crawlers were maintained in soil-filled Styrofoam containers in an environmental chamber at temperatures ranging from 3 to 9 °C and were fed dried oatmeal.

Adults of these 2 species of earthworms were used for test method development for different reasons. Both species have a large number of chemoreceptors concentrated in the prostomium and anterior segments, and the distribution of epidermal tubercles and free nerve endings within and among body segments render them highly sensitive to chemicals in the environment [6]. Both species are highly mobile and tactile. This acute sensitivity to chemicals, coupled with a superior locomotory ability, enables the earthworms to avoid unfavorable environments such as those containing environmental contaminants. Both species belong to the family Lumbricidae of the Class Oligochaeta and are obligatorily amphimictic [7]. The species differ in that *E. fetida* is a gregarious, epigeic organism that selectively feeds on organic material distributed throughout the soil [6]. It tends to be found in gardens, compost, and manure piles associated with human activities [8]. *E. fetida* casts and copulates below ground. *L. terrestris* is widely distributed in Ontario and is found primarily in wheat, corn, or soybean fields; meadows; grasslands; pastures; and golf courses [9]. It is considered to be a nongregarious species and exhibits behavior that suggests it might be territorial in terms of food supply [10]. The organisms are classified as anecic organisms that selectively feed on organic material at the surface of soils but, because they are capable of burrowing deeply into the soil, they tend also to be terrigenous. Although *L. terrestris* will form middens at the soil surface, it primarily casts below ground, and it feeds and copulates at the soil surface.

The behavioral differences (e.g., foraging and feeding behavior and intraspecific interactions) exhibited by these 2 species of earthworm might influence the results of the avoidance test; therefore, separate tests were conducted with both species, and the influence of adding 1, 5, or 10 worms to a test unit was examined. Given the gregarious nature of *E. fetida*, it is possible that individuals might attract each other, forming clumps in certain habitats (e.g., chambers), while *L. terrestris* individuals might interact in such a way that the dominant individuals occupy the most favorable habitats and less dominant individuals are forced to occupy less suitable habitats.

Test soils

The condensate-contaminated soil originated from a site in Western Canada where natural gas was processed. "Condensate" is a term that describes a complex and variable mixture of petroleum hydrocarbons that results from the process of separating natural gas from sour gas. The site and reference soils were collected in 1994 and 1995, respectively, and placed into high-density, polyethylene, 20-l buckets and stored at 4 ± 2 °C. The reference soil was air dried to 20% soil moisture and passed through a 4.75-mm sieve [11]. The physicochemical characteristics of the control soils and the contaminated site soil are presented in Table 7-1.

Two control soils, selected for different reasons, were used in the tests. The artificial soil (AS), which is formulated in the laboratory from constituents, was similar in composition to a naturally occurring agricultural soil in Ontario referred to as "Fox Sand." The

Table 7-1   Physicochemical characteristics of the artificial and reference control soils and the condensate-contaminated site soil

Parameter	Artificial soil	Reference soil	Condensate-contaminated site soil[1]
Phosphorous (mg/kg)	5	7	0.5
Potassium (mg/kg)	20	626	96
Magnesium (mg/kg)	88	407	1900
Calcium (mg/L)	8 to 10	40	520
Sodium (mg/L)	NA	16	4075
Sodium adsorption ratio	0.53	0.6–0.15	16.6
Total carbon (%)	2.15	6.48	1.5
Total nitrogen (%)	0.02	0.62	0.11
Cation exchange capacity (Cmol+/kg)	13.6	41.8	NA
Soil texture	fine sandy loam	clay loam	loam
% sand	76.4	26.6	44
% silt	8.9	42.8	37
% clay	14.8	30.6	19
Organic matter (%)	3–5	11.6–12.2	2.8
Bulk density (g/cm^3)	0.98	0.83	NA[2]
pH (units)	6.4–6.6	6.1–6.3	8.1
Conductivity (mS/cm)	0.08	0.16	21.5
Percent saturation	30–40	30–40	36
Source	Formulated from constituents	Field-collected from Alberta	Field-collected from Alberta

[1] Golder Associates [12]
[2] Not available

artificial soil was also similar to that recommended in the Organization for Economic Cooperation and Development (OECD) earthworm acute toxicity test method [2]; the amount of sphagnum peat in the formulation was closer to 5% than to 10%. The field-collected reference soil (RS) was a clean, uncontaminated soil that was collected in the vicinity of the site soils but was free from the influence of the contamination [11]. The field-collected reference soil had physicochemical characteristics similar to that of the contaminated site soil with the exception of the contaminants.

The 2 control soils were used to dilute the contaminated site soil in order to achieve different exposure dilutions (i.e., % contamination). The physicochemical characteristics of the 2 control soils differed (Table 7-1), particularly with respect to distributions of cations, cation exchange capacity, texture, and organic matter content. The black chernozemic reference soil was a clay-loam soil with a relatively high cation exchange capacity and an organic matter content that was twice that of the artificial soil. The artificial soil was a gray-colored sandy soil with a relatively low cation-exchange capacity.

Preparation of test soils

The preparation of test soils required the following procedures to obtain the desired exposure dilutions. In sublethal, acute, or chronic toxicity tests with the condensate-contaminated soils, a batch of soil, sufficiently large for all the replicates, was proportionally diluted on a dry-weight basis with either the artificial or reference soil to achieve the desired dilution. The worms were presented with different exposure dilutions, ranging from 100 to 0% contamination (i.e., 100% contaminated site soil to 0% contaminated site soil or 100% control soil), depending on the objective of the test. The appropriate amount of water was added to the batch to standardize the test soil to approximately 70% of the water-holding capacity of the mixture, which would equate, on average, to between 30 and 40% moisture content.

The amount of water that was required to achieve this level differed for each dilution and therefore was predetermined by experimentation. The 2 soils (site and diluent) were combined with deionized water in a large stainless steel bowl and mixed with a hand-held, electric mixer until the mixture was visibly uniform in color and texture. The lumps that sometimes formed during this mixing procedure were eliminated by pressing them with the back of a stainless steel spoon. The batch of soil was then randomly allocated on the basis of wet mass to replicate polyethylene-lined test chambers within or among test units or to replicate test units per treatment for the acute and chronic toxicity tests.

Procedures for avoidance-response tests

The surfaces of the chambers in each test unit were lined with polyethylene at the beginning of each test. The test soils were prepared and assigned to the appropriate chamber in the test units. The worms were then placed 1 at a time, or in groups of 5 or 10, into the soil-free, central chamber. A light was placed above or below this central chamber. Because of their sensitivity to light, the worms quickly entered the compartments containing the soil. The worms entered the compartments at random. They usually refused to enter a compartment only when they encountered extremely lethal test soils. They would start to enter the soils in such compartments, then they would reverse direction rapidly, back out of these compartments, and enter another compartment with a more hospitable or less contaminated soil. It usually took 3 to 5 minutes from the time the worms were added to the chamber to the time when the central chamber was empty. The foil-lined lids were placed onto the test units, and the chambers were placed either into a room with a constant temperature of 20 °C for tests with *E. fetida* or into an environmental chamber of 15 to 18 °C for *L. terrestris*.

At the end of a test or a specific exposure duration, the metal plates were positioned to prevent further movement of worms among compartments, and the location of the worms was determined by simply scooping out the soil in each compartment and recording the number of worms present. If a worm was inadvertently sectioned by the metal plate as the plate was put into place, the worm was recorded as being in the compartment that contained the anterior segments of the worm. This happened on only 2 occasions.

Experiment #1: control soil preference

One of the factors that could potentially affect the results of an avoidance test was the nature of the control soil used to dilute the contaminated site soils in order to generate soils with different levels of contamination. A series of tests were performed to present *L. terrestris* with a choice between the AS and the RS by alternating the control soils among the compartments in each of 3 test units. The influence of different loading rates (i.e., the introduction of 1, 5, or 10 worms to each test unit) on the results of the test was also examined by recording the location of the worms in each test unit at the beginning of a test (e.g., t = 0) and after exposure durations of 24 and 72 h.

Experiment #2: autoclaved versus non-autoclaved control soils

A second experiment was conducted whereby *L. terrestris* individuals were presented with a choice between autoclaved and non-autoclaved control soils. Two kilograms of each of the control soils with soil moisture adjusted to 35% by weight were autoclaved (20 min at 121 °C) and about 270 g of the artificial soil dry weight were placed into 3 of the compartments of each replicate test unit, alternating with the non-autoclaved soil, so that 3 compartments had autoclaved soil and 3 had non-autoclaved soil. The earthworms were presented a choice between soils by placing the worms into the soil-free central chamber. The test was repeated with autoclaved and non-autoclaved reference soil.

Experiment #3: avoidance of condensate-contaminated site soils

A third experiment was conducted to assess the avoidance response of the 2 species of earthworm to different dilutions of a contaminated site soil. The objective of these tests was to expose individuals of both earthworm species to an acutely lethal dilution that approximated the 14-d LC50 estimate and to sublethal dilutions that were < the lowest-observed-adverse-effect level of contamination (LOAEL), where the effect was lethality as determined by the acute tests. There were 6 treatments, 0, 0.5, 1, 3, 6, and 12% condensate contamination and 10 worms per test unit. The treatments (i.e., the exposure dilutions) were randomly allocated to compartments within the test unit. The pattern of allocation was the same in all 3 test units for tests with *E. fetida* (Figure 7-5) and *L. terrestris* (Figure 7-6). Observations were made at 0, 24, and 72 h.

The data were analyzed using chi-square procedures [13] that determined whether the distribution of the worms at time 0, 24, or 72 h among the compartments was random.

Procedures for acute toxicity tests

L. terrestris and *E. fetida* were each exposed to 8 exposure dilutions (e.g., treatments) of the condensate-contaminated site soil diluted with both control soils. The treatments ranged from 0 to 100% condensate contamination. There were 4 exposure durations with time-independent replication of 8 treatments, each with 4 replicate test units per treatment. The test units had 270 g test soil dry weight and 5 and 3 test organisms each for *E. fetida* and *L. terrestris*, respectively. At the end of the appropriate exposure duration (e.g., d 4, 7, 14, 21), each replicate test unit for each treatment was examined by opening the jar and dumping the contents onto a tray. The test soils were spread and searched with a probe, and the number of live or dead worms was recorded.

The LC50 values were determined using Trimmed Spearman-Karber or Spearman-Karber methods [14]. The LOAEL and no-observed-adverse-effect level (NOAEL) for lethality were determined using analysis of variance (ANOVA) procedures [15]; treatment means significantly different from the controls at P < 0.05 were identified by a Dunnett's test [15].

Procedures for chronic (reproduction) tests

A test was conducted to examine the effects of the condensate-contaminated soils on reproduction in *E. fetida*. The factorial experiment consisted of 2 types of food (alfalfa and oatmeal), each with 5 exposure dilutions and a control, for a total of 6 treatments. Each treatment had 10 replicate test units. The earthworms used in the test were separated from the bedding material in the laboratory culture 24 h prior to use and placed into a plastic container lined with wet paper towels to allow the individuals to purge the contents of their gut. The test units consisted of a 500-ml glass jar with 270 g test soil dry weight and 2 adult clitellate *E. fetida*. Food (5-ml wet) was added approximately every 2 weeks to the test soils. The adults were removed on d 35, and the test units were processed on d 70. The measurement endpoints included number of juveniles produced per treatment and the wet biomass of non-purged juveniles per treatment.

Data were analyzed using a multivariate general linear model [15]. This hypothesis-testing approach used a model that automatically adjusts for unbalanced experimental data and tests for main effects (e.g., food type and concentration) and interaction effects (food × concentration). The concentration effects were determined for each food type using ANOVA procedures, and Dunnett's posthoc comparison of each treatment with the control was used to determine significant differences from the control treatment at P < 0.05.

Results and discussion

Avoidance-response tests

The following section details the avoidance-response tests.

Experiment #1: control soil preference

L. terrestris individuals were presented with a choice between the AS and the RS. The influence of different loading rates on the preference of one control soil over the other also was determined. The pooled results for the 3 test units are summarized in Figure 7-2 for each of the loading-rate scenarios.

When 1 worm was introduced to a test unit, it would enter a compartment with either AS or RS, but 24 h later the earthworm would be found in a compartment containing the RS (Figure 7-2A). When the loading rate was increased to 5 worms per test unit, with alternating AS and RS soils in each test unit, the earthworms also appeared to prefer the RS over the AS (Figure 7-2B). This preference was consistent with a loading rate of 10 worms per test unit and exposure durations of 24 and 72 h (Figure 7-2C). To summarize, comparative testing demonstrated that *L. terrestris* preferred the field-collected reference soil to that of the artificial soil. At the end of each test, regardless of the loading rates, more worms were located in the reference soil (e.g., 100, 73, and 80%, for Figure 7-2A, 7-2B, and 7-2C, respectively).

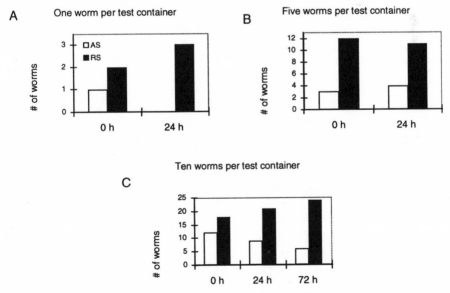

Figure 7-2   A comparison of different loading rates and different exposure durations in avoidance tests with *L. terrestris* exposed to control soils, an AS and a RS. Results from each of the 3 test units have been pooled for each scenario.

Experiment #2: preference of autoclaved versus non-autoclaved control soils

The microbial populations in soils are germane to the chemotactic responses of earthworms [16]. Therefore, *L. terrestris* individuals were presented with a choice between autoclaved and non-autoclaved control soils (Table 7-2). When presented with this choice, the worms tended to enter the compartments at random (t = 0), but by the end of 12 h the earthworms preferred the non-autoclaved soil to that of the autoclaved soil, regardless of the type of soil (i.e., AS or RS). The preference was also exhibited in tests with different loading rates and longer exposure durations (Table 7-2). We concur with the hypothesis that the natural assemblages of microbial organisms in the soil play an important role in chemotaxis of earthworms. To summarize, comparative testing demonstrated that *L. terrestris* preferred a non-autoclaved control soil over an autoclaved control soil and a field-collected reference control soil over an artificial soil formulated in the laboratory from commercially available constituents.

Experiment #3: avoidance of condensate-contaminated soil

Results indicated that at the beginning of the test (e.g., t = 0), the earthworms of both species entered the compartments at random. At the end of 24 h, the *E. fetida* had moved

Table 7-2   Relative preferences of *L. terrestris* for soils with (i.e., non-autoclaved soils) and without (autoclaved soils) microorganisms

Soil type	Loading rate	Time (h)	Nr. *L. terrestris* in autoclaved soil	Nr. *L. terrestris* in non-autoclaved soil
Artificial[1]	1 worm/test unit	t = 0	0	3
		t = 12	0	3
Reference[1]	1 worm/test unit	t = 0	2	1
		t = 12	0	3
Artificial[2]	5 worms/test unit	t = 0	6	9
		t = 72	1	14
Reference[2]	5 worms/test unit	t = 0	4	11
		t = 72	1	14

[1] N = 3
[2] N = 15

out of the compartments with the 2 highest levels of contamination. This suggests that the worms are actively avoiding these treatments (Figure 7-3). It took slightly longer for the *L. terrestris* to move out of the compartments with the highest levels of contamination, but the pattern of avoidance was apparent by 72 h (Figure 7-4).

To summarize, the sublethal behavioral test with avoidance response of earthworms as a measurement endpoint was more sensitive than the acute toxicity test (Table 7-3). Earthworms will avoid soils contaminated to a level equal to or below the NOAEL de-

Test design

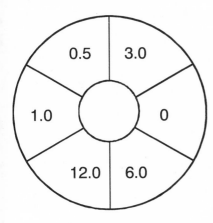

% Contamination

Test result

% condensate	t = 0	t = 24
0	9	14
0.5	2	4
1.0	4	9
3.0	7	3
6.0	4	0
12.0	4	0

Figure 7-3   Pooled results of avoidance-response test in which *E. fetida* (10 worms/test unit; 3 test units) were presented with different dilutions of a condensate-contaminated site soil diluted with the reference control soil for exposure durations of 0 and 24 h. Results are pooled among test units.

Test design

Test result

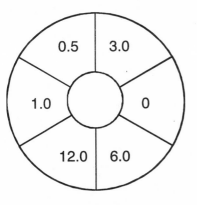

% Contamination

% compensate	t = 0	t = 24	t = 72
0	7	5	7
0.5	6	8	12
1.0	5	4	6
3.0	4	3	4
6.0	4	4	1
12.0	4	6	0

Figure 7-4   Results of avoidance-response test where *L. terrestris* (10 worms per test unit; 3 test units) were presented with different dilutions of a condensate-contaminated site soil diluted with the reference control soil for exposure periods of 0, 24, and 72 h. Results are pooled among test units.

rived from acute toxicity tests, where lethality was the measurement endpoint. This also suggests that using an exposure dilution factor of 0.5 in the acute tests might not be the most appropriate factor for this particular type of contamination problem.

Acute toxicity of condensate-contaminated soils

Results of the acute tests indicated that the dose-response curves for both species were characterized by relatively steep slopes and that the relationship was notably consistent for each exposure duration (Figures 7-5 and 7-6). The 14-d LC50 values for *L. terrestris* were comparable for the control soils (Table 7-3), as were the NOAEL and the LOAEL in which lethality was the endpoint. The toxicity estimates for *E. fetida* indicated that the species was slightly less sensitive to the condensate-contaminated site soils relative to *L. terrestris* (Table 7-3).

The toxicity estimates (e.g., 14-d NOAEL and LOAEL) derived from tests with the RS differed from those in tests with the AS because the dose-response curve was shifted slightly to the right on the x-axis (Figure 7-4), but the LC50 values were comparable (Table 7-3). In summary, the condensate-contaminated site soil was more toxic to *L. terrestris* than to *E. fetida*, and the nature of the control soil had no effect on the toxicity of the site soil to *L. terrestris*, but there were slight differences in tests with *E. fetida*.

Effects of condensate-contaminated site soils on reproduction of *E. fetida*

Two types of food (e.g., alfalfa and oatmeal) were used to determine if

Table 7-3   Toxicity endpoints[1] for 2 species of earthworms exposed to condensate-contaminated site soils diluted with the AS and RS control soils in 14-d acute toxicity tests

Species	Endpoint	AS diluent	RS diluent
L. terrestris	NOAEL	6	6
	LOAEL	12	12
	14-d LC50	10.5	11.8
	LC100	25	25
E. fetida	NOAEL	6	12
	LOAEL	8	25
	14-d LC50	25.6	21

[1] Values for toxicity endpoints are expressed as % condensate in the diluent control soil

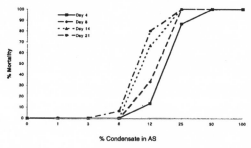

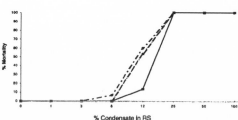

Figure 7-5   Mortality of *L. terrestris* exposed for 4, 8, 14, or 21 d to condensate-contaminated site soils diluted with the AS and RS control soils. Note that data for days 8 and 14 overlap in the lower plots.

there was a significant interaction between the type of food used in an earthworm reproduction test and the effects of the soil contaminants on earthworm reproduction. Previous studies have indicated that food type and quality can have a significant effect on reproduction parameters [4,17]. The results of the reproduction test with *E. fetida* (Figure 7-7) indicated that the condensate-contaminated soils adversely affected earthworm reproduction (e.g., number of juveniles produced) at the highest concentration in both the oatmeal ($P = 0.036$; F-ratio = 2.30) and the alfalfa groups ($P = 0.031$; F-ratio = 5.91). This level could have been predicted from the sublethal test results because the earthworms actively avoided the 12% contamination treatment. The mean wet biomass indicated that, although there were more worms produced in the alfalfa treatments with contamination, they were, on average, smaller than those found in the corresponding oatmeal treatments. The wet biomass of the juveniles produced by adults exposed to the contaminated-site soil and fed oatmeal was significantly reduced in the 2 highest concentrations (6 and 12% contamination) and in the 0.25% contamination treatment relative to that in the control treatment. The wet mass of the juveniles was also significantly reduced in the 2 highest treatments (6 and 12% contamination) for adults that were fed alfalfa. This level could have been predicted from the sublethal test results because the earthworms actively avoided both of these exposure

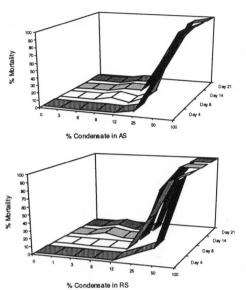

Figure 7-6   Mortality of *E. fetida* exposed for 4, 8, 14, or 21 d to condensate-contaminated site soils diluted with the AS and RS control soils. Note that data for days 14 and 21 overlap in the lower plots.

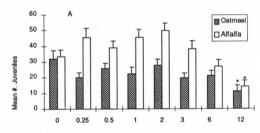

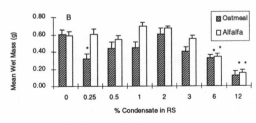

Figure 7-7   Mean number of juveniles (A) and mean wet mass of juveniles (B) produced per treatment in a reproduction test where *E. fetida* were exposed to condensate-contaminated soil diluted with the reference soils and fed alfalfa or oatmeal

levels (Figures 7-5 and 7-6). Juvenile wet mass is a more sensitive measurement endpoint than fecundity (e.g., number of juveniles produced).

There was a significant interaction between the 2 main effects of food and concentration; the progeny of the earthworms fed alfalfa, in the lower levels of contamination, produced significantly more young than those fed oatmeal ($P < 0.0001$; $F = 2.158$). However, this difference was not evident for the wet biomass measurements. In other words, although there were fewer juveniles produced in the oatmeal treatments, they were, on average, larger than the juveniles produced in the alfalfa treatments. This applied only for treatments ranging from 0.5 to 3% contamination.

Conclusions

A short-term behavioral toxicity test that measures the avoidance response of earthworms to sublethal levels of contaminants in soils was predictive of the results of the traditional chronic tests. This test may be useful as a rapid assessment tool for soils where contamination is suspected. The suspected contamination must be such that the contaminants, or presence of contaminants, can be detected by earthworms. Further research is required to examine other factors that might affect the results of the avoidance test and to determine if the predictive capability applies to other types of contamination problems. More data reflecting different types of site-soil contamination are required to impart greater certainty to the relative usefulness of the test.

Acknowledgments - The research on terrestrial test method development is a joint effort on the part of the federal government of Canada (Environment Canada), academia (Universities of Waterloo and Guelph), and industry (Canadian Association of Petroleum Producers and Ecological Services Group). We would like to thank the following individuals for technical assistance in the laboratory with the processing of the acute and chronic toxicity tests: E. Gunby, I.J.C. Middelraad, Y. Um, M. Whelly, L Diener, S. de Solla, N. Koper, C. MacDonald, A. Schulte-Hostedde, L. Beaton, D. Cunnington, K. Welstead, H. Brooks, N. Mackintosh, and C. Laing. We acknowledge the contributions of 3 anonymous reviewers.

References

1. [ASTM] American Society for Testing and Materials. 1997. Standard guide for conducting a laboratory soil toxicity test with lumbricid earthworm *Eisenia fetida*. Draft revision of the E1676-95 ASTM Standard. West Conshohocken PA: ASTM.

2. [OECD] Organization for Economic Cooperation and Development. 1984. OECD guidelines for testing of chemicals: earthworm acute toxicity test. Organization for Economic Cooperation and Development Guideline No. 207. Paris, France: OECD.

3. [ISO] International Standards Organization. 1997. Soil quality: effects of pollutants on earthworms (*Eisenia fetida*). Part 2: Determination of effects on reproduction. Draft 1996 ICS 13.080. 19 p.

4. Gibbs MH, Wicker LF, Stewart AJ. 1996. A method for assessing sublethal effects of contaminants in soils to the earthworm, *Eisenia fetida*. *Environ Toxicol Chem* 15:360–368.

5. Von Wachs B. 1967. Die oligocheaeten-fauna der fliessgewässer unter besonderer berhcksichtigung der beziehungen zwischen der tubificiden-besiedlung und demsubstrat. *Arch Hydrobiol* 63:310–386.

6. Wallwork JA. 1983. Earthworm biology. Studies in Biology No. 161. Southampton, England: Camelot Pr. 58 p.

7. Reynolds JW. 1977. The earthworms (Lumbricidae and Sparganophilidae) of Ontario. The Royal Ontario Museum Publications in Life Sciences. Toronto, Ontario, Canada: Hunter Rose. 141 p.

8. Edwards CA, Lofty JR. 1972. Biology of earthworms. London UK: Chapman and Hall. 283 p.

9. Tomlin AD. 1995. Behavior as a source of earthworm susceptibility to ecotoxicants. In: Greig-Smith PW, Becker H, Edwards PJ, Heimbach F, editors. Ecotoxicology of earthworms. Hants UK: Intercept. 269 p.

10. Tomlin AD, Protz R, Martin RR, McCabe DC, Lagace RJ. 1993. Relationships amongst organic matter content, heavy metal concentrations, earthworm activity, and soil microfabric on a sewage sludge disposal site. *Geoderma* 57:89–103.

11. Komex International, Ltd. 1995. Sampling and shipping of reference soil for the terrestrial soil toxicity method development project. Prepared for Environmental Technology Centre, Environment Canada, Ottawa, Canada.

12. Golder Associates Ltd. 1995. Testing of toxicity-based methods to develop site-specific clean-up objectives: Phase II toxicity identification and evaluation. Report No. 932-7204. 60 p. (appendices)

13. REPFIT. 1996. Software program for calculating exact chi-square probability estimates. Ashton Statistical Laboratory (William Mathes-Sears). Ontario, Canada: University of Guelph.

14. Stephan CE. 1989. Software to calculate LC50 values with confidence intervals using probit, moving averages, and Spearman-Karber procedures. Modified by RG Clements and MC Harrass. Duluth MN: U.S. Environmental Protection Agency

15. SYSTAT. 1994. The system for statistics. Version 5.0 for Windows. Evanston IL: SYSTAT, Inc.16. Satchell JE, Graff O. 1967. Progress in biology. Amsterdam, the Netherlands: North Holland.

17. Stephenson GL, Whelly M, Solomon KR, Scroggins RP. 1998. Effect of food type on reproduction of *E. fetida* in 2 control soils. *Environ Toxicol Chem* (submitted).

Chapter 8
Enchytraeid reproduction test

Jörg Römbke, Thomas Moser, and Thomas Knacker

According to international regulations (e.g., from the European Union [EU]), the environment should be protected against hazardous effects of chemicals. Following accepted rules, the ecotoxicological risk of these substances must be assessed based on test data. For the terrestrial environment, this assessment depends mainly on acute and long-term earthworm tests (*Eisenia fetida, E. andrei*) according to the Organization for Economic Cooperation and Development (OECD) [1] or Biologische Bundesanstalt für Land- und Forstwirtschaft (BBA) [2] guidelines. The 2 *Eisenia* species were selected as test organisms mainly because they are easy to breed and have, in comparison to other Lumbricidae, a short life cycle. Unfortunately, these earthworm species normally do not occur in soils but in organic matter such as compost. Therefore, for risk assessment purposes it is not fully scientifically justified to regard *E. fetida* and *E. andrei* as typical representatives of the soil community [3].

Additionally, parallel to testing strategies for the aquatic medium, there is a growing need for further tests with soil organisms [4]. Species from different trophic levels and different taxonomic, physiological, and/or functional groups should be selected and long-term endpoints should be measured in order to improve the risk assessment of chemicals in soil [5]. In such a tiered test strategy with increasing ecological relevance and complexity, a test battery with, e.g., Collembola, Carabidae, Staphylinidae, or Enchytraeidae (Potworms; Oligochaeta), could be used.

In this chapter, the new Enchytraeid Reproduction Test (ERT), with the potworm, *Enchytraeus albidus*, is presented. Recently, a draft protocol of this test was used as a basis for an international validation study [6]. The most important part of this validation process is the performance of a ring test that started in autumn 1996. Some general aspects of ring test performances will also be presented. Before describing the test and some preliminary results in detail, information on the ecology of these small and inconspicuous animals and their use in ecotoxicology is given. The following issues will be discussed:

1) biology of terrestrial Enchytraeidae,

2) history of various test approaches using enchytraeids,

3) description of the ERT,

4) presentation of selected test results,

5) comparison of the ERT with existing earthworm tests, and

6) status of the international ring test.

Biology of terrestrial Enchytraeidae

Enchytraeids have been known worldwide since the second half of the 19th century. However, the ecological role of these small (usually 1 to 20 mm), inconspicuous, and often whitish worms has hardly been studied. This lack of knowledge is at least partly due to the difficult taxonomy of the family in general. Altogether, several hundred species have been described from terrestrial habitats, but only a relatively small number of species (usually 5 to 25) occur at any given field site. The increasing number of soil biological studies as part of various acid rain projects during the 1970s and 1980s contributed significantly to the understanding of the role of Enchytraeids in the soil ecosystem.

Enchytraeids are important members of the soil biocenosis (at least in temperate regions) in many different habitats, especially where earthworms are rare [7]. At sites with low pH and a high content of organic matter, they can reach annual average densities of more than 100,000 individuals per m² (Table 8-1). Such high densities, e.g., in coniferous forests, are usually caused by few species, whereas the majority of the Enchytraeids prefer slightly acidic to alkaline soils. At many sites (including mull soils), they show compared to the earthworms, the highest biomass of all other invertebrate groups [8]. Additionally, a significant contribution of Enchytraeids to total soil respiration has been found [8]. Despite their small individual size, Enchytraeidae probably influence soil structure evolution by producing excrements, burrowing actively, and transporting of soil particles [7]. Even though some species might act as primary decomposers, other Enchytraeids are known to play a key role affecting the structure or function of the soil microflora [9,10]. Therefore, reduction of these oligochaete populations by chemicals may reduce the decomposition rate even if there is no direct effect on microorganisms.

Enchytraeids prefer to live close to or in soil pore water and consequently are directly exposed to dissolved chemicals. They have species-specific feeding preferences (e.g., fungi, bacteria, algae, protozoa, collembola excrements, and organic material like litter, often in combination with soil particles). Thus, their exposure to chemicals could occur from the surrounding soil water or from the food. From field studies it is known that enchytraeids react sensitively to a wide variety of anthropogenic stressors like heavy metals [11], petroleum [12], or pesticides [13]. However, the reaction of the whole coenosis is often very differentiated, since the various species are differently exposed and affected, depending on their preferred soil depth or food. For example, it seems that species of the genus *Enchytraeus* are more sensitive than species from other genera under field conditions [14].

History of test approaches using enchytraeids

Enchytraeids (especially *Enchytraeus albidus*) have been used in ecotoxicological laboratory tests for more than 30 y (Table 8-2). The first was a late-1960s test with *Enchytraeus albidus* in water to assess the effects of anthelminthics, but it was not until the early 1990s that several groups, working independently from each other, started to develop different test approaches for enchytraeids. In these test approaches, test substrates like water, agar, or soil were used (Table 8-3). However, the behavior of a chemical in agar or

Table 8-1   Abundance, biomass, and number of species of Enchytraeidae in Central Europe [7, 30]

Biotope	Average ± SD [Ind./m² × 10³]	Minimum	Maximum
Moor	45.5 ± 47.4	3	142
Meadow (pH < 5.5)	41.4 ± 20.2	31	74
Meadow (pH > 5.5)	20.6 ± 12.6	5	24
Conif. forest (Mor)	51.0 ± 47.6	10	134
Conif. forest (Raw Humus)	53.5 ± 55.34	4	134
Decid. forest (Mull)	17.3 ± 9.7	6	39
Decid. forest (Mor)	60.9 ± 56.1	5	146
Ruderal places	25.2 ± 21.2	6	56
Arable land	10.5 ± 7.2	2	30
Sewage sludge	≈ 2390	?	?

Biotope	Average ± S.D [g/m²]	Minimum	Maximum
Moor	0.94 ± 1.00	0.09	2.38
Meadow (pH < 5.5)	0.84 ± 0.51	0.53	1.76
Meadow (pH > 5.5)	0.59 ± 0.27	0.34	0.97
Conif. forest (Mor)	1.13 ± 0.53	0.50	1.94
Conif. forest (Raw Humus)	0.82 ± 1.25	0.17	2.69
Decid. forest (Mull)	0.59 ± 0.47	0.23	1.60
Decid. forest (Mor)	0.95 ± 0.62	0.40	1.64
Ruderal places	0.47 ± 0.39	0.13	0.90
Arable land	0.37 ± 0.21	0.11	0.64
Sewage sludge	≈ 206.8	?	?

Biotope	Average ± SD [# of species]	Minimum	Maximum
Moor	9.1 ± 7.3	1	23
Meadow (pH < 5.5)	8.3 ± 8.8	4	28
Meadow (pH > 5.5)	10.7 ± 8.2	3	24
Conif. forest (Mor)	9.5 ± 6.9	2	19
Conif. forest (Raw Humus)	4.3 ± 0.6	1	14
Decid. forest (Mull)	25.7 ± 7.4	18	36
Decid. forest (Mor)	16.1 ± 9.2	7	27
Ruderal places	8.0 ± 3.9	2	13
Arable land	9.1 ± 7.7	3	22
Sewage sludge	2	1	3

water is not comparable to that in soil, and thus the conditions for a worm are quite different compared to its normal habitat. Therefore, for risk assessment purposes only, soil tests are useful because otherwise the extrapolation to soil quality criteria is rather difficult [3]. It is useful to perform tests simultaneously in different substrates. In tests using agar as a substrate, the measurement endpoint of cocoon production can be used, which would be difficult to determine in soil. From an ecological point of view, assessing of the number of those animals that reproduce is more important than assessing the number

Table 8-2   Historical overview on enchytraeid tests

Year	Reference	Substrate, type, species, chemical
1968	Weuffen [31]	Water; acute; *E. albidus*; diff. Anthelminthika
1975	Kaufman [32]	Soil; acute; *E. albidus*; Phenol
1984	Heungens [33]	Litter; acute; *C. sphagnetorum*; pH, salts
1984	Huhta [34]	Soil, litter; acute; *C. sphagnetorum*; pH, nutrients
1989	Römbke [35]	Soil; prolonged; *E. albidus*; brass
1989	Römbke & Knacker [36]	Water; acute; *E. albidus* + 4 other species; 8 chemicals
1989	Westheide et al. [37]	Agar; prolonged; *E. crypticus*; 9 chemicals, liquid manure
1990	Rüther & Greven [18]	Agar; acute, accumulation; *E. buchholzi*; 4 metals
1991	Funke & Frank [38]	Soil; acute; natural cenosis; diff. volatile hydrocarbons
1991	Graefe [39]	Water; acute; *E. minutus, E. lacteus*; soil extracts
1991	Purschke et al. [40]	Agar; prolonged; *E. crypticus*; diff. pesticides
1991	Römbke [16]	Soil; prolonged; *E. albidus, E. crypticus*; diff. chemicals
1991	Westheide & Bethge-Beilfuß [41]	Agar (+ soil); prolonged; *E. crypticus*; 3 pesticides
1991	Westheide et al. [42]	Agar, soil; prolonged; *E. crypticus, E. minutus*; Benomyl
1992	Mothes-Wagner et al. [43]	Soil (microcosm); prolonged; natural cenosis; chemicals
1993	Born [44]	Soil (microcosm); prolonged; natural cenosis; Aldicarb
1993	Elzer [45]	Soil; prolonged; *E. buchholzi*; lead, cadmium
1994	Dirven-van Breemen et al. [17]	Soil; prolonged; *E. albidus, E. crypticus*; Zinc
1994	Notenboom & Posthuma [46]	Soil; prolonged; *E. albidus, E. crypticus*; 4 metals
1994	Römbke et al. [15]	Soil (microcosm); prolonged; natural cenosis; 2 pesticides
1994	Willuhn et al. [47]	Agar, water; acute; *E. buchholzi*; cadmium
1995	Achazi et al. [48]	Soil, agar, water; prolonged; *E. albidus, E. crypticus*; 3 chemicals + contaminated soils
1995	Christensen et al. [49]	Water; acute; *E. bigeminus*; 3 pesticides
1995	Heck [50]	Soil; prolonged; *E. albidus, E. crypticus*; PAKs + PCBs
1995	Kristufek et al. [51]	Agar; prolonged; *E. crypticus*; 7 chemicals
1995	Römbke & Federschmidt [14]	Soil; prolonged; *E. albidus, Fridericia sp.*; Carbendazim
1995	Rundgren & Augustsson [52]	Soil; prolonged; *C. sphagnetorum*; Cu, LAS, Dimethoate

Table 8-3   Advantages and disadvantages of different test substrates

Approach	Advantages	Disadvantages
Agar	Easy, various reproduction parameters measurable	Exposure not comparable to soil or water
Water	Easy, quick, inherent toxicity	Only 1 exposure pathway
Pore water	Same as water	Same as water
Artificial soil	Comparable, standardized test substrate	Different from natural soils e.g., organic content too high
Natural soil	Ecological reality; regional situations accessible	Highly variable; too many different possibilities
Sediment	Same as soil	Same as soil

of cocoons. Most of the tests listed in Table 8-1 were performed in the context of field validation or monitoring studies. The ERT, however, was developed as a standard test designed for the notification and registration of chemicals.

In the laboratory, a high number of stress factors have been tested that show an effect on enchytraeids, including heavy metals, environmental chemicals, pharmaceuticals, and at least 13 pesticides [e.g., 15,16,17], (Table 8-4). Enchytraeids are also useful for bioaccumulation studies, especially with heavy metals [18; Chapter 10, this volume]. Nearly all test methods using soil as a substrate can be modified for soil quality assessment (the ERT is already listed as a possible method for this purpose [19]). However, in this chapter, only the testing of single substances (e.g., as part of the registration process of pesticides) will be discussed.

Table 8-4   Stress factors tested with enchytraeids in the laboratory

Group	Type
Insecticides	Aldicarb, DBCP, dicofol, dimethoat, lindan, parathion, propoxur
Herbicides	2,4,5-T, amitrole, diuron
Fungicides	Benomyl, carbendazim, pentachlorophenol
Environmental chemicals	Benzothiaxole, chloroacetamide, chloroform, potassium dichromate, tetrachloroethene, trichloroethene, TPBS
Heavy metals	Cadmium, copper, lead, zinc
Pharmaceuticals	Various (» 50) anthelminthic acitive ingredients
Other stress factors	Ash, lime, nitrogen, sulphuric acid, soil cultivation

Despite the fact that many different species were examined, only species of the genus *Enchytraeus* are widely used in ecotoxicological tests. This is due to the simple culturing and handling of many but not all species of this genus. Species like *E. albidus, E. buchholzi, E. coronatus*, or *E. crypticus*, which can reach enormous densities in laboratory cultures, can be well fed with rolled oats [20]. All other enchytraeid species, probably with the exception of *Cognettia sphagnetorum* [21], are too difficult to breed in mass cultures, and therefore cannot be used as standard test organisms.

Within the genus *Enchytraeus*, various species have been proposed for testing purposes. Test results are available for *E. albidus* and for some species of the *E. minutus/E. buchholzi* complex. *E. albidus* has been promoted mainly because of its size (adult animals have a length of approximately 10 to 15 mm) (Figures 8-1 and 8-2) and the simple breeding method of this species, which has been kept in culture as fish food for decades [22]. Difficulties with the handling of small species like *E. crypticus* can be avoided when staining is applied instead of wet extraction for the purpose of counting juveniles. Still there is the question of ecological relevance: *E. albidus* is distributed worldwide in various biotopes, whereas the origin of *E. crypticus* is not known (the species was described from a compost heap). However, other recently extracted species of the *E. minutus* complex are abundant in German meadow soils and can be bred easily, too. In fact, including various species in the annex of the ERT guideline is planned, since the test performance is not affected by the choice of species. The same procedure was used in other recently finalized Organiza-

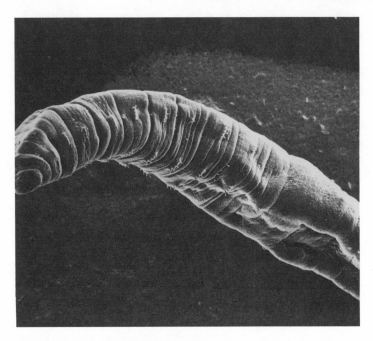

Figure 8-1   Scanning electron microscope picture of the anterior body of *Enchytraeus albidus* (Enchytraeidae, Oligochaeta); magnification factor: 25. The characteristic bundles with 4 setae are clearly visible.

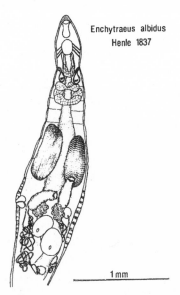

Figure 8-2   Drawing of internal organs of an adult *Enchytraeus albidus* (Enchytraeidae, Oligochaeta). Note the extraordinary long and coiled sperm duct in the lower half of the body cavity behind the clitellum.

tion for Economic Cooperation and Development (OECD) guidelines (e.g., the new guideline on fish bioaccumulation, which is currently discussed as a final draft).

Materials and methods

Description of the Enchytraeid reproduction test

The actual version of the ERT protocol, used in the recently finished international ring test, is described briefly. Further details of the test development and results of tests with various plant protection products and environmental chemicals (partly with slightly different versions) can be found in [16] or [17]. Listed below are the criteria details of the protocol:

- Guideline: Draft guideline according to OECD standards

- Test species: *Enchytraeus albidus; Enchytraeus sp.*; 10 adult worms (as identified by visible eggs in the clitellum region)

- Substrate: Artificial soil according to OECD [1]

- Parameter: Range-finding test: mortality, behavior
  Reproduction test: number of juveniles

- Performance: 4 replicates; room temperature (20 ± 2 °C); weekly feeding oats strewn onto the soil surface); removal of the adult worms after 3 weeks; counting of the juveniles hatched after 3 more weeks after extraction by a wet funnel method (however, staining with Bengal red seems to be more efficient and time-saving, as proposed by W. de Coen, University of Ghent, Belgium, and used, e.g., by RIVM Bilthoven [17])

- Concentration: Range-finding test: 0.1, 1, 10, 100, 1000 mg/kg

- Reproduction Test: No-observed-effects concentration (NOEC) approach: at least 5 concentrations (4 replicates)

  EC_x-approach: at least 12 concentrations (2 replicates)

- Reference substance: Carbendazim (formulation Derosal)

- Validity: In the control: after 3 weeks mortality < 10%; after 6 weeks number of juveniles > 25 per vessel

- Evaluation: LC50: Probit analysis; NOEC estimation: analysis of variance (ANOVA)

  EC_x: Regression analysis or similar methods

Results

Presentation of selected test results

Two tests were performed at ECT GmbH according to the ERT draft guideline in autumn of 1996. The fungicide benomyl (a.i. benomyl: 50%) was selected because it is metabolized quickly in soil to the proposed reference chemical carbendazim. There was nearly no mortality observed in the NOEC approach (mean survival rate: 97.3 ± 3.0%) as well as in the EC_x approach (97.6 ± 3.8%). Due to the low number of dead worms, no statistical evaluation was possible. The measured NOEC for reproduction is 4.04 mg/kg (multiple sequentially U-test after Bonferroni). For the EC_x approach, a dose-response relationship is not recognizable (Figure 8-3).

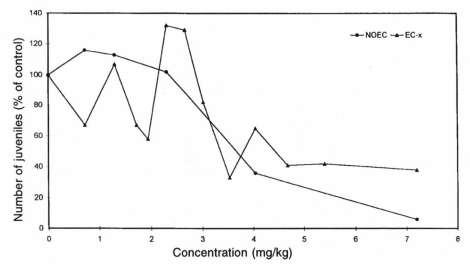

Figure 8-3   Effects of benomyl (a.i. benomyl: 50%) on the reproduction of *E. albidus* (ERT: Comparison of NOEC and EC_x approach). All data are given in mg/kg dry weight artificial soil.

In the second test, the fungicide Derosal (a.i. carbendazim: 36%) was used. The results of the 2 approaches show reasonable consistency. According to the validity criteria of the ERT draft guidelines, the test can be considered valid. The NOEC of the mortality data is 5.6 mg/kg (Williams-Test). The mean survival rate in the control and in the 4 lowest concentrations of the NOEC approach is 96.2 ± 2.5%; in the LOEC (10.0 mg/kg), the survival rate is 87.0 ± 0.5%. EC_x determination was impossible because a dose-response relationship was not observed and the average mortality was low (mean survival rate: 97.4 ± 3.8%). To calculate the NOEC of the reproduction data, the second concentration was excluded because of the increase in the number of juveniles (Figure 8-4). The determined NOEC for reproduction is 3.2 mg/kg (ANOVA). In a first statistical approach used to evaluate the EC_x, all concentrations producing more juveniles than the control had to be

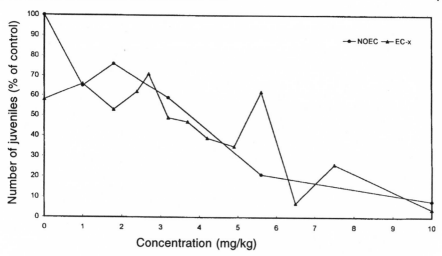

Figure 8-4   Effects of Derosal (a.i. carbendazim: 36%) on reproduction of *E. albidus* (ERT: Comparison of NOEC and EC_x approaches). All data are given in mg/kg dry weight artificial soil.

excluded. An increase in the measured endpoint (# of juveniles) at low concentrations, i.e., a hormesis effect, was originally described from plant tests but seems to be very common in many different botanical and zoological test systems. Therefore, only 8 concentrations were used for the determination of the EC10: 2.5 mg/kg (confidence limits [95%]: 1.0 to 3.4 mg/kg).

The test results for Derosal obtained so far by other participants are within the same order of magnitude. The average mortality of adult worms was low (mean survival rate: 90.7 ± 6.9%). Additionally, the mortality in the tested range of sublethal concentrations (1.0 to 10.0 mg/kg Derosal) was low (LC50 > 10.0 mg/kg). The NOEC and EC10 values varied between 1.0 and 6.3 mg/kg.

These results are within the range that could be expected from data in the literature (e.g., [14]). For example, LC50 values for enchytraeids in laboratory tests can vary between approximately 20 and 30 mg/kg and NOEC values between 1 and 10 mg/kg Derosal, depending on the test species and soil type. LC50 or NOEC values for Derosal from acute and long-term tests with earthworms according to OECD or BBA guidelines are in a similar order of magnitude when recalculated to active ingredient concentrations [23].

Discussion

Comparison of the ERT with existing earthworm tests
In order to assess the applicability of the existing earthworm test protocols and the ERT draft guideline, their pros and cons should be discussed. Concerning the sensitivity of the test species toward chemicals, there is no clear answer. It can be assumed that for each

chemical a different species is the most sensitive. For example, some mites are extremely affected by Cd, whereas earthworms seem to be very sensitive to Cu [24]. After completion of the ring test, a detailed comparison of Enchytraeid and earthworm sensitivity toward chemicals will be presented, which will show not only single test results for a specific chemical but also ranges of sensitivity for some chemicals.

The decision process for 1 of the 2 test methods using *Enchytraeus sp.* or *E. fetida* and *E. andrei* should be based on more than just sensitivity, and these criteria might include the following aspects:

Ecological relevance: Some soil-inhabiting species of the genus *Enchytraeus* (mainly some small species, but also *E. albidus)* are ecologically more relevant than *E. andrei* and *E. fetida*, since the latter occur almost exclusively in compost heaps [3]. A soil-inhabiting lumbricid species like *Aporrectodea caliginosa* would be, with regard to ecological relevance, a better choice of species for the earthworm test. However, tests performed with true soil-living earthworm species revealed breeding problems and long generation cycles [25]. It is not clear whether a fragmenting (i.e., genetically uniform) *Enchytraeus* species could be useful for test purposes.

Practicability: Many *Enchytraeus* species are very easy to handle and to breed. Moreover, their generation time is significantly shorter than for earthworms (test duration depending on the species: 4 to 6 weeks compared to 12 weeks). Due to the small amount of soil necessary for the test (20 g in 1 test vessel, compared to 500 g), only small amounts of contaminated soil are produced in the laboratory. Overall, the enchytraeid test performance is less expensive because of the short test period, the minimum of space required, and except for the start of the test, the small amount of work.

Data availability: Since the earthworm test guidelines were established in 1984 (acute) and 1994 (long-term or chronic), much experience and data have become available. Based on this experience and despite the fact that only a limited number of studies have been performed with enchytraeids so far, the standardization according to OECD rules for the enchytraeid reproduction test has reached a high level in a short period of time.

Extrapolation to semi-field and field conditions: The comparison of laboratory test results with semi-field and field conditions by performing long-term microcosm and field studies within tiered risk assessment schemes [15] can be achieved much easier with Enchytraeids than with earthworms because the same or very similar (same genus) species can be used at all tiers. For earthworms, at the moment, the species recommended for the acute test (Tier 1) and the long-term (reproduction) test (Tier 2) belong to different genera from different ecological groups (epigees, endogees, and aneciques). To solve this problem, an earthworm reproduction test is listed as a possibility for Tier 1 testing in a recent proposal for an ecotoxicological test strategy [5]. In the same proposal, the ERT is mentioned as an option (i.e., as part of a battery of standardized tests that are required only in special situations) for Tier 2. Further experience from the ongoing ring test and a recently started microcosm test should show whether the obvious advantages of the ERT in a tiered test approach in combination with a tiered risk-assessment scheme are important enough to eventually replace the existing earthworm laboratory tests.

Status of the international ring test

The aim of the ring test is to provide data to assess the suitability of the ERT draft guideline for the risk assessment of chemicals. Concerning legal requirements, the ERT is already included in the appendix of the European Union Technical Guidance Document 93/67/EEC + 1488/94 for the notification of existing and new environmental chemicals. In general, the development and validation of a new test guideline can be classified in the following steps [26] (the schedule for ERT is shown in parentheses):

1) test development (1988 to 1991);

2) prevalidation (informal studies in several labs; 1991 to 1995);

3) validation (currently performed and sponsored by German Federal Environment Agency [UBA] and European Chemical Bureau (ECB);

4) independent assessment of study results and protocols (planned for 1998); and

5) progression toward regulatory acceptance, e.g., finalization as an OECD guideline (planned for 1998).

About 25 institutions have agreed to participate in the ring test. Approximately 10 other institutions are interested in conducting at least some test runs or would like to contribute otherwise to the success of the ring test. Universities, governmental institutions, industry, and contract research laboratories joined the group. Geographically, Germany is represented very well, but nearly all other nations of the EU are also participating. Additionally, laboratories from the Czech Republic, Russia, Canada, the United States, Japan, and South Africa are interested in the ring test.

For the practical work, the fungicide carbendazim (formulation Derosal) and the industrial chemical 4-nitrophenol were selected. Some participants also tested potassium chloride (KCl), which was recently proposed as a reference chemical for earthworm tests [27]. The actual ring test is performed using 2 different statistical approaches, the NOEC- and the EC_x-test design (5 concentrations $\times$ 4 replicates and 12 concentrations $\times$ 2 replicates, respectively). For the NOEC approach, 28 test vessels are required, while for the EC_x approach, 30 test vessels are used; the number of control vessels is 8 and 6, respectively. The background for this laborious way of performing the ring test is related to the recent discussion about whether the well-known NOEC or a somehow defined and calculated EC_x (e.g., EC10, EC25) value is better suited for risk assessment purposes [25,28].

Based on the experiences gained in the ring test, the ERT draft guideline has been already improved, in particular by reduction in the amount of food in order to avoid molding by fungi and use of staining methods (e.g., Bengal red) to facilitate the counting of juveniles.

However, further improvements are necessary in order to reduce the variability of the test results. Besides the work aimed at upgrading the draft guidelines and completing the tests with the second test chemical, some participants agreed to do further investigations. For example, other species of the genus *Enchytraeus* and other test substrates, especially natural soils (e.g., LUFA standard soil 2.2) will be tested in order to improve the extrapolation of test data from laboratory to field conditions. Finally, all experience gained in this validation process will be published in a way comparable to those data from other recent ecotoxicological ring tests, e.g., with fish [29].

Acknowledgments - We thank the German Federal Environmental Agency (UBA) and the European Chemical Bureau (ECB) for funding and supporting the ERT ring test. Additionally, we wish to thank all participants in the ring test for their dedicated work, especially Dr. H-T. Ratte (RWTH Aachen) for advice on statistical issues and Ms. E. Leweke for improving the English. Acknowledgments are also due to the members of the task force: Mr. Didden (Agricultural University, Wageningen), Ms. Martin (UBA, Berlin), Mr. Ratte (RWTH, Aachen), Mr. Riego Sintes (ECB, Ispra), and Mr. Riepert (BBA, Berlin).

References

1. [OECD] Organization for Economic Cooperation and Development. 1984. Organization for Economic Cooperation and Development guidelines for testing of chemicals: earthworm acute toxicity test. OECD Guideline No. 207. Paris, France: OECD.

2. [BBA] Biologische Bundesanstalt für Land- und Forstwirtschaft. 1994. Richtlinienvorschlag für die Prüfung von Pflanzenschutzmitteln (Nr. VI, 2-2): Auswirkungen von Pflanzenschutzmitteln auf die Reproduktion und das Wachstum von *Eisenia fetida/Eisenia andrei*.

3. Van Gestel CAM, van Straalen NM. 1994. Ecotoxicological test systems for terrestrial invertebrates. In: Eijsackers H, Donker MH, Heimbach F, editors. Ecotoxicology of soil pollution. Chelsea MI: Lewis. p 205–228.

4. [VKI] Water Quality Institute. 1994. Discussion paper regarding guidance for terrestrial effects assessment. Report for the OECD. 63 p.

5. Römbke J, Bauer C, Marschner A. 1996. Hazard assessment of chemicals in soil— proposed ecotoxicological test strategy. *Environ Sci Pollut Res* 3(2):78–82.

6. Römbke J. 1996. Recent advantages of the Enchytraeid reproduction test. *Newsletter on Enchytraeidae* 5:73–81.

7. Didden WAM. 1993. Ecology of terrestrial Enchytraeidae. *Pedobiologia* 37:2–29.

8. Beck L. 1993. Zur Bedeutung der Bodentiere für den Stoffkreislauf in Wäldern. *Biologie in unserer Zeit* 23:286–294.

9. Williams BL, Griffiths BS. 1989. Enhanced nutrient mineralization and leaching from decomposing Sitka spruce litter. *Soil Biol Biochem* 21:183–188.

10. Förster B, Römbke J, Knacker T, Morgan E. 1994. Microcosm study of the interactions between microorganisms and enchytraeid worms in grassland soil and litter. *European J Soil Biol* 31:21–27.

11. Bengtsson G, Rundgren S. 1982. Population density and species number of enchytraeids in coniferous forest soils polluted by a brass mill. *Pedobiologia* 24:211–218.

12. Pirhonen R, Huhta V. 1984. Petroleum fractions in soil: Effects on populations of Nematoda, Enchytraeidae and Microarthropoda. *Soil Biol Biochem* 16:347–350.

13. Römbke J. 1994. Die Auswirkungen von Umweltchemikalien auf die Enchytraeidae (Oligochaeta) eines Moder-Buchenwalds. *Mitt Hamburg Zool Mus Inst 89 Ergbd* 2:187–197.

14. Römbke J, Federschmidt A. 1995. Effects of the fungicide Carbendazim on Enchytraeidae in laboratory and field tests. *Newsletter on Enchytraeidae* 4:79–96.

15. Römbke J, Knacker T, Förster B, Marcinkowski A. 1994. Comparison of effects of two pesticides on soil organisms in laboratory tests, microcosms and in the field. In: Eijsackers H, Donker MH, Heimbach F, editors. Ecotoxicology of soil pollution. Chelsea MI: Lewis. p 229–240.

16. Römbke J. 1991. Entwicklung eines Reproduktionstests an Bodenorganismen - Enchytraeen. Bericht im Auftrag des Umweltbundesamts, F+E-Vorhaben Nr. 106 03 051/01. Frankfurt, Germany: Battelle Institut Frankfurt.

17. Dirven-Van Breemen E, Baerselmann R, Notenboom J. 1994. Onderzoek naar de Geschiktheid van de Potwormsoorten *Enchytraeus albidus* en *Enchytraeus crypticus* (Oligochaeta, Annelida). Bodemecotoxicologisch Onderzoek. RIVM Rapport Nr. 719102025. 46 p.

18. Rüther U, Greven H. 1990. The effect of heavy metals on Enchytraeids. 1. Uptake from an artificial substrate and influence on food preference. *Acta Biol Benrodensis* 2:125–131.

19. Dott W. 1995. Bioassays for soils. DECHEMA-Fachgespräche Umweltschutz, 4. report, DECHEMA e.V. Frankfurt. 45 p.

20. Reichert A, Mothes-Wagner U, Seitz K-A. 1996. Ecohistological investigation of the feeding behaviour of the enchytraeid *Enchytraeus coronatus* (Annelida, Oligochaeta). *Pedobiologia* 40:118–133.

21. Sjögren M, Augustsson A, Rundgren S. 1995. Dispersal and fragmentation of the enchytraeid *Cognettia sphagnetorum* in metal polluted soil. *Pedobiologia* 39:207–218.

22. Geyer H. 1913. Rationelle Enchytraeenzucht. *Blätter zur Aquarien- und Terrarienkunde* 24:404–406.

23. Van Gestel K. 1991. Earthworms in ecotoxicology. [Ph.D. Dissertation.] Utrecht, the Netherlands: Rijksuniversiteit te Utrecht.

24. Van Straalen NM, Leeuwangh P, Stortelder PBM. 1994. Progressing limits for soil ecotoxicological risk assessment. In: Eijsackers H, Donker MH, Heimbach F, editors. Ecotoxicology of soil pollution. Chelsea MI: Lewis. p 397–410

25. Kula H. 1994. Species-specific sensitivity differences of earthworms to pesticides in laboratory tests. In: Eijsackers H, Donker MH, Heimbach F, editors. Ecotoxicology of soil pollution. Chelsea MI: Lewis. p 241–250.

26. Balls M et al. 1995. Practical aspects of the validation of toxicity test procedures. The Report and Recommendations of the ECVAM Workshop 5. *ATLA* 23:129–147.

27. Yeardley RB, Lazorchak JM, Pence MA 1995. Evaluation of alternative reference toxicants for use in the earthworm toxicity test. *Environ Toxicol Chem* 14:1189–1194.

28. Joermann G, Koepp H, Kula C. 1997. Meeting on statistics in ecotoxicology. *Reports from the Federal Biological Research Centre for Agriculture and Forestry* 17:1–34.

29. Mallett MJ, Grandy NJ, Lacey RF. 1997. Interlaboratory comparison of a method to evaluate the effects of chemicals on fish growth. *Environ Toxicol Chem* 16:528–533.

30. Römbke J, Beck L, Förster B, Fründ C-H, Horak F, Ruf A, Rosciczewski K, Scheurig M, Woas S. 1996. Fortführung der Literaturstudie: Bodenfauna und Umwelt. Bericht der ECT GmbH für die Landesanstalt für Umweltschutz Baden-Württemberg (Karlsruhe). Flörsheim am Main, Germany.

31. Weuffen W. 1968. Zusammenhänge zwischen chemischer Konstitution und keimwidriger Wirkung. 20. *Arch Exp Vet Med* 22:127–132.

32. Kaufman ES. 1975. Certain problems of phenol intoxication of *Enchytraeus albidus* from the view-point of stress. *Hydrobiol J* 11:44–46.

33. Heungens A. 1984. The influence of some acids, bases and salts on an enchytraeid population of a pine litter substrate. *Pedobiologia* 26:137–141.

34. Huhta V. 1984. Response of *Cognettia sphagnetorum* (Enchytraeidae) to manipulation of pH and nutrient status in coniferous forest soil. *Pedobiologia* 27:245–260.

35. Römbke J. 1989. *Enchytraeus albidus* (Enchytraeidae, Oligochaeta) as a test organism in terrestrial laboratory systems. *Arch Toxicol* 13:402–405.

36. Römbke J, Knacker T. 1989. Aquatic toxicity test for enchytraeids. *Hydrobiologia* 180:235–242.

37. Westheide W, Bethke-Beilfuss D, Hagens M, Brockmeyer V. 1989. Enchytraeiden als Testorganismen – Voraussetzungen für ein terrestrisches Testverfahren und Testergebnisse. *Verh Ges Ökol* 17:793–798.

38. Funke W, Frank H. 1991. Auswirkungen leichtflüchtiger Halogen-Kohlenwasserstoffe auf den Gesundheitszustand von Waldökosystemen. Ber. für das Bay. Staatsmin. f. Landesentwicklung und Umweltfragen Nr. 6487-953-127147.

39. Graefe U. 1991. Ein Enchytraeentest zur Bestimmung der Säure- und Metalltoxizität im Boden. *Mitt Dt Bodenkdl Ges* 66:487–490.

40. Purschke G, Hagens M, Westheide W. 1991. Ultrahistopathology of enchytraeid Oligochaetes (Annelida) after exposure to pesticides: a means of identification of sublethal effects? *Comp Biochem Phys* 100 C:119–122.

41. Westheide W, Bethge-Beilfuss D. 1991. The sublethal enchytraeid test system: Guideline and some results. In: Esser G, Overdieck D, editors. Modern ecology: basic and applied aspects. Amsterdam, the Netherlands: Elsevier. p 497–508.

42. Westheide W, Bethge-Beilfuss D, Gebbe J. 1991. Effects of benomyl on reproduction and population structure of enchytraeid Oligochaetes (Annelida)—sublethal tests on agar and soil. *Comp Biochem Physiol* 100 C:221–224.

43. Mothes-Wagner U, Reitze HK, Seitz KA. 1992. Terrestrial multispecies toxicity testing. 1. Description of the multispecies assemblage. *Chemosphere* 24:1653–1667.

44. Born H. 1993. Die Sukzession der Enchytraeen-Synusie (Annelida, Oligochaeta) eines Ruderal-Ökosystems unter natürlichen und anthropogenen Einflüssen. Dissertation Universität Bremen. 170 p.

45. Elzer U. 1993. Der Einfluss von Kupfer, Blei und Cadmium auf Mortalität, Wachstum und Reproduktionserfolg von *Enchytraeus buchholzi* (Oligochaeta, Annelida) unter Laborbedingungen. Diplomarbeit FU Berlin. 82 p.

46. Notenboom J, Posthuma L. 1994. Validatie Toxiciteitsgegevens en Risikogrenzen Bodem: Voortgangsrapportage 1993. RIVM-Rapportnr. 719102029. 70 p.

47. Willuhn J, Schmitt-Wrede HP, Greven H, Wunderlich F. 1994. cDNA cloning of a cadmium-inducible mRNA encoding a novel cysteine-rich, non-metallothionein 25-kda protein in an enchytraeid earthworm. *J Biol Chem* 269:24688–24691.

48. Achazi RK, Chroszcz G, Düker C, Henneken M, Rothe B, Schaub K, Steudel I. 1995. The effect of fluoranthene (Fla), benzo(a)pyrene (BaP) and cadmium (Cd) upon survival rate and life cycle parameters of two terrestrial annelids in laboratory test systems. *Newsletter on Enchytraeidae* 4:7–14.

49. Christensen B, Jensen CO. 1995. Toxicity of pesticides to Enchytraeus bigeminus. In: Danish National Environmental Research Institute, Løkke H. editor. Effects of pesticides on meso- and microfauna in soil. Nr.8. Danish Environmental Protection Agency.185 p.

50. Heck M. 1995. Enchytraeidenzönosen als Indikatoren belasteter Flächen in der Region Berlin. *Newsletter on Enchytraeidae* 4:69–77.

51. Kristufek V, Hallmann M, Westheide W, Schrempf H. 1995. Selection of various *Streptomyces* species by *Enchytraeus crypticus* (Oligochaeta). *Pedobiologia* 39:547–554.

52. Rundgren S, Augustsson AK. 1996. Sublethal toxicity test with the enchytraeid worm *Cognettia sphagnetorum*. In: Løkke H, van Gestel CAM, editors. SECOFASE Final Report. EU contract Nos. EV5V-CT92 u. ERB-CIPD-CT93-0059.

Processes important to earthworm ecotoxicology

Chapter 9

Combined effects of Cu, desiccation, and frost on the viability of earthworm cocoons*

Martin Holmstrup, Birgitte Friis Petersen, Martin Mørk Larsen

The effects of heavy-metal pollution on earthworms have been extensively studied, but no studies have examined how earthworms react if they are simultaneously exposed to metal pollution and climatic stress. This question has been addressed in a laboratory study where cocoons of *Aporrectodea caliginosa* and *Dendrobaena octaedra* were initially exposed to Cu in aqueous solutions of Cu chloride and thereafter exposed to realistic degrees of either desiccation or frost. Earthworm embryos absorbed Cu in amounts (up to approx. 200 mg/kg dry tissue weight) comparable to concentrations found in various tissues of earthworms from metal-polluted soils. Desiccation and Cu exposure in combination had synergistic effects on survival rates for both species. For example, at full saturation, the no-observed-effects concentration (NOEC), the highest tested concentration with no statistically significant effect, for Cu of *A. caliginosa* was 12 mg/L, whereas at 97% relative humidity, it was only 6 mg/L. Frost and Cu exposure in combination also showed synergistic effects in some experiments. No cocoons of *A. caliginosa* exposed to 20 mg Cu/L were viable after exposure to –3 °C, but at 0 °C, viability was as high as 95%. The same tendency was seen in *D. octaedra* but not as clearly as in *A. caliginosa*. A change of the environmental conditions (moisture, temperature) to increasing severity caused a shift in the statistically derived NOEC toward lower critical values of Cu. The involvement of combination effects in ecotoxicological tests could therefore improve risk assessment of soil-polluting compounds.

Annual and seasonal fluctuations in earthworm populations are natural and normal phenomena. The amplitude and duration of these fluctuations depend first of all on variations in soil humidity and temperature, and earthworms therefore from time to time encounter environmental bottlenecks such as severe drought in summer or extreme and sudden frosts in autumn or spring when no insulating snow cover is present. Such events may drastically reduce populations or even cause local extinction of species [1]. Heavy metals such as Cu have been shown to cause significant reductions in earthworm populations [2,3]. However, no studies have emphasized the combined effects of metal pollution and climatic stress on earthworms.

Earthworms possess physiological adaptations that enable them to populate a variety of habitats except for deserts and areas permanently covered with ice. Drought and frost are important limiting factors for earthworm populations. If soils become too dry or too cold, earthworms may move deeper in the soil or to microsites where conditions are more favorable [4]. However, earthworm cocoons, mostly present in the upper soil layers [5], depend on tolerance to drought and frost. Field and laboratory investigations

---

* This chapter is reprinted from *Environmental Toxicology and Chemistry* 17:897–901.

suggest that earthworm cocoons are much more tolerant of drought and frost than are postembryonic individuals [1,6–8]. Thus, earthworm populations may, under extreme conditions, survive mainly as cocoons. It is not known whether earthworm cocoons take up Cu from soil pore water, but it has been shown that some Cu taken up by adults is transferred to the cocoons they produce [9]. Consequently, it seems important to investigate viability of cocoons subjected to both climatic stresses and stresses from metal pollution. This question has been addressed in a laboratory study where cocoons of 2 common earthworm species, *Aporrectodea caliginosa* Savigny and *Dendrobaena octaedra* Savigny, were initially exposed to Cu in aqueous solutions and thereafter exposed to realistic degrees of either desiccation or frost.

Materials and methods

Biological material
Cocoons of *A. caliginosa* and *D. octaedra* were obtained from laboratory cultures as earlier described [10]. The cocoons were sampled by washing and sieving the culture soil through a 1-mm mesh after 3 weeks of cocoon production. The cocoons were incubated at 20 °C in petri dishes with moist filter paper until the embryos had developed to an early differentiated stage [11]. Only cocoons with developing embryos were used to ensure a high survival rate in controls, and the experiments were conducted with cocoons of the same stage of development.

Exposure to Cu
Cocoons were exposed to Cu in aqueous solutions of $CuCl_2 2H_2O$ with concentrations ranging from 0 to 20 mg Cu/L. For each concentration, about 90 cocoons were submerged in 25 ml solution in a petri dish and kept at 10 °C (desiccation experiments) or 2 °C (frost experiments) for 14 d. Keeping the cocoons at these temperatures was necessary in order to slow down the development of the embryos and to acclimatize them to cold before frost experiments [12]. For determination of Cu concentrations in embryo tissues, 3 cocoons were sampled from each concentration (10 °C; desiccation experiment) after their exposure. The cocoons were dried with filter paper, placed in a watch glass, and cut open with needles and fine tweezers. The embryos were carefully removed from the albuminous fluids and placed in an acid-rinsed glass vial, dried for 24 h at 80 °C, and stored in a freezer for later analysis.

Desiccation and frost experiments
After exposure to Cu solutions, the cocoons were gently dried with filter paper before they were used in desiccation or frost experiments. Exposure to desiccation was done by placing cocoons over either pure water or NaCl solutions (19.5, 43.5, or 61.0 g/L) in closed 150-ml plastic beakers, producing relative humidities (RH) of 100, 98.9, 97.5, and 96.5%, and exposing them at 20 °C for 14 d [7], a period corresponding to approximately 1/3 of the incubation time at 20 °C. This range of RH represents realistic humidity conditions in soil. For example, 98.9% RH corresponds to the soil water potential at the permanent wilting point for plants, pF 4.2 [13]. Cocoons of *A. caliginosa* have a deeper

distribution in the soil profile and are less drought and cold tolerant than *D. octaedra* cocoons [7,12]. Therefore, the RH levels and temperatures used were different for the 2 species. For each level of Cu and drought, 3 replicates with 10 to 15 cocoons each were set up. After drought exposure, the cocoons were placed in petri dishes with moist filter paper, incubated at 20 °C, and assessed for hatchability [10].

Cold tolerance was investigated by use of a previously described method in which cocoons were buried midway in soil in small petri dishes [12]. For *A. caliginosa*, the exposure temperatures were 0 (control), –3, and –6 °C, and for *D. octaedra* they were 0, –3, and –8 °C. Cocoons exposed to subzero temperatures were initially held at –3.0 ± 0.1 °C for 14 d and then cooled to the final exposure temperature (± 0.1 °C) and held there for 14 d. For each Cu concentration and exposure temperature, 3 replicates with 10 to 15 cocoons each were set up.

After thawing at 2 °C for 2 d, the cocoons were collected by washing and sieving the soil, incubated at 20 °C, and assessed for hatchability as previously described.

Determination of Cu concentration in embryo tissue

Apparatus
An MDS 81d microwave oven (CEM, Indian Trail, NC, USA) with 0 to 100% of full power (630 W) adjustable in 1% steps and a pressure sensor for 1 of 12 digestion vessels were used for digestion of embryos. The 100-ml lined digestion vessels of CEM (Teflon PFA-lined Ultem polyetherimide vessels) were used with 10 ml of deionized water with a 5-ml Teflon cup (Savillex) inserted, holding the actual sample and digestion acid. The energy program utilized was 5 min at 50% energy, 5 min at 70%, and 10 min at 100%, with a maximum pressure of 170 psi in the control vessel [14]. The determination of Cu was carried out using the stabilized temperature platform furnace (STPF) technique and Zeeman background correction on a Perkin-Elmer 5100pc graphite furnace (Norwalk, CT, USA) [15].

Reagents
Suprapure nitric acid was used for digestion of tissue (Merck, Darmstadt, Germany). The palladium and magnesium nitrate matrix modifier (Perkin-Elmer) was 10 g/L of $Pd(NO_3)$ and $Mg(NO_3)$, respectively. The modifier solution was prepared by mixing 3 ml Pd, 2 ml Mg, and 5 ml deionized water without further purification.

Standard material was made up from 100 mg/L multielement standards (VHG lab SM-20 and SM-50) by dilution with 1.5% nitric acid. Calibration against acidified standard solutions (5 to 40 µg Cu/L) was done, but standard additions were also used during method development.

Procedure
The embryos were weighed with 0.01 mg resolution into a Savillex cup (0.07 to 0.55 mg dry weight for *D. octaedra*; 0.29 to 4.66 mg for *A. caliginosa*). Five reagent blanks were

carried through the same procedure as the samples. Concentrated HNO_3 (Merck) was added (0.5 ml), and vial tops were loosely secured. The samples and blanks were carried through the oven program (10 vessels at a time), and after cooling to room temperature, 1.0 ml of deionized water was added, and the Savillex cup was weighed for determination of final weight of sample plus water.

Digested samples (1.5 ml) were transferred to acid-rinsed polyethylene cups and analyzed on graphite furnace with Perkin-Elmer pyrocoated, integrated platform graphite tubes using 30 µl sample, 5 µl 1.5% nitric acid, and 5 µl modifier. For samples with high Cu content, on-line dilution was made with 3 µl of sample. Some of the high results for Cu were checked using a 10- to 100-µg Cu/L standard curve and 10 µl of sample plus 5 µl modifier in fork platform graphite tubes. The detection limit for *A. caliginosa* was estimated to 15 mg/kg dry weight from 3 × the standard deviation of 7 embryos from uncontaminated culture. Average dry weight of the 7 embryos was 1.14 mg. As the dry weights of *D. octaedra* embryos were significantly lower (average 0.2 mg), the detection limit for these is estimated to be twice the detection limit of *A. caliginosa* (30 mg/kg).

Results

Copper had only a minor effect on the viability of *A. caliginosa* cocoons if they were not stressed by desiccation or frost; only the highest concentration (20 mg/L) drought experiment caused a reduced viability of about 30% (Figure 9-1 and Table 9-1). On the other hand, viability of *D. octaedra* cocoons was drastically reduced by Cu at concentrations above 2 mg/L (drought experiment) or 4 mg/L (cold experiment). Embryos that died from Cu exposure turned dark and apparently disintegrated in the cocoon. Embryo tissue concentration of Cu in *A. caliginosa* increased almost linearly from below the detection limit (control) to about 200 mg/kg at the highest exposure concentration, for those exposed at 10 °C (Figure 9-2). In *D. octaedra*, a similar increase in embryo tissue concentration was observed although the variation of the data in this species was somewhat larger, probably due to the lower mass of the embryos. At each exposure concentration, there was a tendency to higher internal concentrations in *D. octaedra* as compared to *A. caliginosa*.

With low concentrations of Cu (up to 6 mg/L), desiccation had no effect on viability of *A. caliginosa*, but combined effects of higher concentrations and 97.5% RH decreased viability much more than the sum of the separate effects of Cu and desiccation (Figure 9-1). Also *D. octaedra* was unaffected by desiccation in controls, but combined effects from Cu and desiccation caused reduced viability relative to cocoons at full saturation. Exposure in frozen soil decreased viability by 10 to 30% in both species when cocoons had not been exposed to Cu, but this effect was statistically significant only for *D. octaedra* at −8 °C (Figure 9-1). For *A. caliginosa* combined effects of Cu concentrations above 4 mg/L and frost reduced viability significantly and more than the added effects of the separate stressors. This tendency was also observed for *D. octaedra* at −3 °C but not at −8 °C (Figure 9-1).

The highest-tested concentration of Cu with no statistically significant effect (NOEC) as well as the lowest tested concentration causing a statistically significant effect (LOEC) were calculated by Dunnett's test. Addition of cold or desiccation stress to toxic stress from Cu generally caused a shift in NOEC and LOEC values toward lower Cu concentrations (Table 9-1). For example, NOEC for *A. caliginosa* at 20 °C decreased from 12 mg/L at full saturation to 6 mg/L at 97.5% RH.

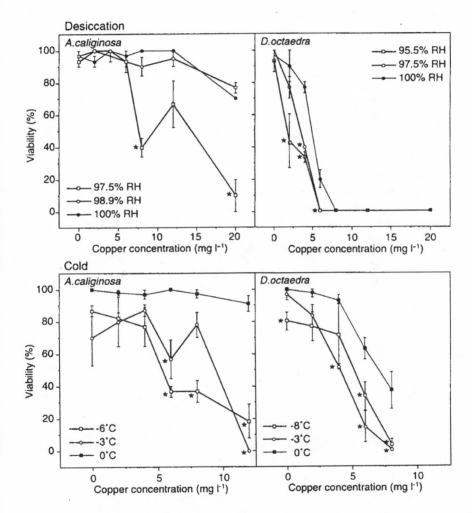

Figure 9-1   Combined effects of Cu exposure and desiccation on viability of earthworm cocoons and combined effects of Cu and frost. Legend: Each point represents the mean viability (± SEM) of 3 replicates with 10 to 15 cocoons each. Asterisks indicate a statistically significant (*P* < 0.05) reduction of the viability in comparison with cocoons not stressed by desiccation or frost, respectively.

Table 9-1   Highest tested concentration of Cu with no statistically significant
             effect (NOEC) and lowest tested concentration causing a statistically
             significant effect (LOEC)

Species	Treatment	NOEC (mg Cu/L)	LOEC mg Cu/L)
A. caliginosa	100% RH/20 °C (control)	12	20
	98.9% RH/20 °C	12	20
	97.5% RH/20 °C	6	6
	0 °C (moist soil) (control)	≥12	>12
	−3 °C (frozen soil)	8	12
	−6 °C (frozen soil)	8	12
D. octaedra	100% RH/20 °C (control)	2	4
	97.5% RH/20 °C	0	2
	96.5% RH/20 °C	0	2
	0 °C (moist soil) (control)	4	6
	−3 °C (frozen soil)	2	4
	−8 °C (frozen soil)	4	6

Discussion

The combined effects of climatic and chemical stressors seems to be an overlooked research area in terrestrial ecotoxicology. Two common species of earthworms were chosen as model organisms because of their importance in terrestrial ecosystems and because their cocoons are regularly subjected to harsh temperature and humidity conditions in the soil. In the present study, earthworm cocoons were exposed to aqueous solutions of $CuCl_2$ where the concentration of Cu was higher than would normally be found in the pore water of metal-polluted soils. Other studies show that the Cu concentration in soil pore water (sampled by centrifugation of moist soil) of natural soils containing from 100 to 800 mg Cu/kg dry soil may be in the range 0.2 to 4.4 mg Cu/L [16,17]. It could be argued that exposure via soil would have been more relevant, but it was decided to investigate more fundamental aspects and to use a simple and easily reproducible exposure method.

A large proportion of the Cu entering the cocoons (by diffusion) was probably adsorbed to the cocoon membrane and proteinaceous material of the albuminous fluids. But embryos also absorbed Cu in amounts (up to ~200 mg/kg; Figure 9-2) comparable to concentrations found in various tissues of earthworms from metal-polluted soils. Bengtsson et al. [2] measured from 100 to 280 mg/kg in tissues of field-collected D. octaedra and A. caliginosa and Bengtsson et al. [9] reported Cu concentrations up to 1600 mg/kg in muscle tissue of Dendrobaena rubida exposed to Cu-contaminated soil in laboratory studies. The same authors found 77 to 240 mg/kg (average 134 mg/kg) in D. rubida cocoons and 200 to 300 mg/kg in hatchlings. Streit [18] found that a Cu concentration higher than 240 mg/kg (whole animals) was lethal to adult Octolasion cyaneum, whereas Bengtsson et al. [9] reported only sublethal effects (i.e., reduced reproduction) in D. rubida with Cu concentrations from 370 to 3800 mg/kg (pharynx, muscles, seminal vesicles, and cerebral ganglion). There seem to be differences in sensitivity between species that were also shown in the present study, where *D. octaedra* was clearly more sensitive than *A. caliginosa*. Previous experiments (M. Holmstrup, unpublished study) revealed that exposure to NaCl solutions with concentrations up to at least 200 mg/L had no negative effect on viability of *A. caliginosa* cocoons that were subsequently exposed to the same levels of desiccation as used

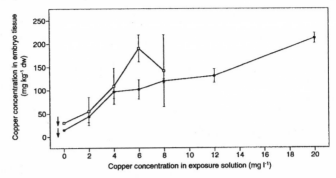

Figure 9-2   Mean (± SEM) Cu concentration of embryo tissues after 4-d exposure in aqueous solutions of $CuCl_2.2H_2O$ at 10 °C. Legend: Closed circles, *A. caliginosa*; open squares, *D. octaedra*. Arrows pointing downward indicate values that were below the detection limit.

in this study. It is therefore most likely that observed effects from $CuCl_2$ are due to Cu and not chloride.

When cocoons were subjected to desiccation without prior exposure to Cu, survival at any desiccation treatment was largely unaffected. These findings are in accordance with results reported by Holmstrup and Westh [7]. Exposure to subzero temperatures in frozen soil (without previous Cu exposure) caused a moderate reduction (10 to 30%) in viability of *A. caliginosa* and *D. octaedra* cocoons, a result that is similar to the viability reported by Holmstrup [12] for these species under the same experimental conditions (7 to 27% reduced viability).

Exposure to Cu, desiccation, or frost had almost no effects on survival in *A. caliginosa*. However, some combinations of sublethal stress from Cu and desiccation (or frost) resulted in an unexpected, high mortality. Such synergistic effects are seen when the combination effect is larger than the added effects of each single stressor as was seen in *A. caliginosa* cocoons subjected to 97.5% RH or frost (−3 and −6 °C). For *D. octaedra*, synergistic effects were evident in the desiccation treatments but negligible in the frost experiment, probably because the subzero temperatures used (−3 and −8°) were somewhat above the lower lethal temperature for this species [8].

Cu is an essential element despite being as inherently toxic as nonessential heavy metals. If animals are exposed to high concentrations, ionic Cu in the body fluids may accumulate in excess of a level that can be incorporated in normally occurring Cu proteins [19]. Cu toxicity seems in large part to be due to free Cu ions combining with proteins and altering their physiological functions. Moreover, it has been shown that excessive amounts of cupric ions can react with thiol groups of membrane proteins and thereby labilize cellular membranes [19,20]. Earthworm cocoons that are subjected to desiccation or frost dehydrate substantially due to evaporative water loss from the very permeable cocoon membrane [8]. Because of a low initial osmolality of cocoon and embryo fluids (~130 mOsm) and the fact that cocoons dehydrate until equilibrium between vapor pressure of cocoon fluids and ambient air is attained, they may lose as much as 80% of the initial water content when in equilibrium with air of 97.5% RH [7]. Similar levels of water loss are reached in frozen soil at temperatures below –3 °C [21]. Thus, the cells of embryos will become dehydrated and shrunken during desiccation and frost exposure. Upon rehydration, embryo cells will encounter a hypoosmotic shock, because the dehydrated cocoons (in dry or frozen soil) after rain or thaw will abruptly be surrounded by free water. Due to dehydration, the intracellular osmotic pressure is high, and the cells will quickly absorb water and swell. Under nontoxic conditions, earthworm embryos are obviously able to tolerate such a hypoosmotic shock. Because Cu has the potential to destroy or negatively affect membrane functioning, it is possible that tolerance to the hypoosmotic shock upon thawing after frost exposure or rehydration after desiccation is reduced (or lost) in embryos previously exposed to Cu. Another explanation (that does not rule out the previous) is that particular enzymes or processes important to dehydration survival are damaged by Cu.

The environmental bottlenecks artificially created in this study (desiccation and frost) may well be encountered in nature by the species investigated. Both species are common in temperate climatic zones where severe droughts and extreme winter frosts frequently occur. Though the method used to expose cocoons to Cu may not be very realistic, the Cu analysis of whole embryos suggests that internal concentrations were not much higher than they are in earthworms from Cu-polluted areas. It is not known if the Cu levels in field-collected cocoons (no data available) are high enough to decrease drought and frost tolerance as shown in this study. However, the potential for synergistic effects may be present under field conditions, not only for earthworms but also for other soil animals in polluted areas. This is probably most important in extreme climatic zones (e.g., the subarctic) where the narrowness and frequency of bottlenecks is highest. The ecological implications of this could be that pollution in such regions may play a much more dramatic role for the local extinction or geographical boundaries of species than in more benign climates. Along these lines, it should be pointed out that in many cases ecotoxicological risk assessment is based on laboratory studies in which the test organisms have optimal temperature and moisture conditions. As shown in Table 9–1, a change of the environmental conditions to increasing severity caused a shift in the statistically derived NOEC toward lower critical values of Cu. The involvement of combination effects in ecotoxicological tests could therefore improve risk assessment of soil-polluting compounds and play a role in the extrapolation from laboratory experiments to the field.

Acknowledgments - This research was performed at The Danish Center for Ecotoxicological Research, the Danish Environmental Research Program. We thank Karen K. Jacobsen for technical assistance and Paul Henning Krogh, Niels Elmegaard, Hans Løkke, and 3 anonymous reviewers for their useful comments on earlier versions of the manuscript.

References

1. Huhta V. 1980. Mortality in enchytraeid and lumbricid populations caused by hard frosts. In: Solokov VE, editor. Animal adaptations to winter conditions. Moscow, Russia: Nauka. p 141–145.

2. Bengtsson G, Nordström S, Rundgren S. 1983. Population density and tissue metal concentration of lumbricids in forest soils near a brass mill. *Environ Pollut A* 30:87–108.

3. Ma W. 1988. Toxicity of Cu to lumbricid earthworms in sandy agricultural soils amended with Cu-enriched organic waste materials. *Ecol Bull* 39:53–56.

4. Rundgren S. 1975. Vertical distribution of lumbricids in southern Sweden. *Oikos* 26:299–306.

5. Gerard BM. 1967. Factors affecting earthworms in pastures. *J Anim Ecol* 36:235–252.

6. Hopp H. 1947. The ecology of earthworms in cropland. *Proc Soil Sci Soc Am* 12:503–507.

7. Holmstrup M, Westh P. 1995. Effects of dehydration on water relations and survival of lumbricid earthworm egg capsules. *J Comp Physiol B* 165:377–383.

8. Holmstrup M, Zachariassen KE. 1996. Physiology of cold hardiness in earthworms. *Comp Biochem Physiol A* 115:91–101.

9. Bengtsson G, Gunnarsson T, Rundgren S. 1986. Effects of metal pollution on the earthworm *Dendrobaena rubida* (Sav.) in acidified soils. *Water Air Soil Pollut* 28:361–383.

10. Holmstrup M, Hansen BT, Nielsen A, Østergaard IK. 1990. Frost tolerance of lumbricid earthworm cocoons. *Pedobiologia* 34:361–366.

11. Holmstrup M. 1992. Cold hardiness strategy in cocoons of the lumbricid earthworm *Dendrobaena octaedra* (Savigny). *Comp Biochem Physiol A* 102:49–54.

12. Holmstrup M. 1994. Physiology of cold hardiness in cocoons of five earthworm taxa (Lumbricidae: Oligochaeta). *J Comp Physiol B* 164:222–228.

13. Hillel D. 1971. Soil and water. New York: Academic.

14. Bettinelli M, Baroni U, Pastorelli N. 1989. Microwave oven sample dissolution for the analysis of environmental and biological materials. *Anal Chim Acta* 225:159–174.

15. Perkin-Elmer. 1984. Analytical techniques for graphite furnace atomic absorption spectrometry. Überlingen, Germany. Perkin-Elmer Publication B332.

16. Marinussen MPJC, van der Zee SEATM, de Haan FAM, Bouwman LM, Hefting MM. 1997. Heavy metal (Cu, lead and zinc)accumulation and excretion by the earthworm, *Dendrobaena veneta. J Environ Qual* 26:278–284.

17. Pedersen MB, Temminghoff EJM, Marinussen MPJC, Elmegaard N, van Gestel CAM. 1997. Cu accumulation and fitness of *Folsomia candida* Willem in a Cu contaminated sandy soil as affected by pH and soil moisture. *Appl Soil Ecol* 155:135–146.

18. Streit B. 1984. Effects of high Cu concentrations on soil invertebrates (earthworms and oribatid mites): Experimental results and a model. *Oecologia* 64:382–388.

19. Scheinberg H. 1992. Cu. In: Merian E, editor. Metals and their compounds in the environment. Weinheim, Germany: VCH Verlagsgesellschaft. p 893–907.

20. Aaseth J, Norseth T. 1986. Cu. In: Friberg L, Nordberg GF, Vouk V, editors. Handbook on the toxicology of metals. 2nd edition. Amsterdam, the Netherlands: Elsevier. p 33–254.

21. Holmstrup M, Westh P. 1994. Dehydration of earthworm cocoons exposed to cold: A novel cold hardiness mechanism. *J Comp Physiol B* 164:312–315.

Chapter 10

Soil acidity as major determinant of zinc partitioning and zinc uptake in 2 Oligochaete worms exposed in contaminated field soils

Leo Posthuma, Jos Notenboom, Arthur C. De Groot and, Willie J.G.M. Peijnenburg

Soil characteristics are known to influence uptake of chemicals by soil organisms, but risk assessment procedures are not yet attuned to this. An approach to correct for bioavailability differences among soils in risk assessment procedures is presented. Data on Zn partitioning in soil and Zn uptake in 2 oligochaete worm species illustrate the approach. Solid/liquid partition coefficients (K_p) for Zn were determined for 20 moderately contaminated Dutch field soils. Zinc biota-to-soil factors (BSAFs) and rates (BSARs) were determined in the same soils in constant laboratory conditions for *Eisenia andrei* and *Enchytraeus crypticus*. Stepwise regression was used to derive empirical multivariate functions that describe K_p, BSAF, and BSAR as a function of soil characteristics. Soil acidity, measured as pH(CaCl$_2$) of the solid phase, appeared to be the primary descriptor explaining variation in K_p, BSAF, and BSAR. Aluminum oxyhydroxides and clay content contributed to variation in BSAF and BSAR only and were identified as secondary descriptors. These observations suggest that soil acidity is associated with both Zn partitioning in soil and uptake in oligochaete worms. This supports a risk assessment procedure for Zn in soil in which the bioavailable fraction is calculated based on total Zn concentrations, partition coefficients, and simple soil characteristics. The results on Zn in Oligochaete worms should be substantiated by future work in soil chemistry and biology, using other metals and organisms with different biological features (e.g., physiology, exposure route). The dynamics of the free metal ion activity in the soil solution will play a key role in the future. Future work on uptake quantification should concentrate on toxicokinetic modeling. Both are necessary to acknowledge that bioavailability research should encompass both supplies and demands of different organisms in a dynamic way. This goal can be reached only by close collaboration between soil chemists and soil biologists.

Ecotoxicological risk assessment and soil quality criteria for metals are based on easily measurable total metal concentrations in soil in many countries. A common risk assessment approach is to express risk as the ratio of the predicted environmental concentration (PEC) to the no-effect concentration (NEC), both usually expressed by total substrate concentrations. Such procedures are flawed, since total concentrations do not accurately predict the magnitude of effects [1]. Soil organisms do not assimilate metals from the total concentration in the soil. Rather, metals are taken up from a complex system of equilibria between a variety of metal species and soil constituents. Therefore, total metal concentrations in soil should be dropped as a basis for soil quality standards.

The free metal ion activity in soil solution has been proposed as an alternative to the total metal concentration; it takes the impact of soil characteristics upon bioavailability into account [2]. According to the Free-Ion-Activity Model (FIAM), exposure of organisms is mechanistically envisaged as a process of uptake of metal ions from the soil solution. The FIAM explicitly emphasizes the importance of both the ions already present in the pore water plus the labile sorbed metal quantity that is readily available for replenishment of assimilated ions. Following this principle, present research in the Netherlands strives to express bioavailability through the ion activities in soil solution and the labile sorbed fraction [3].

Application of the mechanistically based FIAM approach in risk assessment procedures for metals in soils is difficult. Although proper measurement techniques have emerged [4], they are not applicable in daily practice of soil protection. In anticipation of future developments in the FIAM approach, an empirical approach was followed in this study, one that is applicable in daily practice and that considers both abiotic soil processes and uptake in organisms. This empirical approach should 1) encompass state-of-art knowledge on exposure quantification, 2) be applicable in contemporary risk assessment procedures in the short term, and 3) should be in line with the FIAM approach.

"Bioavailability" is a composite concept rather than a clearly defined phenomenon [5]. Three principal processes can be recognized, namely
   1) the chemical supply, the physicochemical-driven distribution of metals over soil constituents, which was called "environmental availability,"
   2) the biological demand, the toxicokinetic behavior of metals during redistribution over soil and exposed organism, which was called "environmental bioavailability," and
   3) the toxicodynamic behavior of metals inside living tissues, eventually leading to intoxication at target sites, which was called "toxicological (bio)availability."

In this chapter, the outline is given of an empirical and pragmatic approach to correct for bioavailability differences among soils in risk assessment. The outline is illustrated with data on Zn partitioning in different soils (refer to #1 above) and uptake (refer to #2 above) in oligochaete worms. The toxicological (bio)availability (refer to #3 above) is not addressed, although an effect-oriented perspective is eventually needed, since risk concerns toxic effects rather than concentrations alone. Advantages and drawbacks are discussed. The approach is not limited to the bioavailability of Zn, but it is also applicable to other contaminant-soil-organism interactions. The ability to measure body residues is a constraint.

Justification of the approach

The working hypothesis was that interactions in soil modify the distribution of Zn species between soil phases (solid, liquid) and that this indirectly changes Zn uptake and effects on organisms (Figure 10-1). When Zn is brought in contact with soil, a series of complex equilibria is established that comprises the various possible reactions between Zn species and different soil constituents (precipitation, ion exchange, and sorption). Free-metal ion concentrations and replenishment fluxes depend on the sorption affinities

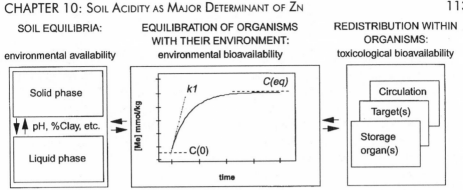

SOIL EQUILIBRIA: environmental availability

EQUILIBRATION OF ORGANISMS WITH THEIR ENVIRONMENT: environmental bioavailability

REDISTRIBUTION WITHIN ORGANISMS: toxicological bioavailability

K_p

Toxicokinetic parameters

Toxicodynamic parameters

Figure 10-1   Scheme of the 3 principal processes involved in the concept of bioavailability. Legend: Left: the soil system, which is related to the environmental availability, in this chapter summarized by the partition coefficient between solid and liquid phase. Soil characteristics are hypothesized to influence the equilibria. Middle: the interface between soil and toxicity is the environmental bioavailability, in this chapter summarized by toxicokinetic parameters. Chemicals can be taken up by the organism from all soil phases. Right: the redistribution of assimilated chemicals in the organisms' body, through the circulation system to the target sites of toxic action, storage organs, and so forth (not addressed in this chapter). C = concentration, k_1 = uptake rate constant, eq = equilibrium

and the proportional dominances of the solid-phase constituents and on liquid-phase characteristics. Following the above hypothesis, it should be possible to quantify the bio-availability of Zn as a function of soil characteristics. This function may differ between organisms, as demonstrated by the differences of body concentrations among different organisms from a single contaminated site (e.g., [6]).

As a gross approximation of the environmental availability of metals, the solid/liquid phase partition coefficient (K_p) might be sufficient to quantify in a static and simple way the net effect of all complex interactions. Note that the use of partition coefficients represents a static approach, thus neglecting the dynamic aspects of metal speciation in soils. Since only circumstantial evidence is available to quantify the kinetics of the various sorption and desorption processes in soil, the equilibrium approach was chosen mainly for pragmatic reasons. This is justified in view of the different time frames envisaged for soil processes that are likely to influence bioavailability (seconds to hours) and biological equilibration processes (for soil invertebrates, usually weeks to months).

The investigation considered whether 1) solid-liquid phase partition coefficients for Zn can be described as multivariate functions of simple soil characteristics and 2) such functions are predictive for Zn uptake in soil-dwelling species.

The basic premise is that the use of partition coefficients would be highly practicable but that their use in risk assessment systems should be validated by evidence collected from exposure experiments.

Materials and methods

Soil sampling, soil characteristics, and partition coefficients

Contaminated field soils, rather than artificially contaminated soils, were sampled to increase field relevance of the measurements. Twenty moderately metal-contaminated field soils were sampled throughout the Netherlands. The soils were sampled to represent the range of soil characteristics present in the field. For the field soils, it is reasonable to assume that Zn species in pore water are in equilibrium, except for the Zn fraction that is continuously but slowly incorporated further in the mineral lattice. Replenishment of Zn to the soil solution from this fraction is assumed to be negligible.

Soil characteristics and Zn partition coefficients were determined following the methods detailed in Janssen et al. [7]. In summary, soils were sampled in the field, and pore water was collected by centrifugation. All soils were characterized with respect to the soil constituents that may influence sorption and speciation in the solid and liquid phases (Table 10-1). The measured soil characteristics were chosen on the basis of shown or suspected influence of metal partitioning (expert judgment, but see also [3]). Total Zn concentrations were determined with flame-atomic absorption spectrophotometry (AAS) after nitric acid digestion of solids or directly in the pore water. Hydrogen fluoride (HF) digestion was not applied, since the Zn fraction incorporated in the mineral lattice will not become available for organisms within the time frame relevant for bioavailability.

Partition coefficients between soil solid and liquid phase (K_p) were calculated as follows:

$$K_P = \frac{C_S}{C_L} \quad \text{(L/kg)} \qquad \qquad \text{Equation 10-1,}$$

with K_p = partition coefficient, C = concentration (µmol/L or mmol/kg dry weight), S = solid phase, and L = liquid phase (pore water).

Exposure experiments

Zinc uptake was studied in 2 oligochaete species (*Eisenia andrei* and *Enchytraeus crypticus*) exposed in unmodified subsamples of the soils under laboratory conditions (20 °C). Animals were taken from mass cultures that have been reared in the laboratory for many generations on horse dung and agar.

Data for *E. andrei* were collected as detailed in Janssen et al. [8]. The fresh weight of the animals was determined before the assay. Subsequently, animals were exposed in the soils for 21 d. After recapture, animals voided their gut contents overnight on wet filter paper, and their fresh weight was determined again. Then they were frozen, lyophilized, their dry weight determined, and the body concentration of Zn measured.

The experimental procedure for *E. crypticus* was similar to that of *E. andrei*, except that animals were sampled at various intervals after the start of exposure until d 35.

Table 10-1   Ranges of variation of the soil characteristics of 20 moderately metal-contaminated Dutch field soils

Characteristics[1]	Min.	Variation Median	Max.	Unit
Solid phase				
Zn	0.08	2.8	47.6	mmol.kg^{-1}
pH(CaCl$_2$)	3.0	6.1	7.2	
OM	2.0	5.5	21.8	%
Clay	0.8	6.5	33.8	%
CEC	1.7	10.6	41.8	cmol.kg^{-1}
Fe$_{ox}$	3.6	88.6	297	mmol.kg^{-1}
Al$_{ox}$	3.5	26.1	128	mmol.kg^{-1}
Liquid phase				
Zn	0.9	3.1	117	µmol.L^{-1}
pH	3.8	6.4	7.9	
EC (20°C)	0.1	0.5	2.4	ms.cm^{-1}
DOC	1.7	5.4	12.4	mmol.L^{-1}
HCO$_3$	< 0.015	0.5	7.0	mmol.L^{-1}
CO$_3$-tot	0.02	0.7	7.0	mmol.L^{-1}
Cl	< 0.002	0.8	4.7	mmol.L^{-1}
NO$_3$	< 0.001	0.2	1.3	mmol.L^{-1}
PO$_4$	0.55	7.1	45	µmol.L^{-1}
SO$_4$	< 0.001	1.0	14.5	mmol.L^{-1}
K	0.04	0.2	1.3	mmol.L^{-1}
Na	0.1	0.8	4.6	mmol.L^{-1}
Ca	0.04	1.2	13.1	mmol.L^{-1}
Mg	0.01	0.3	2.6	mmol.L^{-1}
Fe	1.8	17.9	92	µmol.L^{-1}
Mn	0.2	0.8	15.6	µmol.L^{-1}
Al	3.4	32.2	204	µmol.L^{-1}

[1]Measured solid-phase characteristics were pH(CaCl$_2$), percentage organic matter (OM), percentage clay, cation exchange capacity (CEC), and amount of poorly crystallized and amorphous Al- and Fe oxyhydroxides (Al$_{ox}$ and Fe$_{ox}$). Porewater characteristics, determined after 2.5 µm filtration, were pH, conductivity, anions (Cl$^-$, NO$_3^-$, HCO$_3^-$ and CO$_3^{2-}$, PO$_4^{3-}$), cations (Ca^{2+}, Mg^{2+}, Na$^+$, K$^+$, Fe$^{2+/3+}$, Mn^{4+}, Al^{3+}), and dissolved organic carbon (DOC). Zn concentrations were determined with flame-AAS after nitric acid digestion or directly in the pore water. Details on analytical techniques are given by [7].

Biota-to-soil accumulation factors and rates
For *E. andrei*, BSAFs were calculated as follows:

$$BSAF_1 = \frac{C_W}{C_S}$$

Equation 10-2,

with t = exposure time, and C_W = concentration in worm (mmol/kg dry weight). Note that the BSAF may change over time until equilibrium is reached, although a 21-d equilibration period is plausible for this species [9].

For *E. crypticus*, a 1-compartment model was used to analyze the dynamic change of body concentration over time, following the generalized formula

$$\frac{dC_W}{dt} = k_1 C_S - k_2 C_W$$

Equation 10-3,

with k_1 = uptake rate constant ($kg_s.kg_w^{-1}.d^{-1}$, further expressed as d^{-1}), k_2 = excretion rate constant (d^{-1}), t = time (d). The model was fitted to the data in its integrated form:

$$C_W(t) = C_W(0) + \frac{k_1 C_S}{k_2}(1 - e^{-k_2 t})$$

Equation 10-4.

$C_W(0)$ was a measured value, 1.38 $mmol.kg_w^{-1}$, which was present in animals from the mass culture and was assumed to be an inert pool of constant magnitude. The model was fitted to the data for each soil using GraphPad Prism 2.0.

To judge bioavailability in the different soils, the focus was on 2 different aspects of the uptake curve, i.e., the equilibrium concentration and the initial uptake rate.

The body concentration in equilibrium is estimated to be the following:

$$C_W(eq) = C_W(0) + \frac{k_1 C_S}{k_2}$$

Equation 10-5,

with *eq* = equilibrium. After that, the BSAF in equilibrium was calculated using Equation 10-2. Note that C_W in Equation 10-2 was the measured Zn concentration in *E. andrei* after 21 d and that $C_W(eq)$ in *E. crypticus* is an estimated value. Note further that, in the modeling approach, $C_W(0)$ can be distinguished as a separate quantity within the total body concentration, whereas this is impossible when only a single measurement of the body concentration is made.

The initial uptake phase was quantified with the estimated parameter k_1.

To emphasize similarity in the further use of the BSAF and the initial uptake rate for analyzing bioavailability differences among soils, k_1 was referred to as the BSAR, i.e., a dynamic aspect of bioavailability concept.

Stepwise regression of K_p, BSAF, and BSAR on soil characteristics

Forward stepwise regression was used to identify the crucial soil characteristics that explain most of the observed variation of K_p, BSAF, and BSAR and to compose functions that predict the value of these 3 parameters as a function of soil characteristics. Moreover, the predictability of the parameters was also judged by using soil acidity (based on pH[CaCl$_2$]) only to assess a more practical application.

Stepwise regression yields functions of the general format

$$\log(Y) = a\log(A) + b\log(B) + ...c$$

Equation 10-6,

with $Y = K_p$, BSAF, or BSAR; A, B, and C = significant descriptors (soil characteristics, P < 0.05); and a, b, and c = coefficients. Significant descriptors are arranged in decreasing order of importance (A, B, and so forth). Nonsignificant descriptors were not incorporated in the formulas. The significance of the whole formulas was judged by R^2_{adj}. The robustness of the formulas was determined by randomly removing part of the data (up to 6 of the sampled soils) and recalculating the regression, yielding Q^2. The regression equation is robust if Q^2 (the internally cross-validated R^2_{adj}) is approximately similar to R^2_{adj}; if Q^2 is considerably smaller, then the equation is sensitive for outliers.

Eventually, a comparison was made between the formula obtained with K_p and with BSAF + BSAR. The similarity of descriptors for K_p and BSAF + BSAR suggests that abiotic partitioning on the one hand and biotic uptake on the other hand are governed by the same soil characteristics. Opposite signs for the coefficients of the significant descriptors suggest that an increase in sorption in relation to a soil characteristic is associated with decreased uptake in the exposed organism.

Some aspects of the present approach should particularly be noted, such as the following:

1) the bioavailability assessment bears directly on contaminated field soils rather than on extrapolation from artificially contaminated soils to field soils (see also [10]),

2) a large series of soil characteristics is evaluated with respect to their quantitative importance in modulating Zn partitioning and exposure of organisms, and

3) both equilibrium states and fluxes are addressed (at least in the biotic part of the approach), which is in line with the FIAM hypothesis.

Results

Partition coefficients of Zn

Solid- and liquid-phase characteristics and total soil and porewater Zn concentrations differed considerably among soils (Table 10-1). As an example, Figure 10-2 shows the variation of pH(CaCl$_2$) among the field soils, suggesting that all pH classes were evenly represented. A similar pattern was found for most other measured variables (Table 10-1),

which implies that the statistical relationships are unlikely to be sensitive for
outliers. Zinc sorption differed among
soils, as expected from the differences in
soil composition: the K_p of Zn ranged
from 6 to > 6000 L.kg^{-1}.

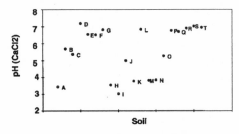

Figure 10-2   Soil acidity (measured as
pH[CaCl$_2$]) of the 20 field soils
Legend: A - T are soil codes.

The quantitative relationship that best
describes the K_p for Zn as a function of
solid phase characteristics is given in
Table 10-2. The pH(CaCl$_2$) was the only
soil characteristic significantly associated
with Zn partitioning. The regression equation is highly significant and robust. This
means that porewater concentration of Zn in a moderately contaminated Dutch field soil
can be predicted with relatively high accuracy based on the total soil concentration of Zn
and the acidity of the soil. Note that this does not imply that the other soil parameters do
not (mechanistically) influence sorption, but that they are statistically insignificant to explain variation in Zn sorption in the study set.

Uptake of Zn

Eisenia andrei
In one soil, all animals died; this was probably due to a biological agent, since the effect
was absent after freezing the soil in a separate observation. Hence, only 19 cases were
used in the statistical analyses. Body concentrations of Zn varied between 1.45 and 5.57
mmol/kg dry weight, and BSAFs ranged from 0.1 to 18. In comparison to the range of K_p
values, which differed by a factor of 1000, the range of variability of BSAFs is small (a
factor of 180).

The quantitative relationship that best describes the BSAF of Zn for *E. andrei* as a function of characteristics of the soil-solid phase is given in Table 10-2. The regression equation is highly significant and robust. The pH(CaCl$_2$), and to a lesser extent Al$_{ox}$ and clay
content, was identified as a solid-phase characteristic that contributed significantly to
the variation of BSAF.

The coefficients of pH(CaCl$_2$) in the formulas for log(K_p) and log(BSAF) have opposite
signs. Thus, an increase of pH(CaCl$_2$) is associated with an increased sorption (K_p) and a
decreased uptake of Zn (BSAF). The coefficients are, however, not exactly opposite,
which probably results from the additional soil factors that were found for BSAF only.
The second important soil characteristic for BSAF was Al$_{ox}$. Judged by its coefficient, an
increase of the Al$_{ox}$ concentration in the soil was associated with a reduction of the BSAF
in the study set. For clay, the factor with the smallest quantitative importance, the coefficients imply the opposite. This is interpreted as a consequence of the empirical approach and should not be identified as a causal relationship.

Table 10-2 Multivariate regression formulas describing the quantitative relationship among K_p, BSAF, and BSAR of Zn in *Eisenia andrei* or *Enchytraeus crypticus* as a function of soil solid-phase characteristics.

Species	Analysis	Regression equation obtained for Zn	Statistics
-	1, 2	$\log(K_{P,Zn}) = 0.61 pH_{(CaCl_2)} - 0.65$	$R^2_{adj} = 0.85, n=20, F=111.2, P<0.0001, Q^2=0.81$
E andrei	1	$\log(BCF_{Zn}) = -0.39 pH_{CaCl_2} - 1.06\log(Al_{OX}) + 0.73\log(Clay) + 3.04$	$R^2_{adj} = 0.87, n=19, F=41.3, P<0.0001, Q^2=0.83$
	2	$\log(BCF_{Zn}) = -0.39 pH_{CaCl_2} + 2.10$	$R^2_{adj} = 0.69, n=19, F=40.8, P<0.0001, Q^2=0.61$
E crypticus	1	$\log(BCF_{Zn}) = -0.52 pH_{CaCl_2} - 1.0\log(Al_{OX}) + 0.71 l\log(Clay) + 3.95$	$R^2_{adj} = 0.84, n=19, F=32.8, P<0.0001, Q^2=0.79$
	2	$\log(BCF_{Zn}) = -0.51 pH_{CaCl_2} + 3.04$	$R^2_{adj} = 0.75, n=19, F=54.5, P<0.0001, Q^2=0.61$
E crypticus	1	$\log(BCR_{Zn}) = -0.37 pH_{CaCl_2} - 0.70\log(Al_{OX}) + 2.18$	$R^2_{adj} = 0.74, n=20, F=27.7, P<0.0001, Q^2=0.68$
	2	$\log(BCR_{Zn}) = -0.45 pH_{CaCl_2} + 1.60$	$R^2_{adj} = 0.67, n=20, F=40.1, P<0.0001, Q^2=0.59$

Enchytraeus crypticus

Mortality during the exposure period was negligible; all soils were used in the statistical analyses. However, animals lost weight in all soils. Weight changes and the toxicokinetics of Zn are illustrated in Figure 10-3 for 2 soils. Soil A contained a total Zn concentration of 0.24 mmol Zn/kg and yielded a BSAF of 9.5. Soil B contained a total Zn concentration in soil of 47.5 mmol Zn/kg and yielded a BSAF of 0.11. In accordance with the regression equations, the low BSAF in soil B compared to soil A can be attributed to the differences in pH(CaCl$_2$), which were 3.1 and 6.3, respectively.

Significant weight loss was observed first after 14 d of exposure (with 1-way ANOVA and a posteriori multiple comparisons, with time as the independent variable). In most cases, weight loss coincided with discontinuities in the uptake pattern of Zn. This suggests that

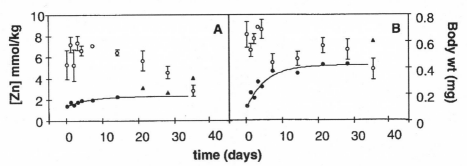

Figure 10-3   Examples of Zn accumulation (observations and fitted curves) and fresh-weight change of *Enchytraeus crypticus* exposed in 2 moderately contaminated Dutch field soils Legend: Please note that the Y-axes (left: body concentration; right: body weight) apply to both figures. Black dots are Zn body concentrations of worms without significant weight loss (individual measurements or average of 2 individuals), black triangles are Zn body concentrations measured in animals showing weight loss. White dots are bodyweight data, bars indicate standard errors.

weight loss did not coincide with the loss of an equivalent amount of Zn. Therefore, only observations for sample dates between 0 and 14 d were used in the analyses.

The toxicokinetic model was fitted to the data that were not biased by weight loss. For one soil, the model converged with a negative estimate of k_1; this soil was further excluded for BSAF calculations, since $C(eq)$ could not be determined. Assimilation rate constants (k_1) ranged from 0.11 to 1.28 d^{-1}, elimination rate constants (k_2) from approximately 0.02 to 0.85 d^{-1}. The equilibrium concentrations calculated from these estimates ranged from 1.9 to 9.3 mmol.kg$_w^{-1}$, the BSAFs from 0.1 to 53, and the BSARs from 0.01 to 2.85 d^{-1}. Although the BSAFs were more variable than in *E. andrei*, the range of BSAF values was smaller than the range of K$_p$ values.

If laboratory-contaminated soils had been used, the optimal strategy to investigate bioavailability differences would have been to add a similar total concentration of Zn to all soils, since BSAFs and BSARs may depend on exposure level. For the field soils, however, this was impossible, and the bias of BASFs and BSARs by exposure level was tested in a separate experiment. A clean field soil was artificially contaminated with 2.8 and 4.9 mmol Zn/kg dry weight, and enchytraeids were exposed therein, using the dynamic approach. The equilibrium concentrations in *E. crypticus* appeared to differ and were 2.4 and 4.6 mmol/kg respectively. BSAFs were 0.8 and 0.9; BSARs were both 0.1. This suggests that BSAF and BSAR are insensitive to the undesired variation introduced by different contamination levels, as assumed when the toxicokinetic parameters are used to analyze bioavailability differences among soils.

The quantitative relationship that best describes the BSAF of Zn in *E. crypticus* as a function of soil solid-phase characteristics is given in Table 10-2. The equation for *E. crypticus* is similar to that for *E. andrei*: soil acidity is the major factor influencing Zn uptake in these species. Note that the coefficients for the BSAF formulas differ slightly between both worm species. Figure 10-4 illustrates the relationship between observed and predicted BSAFs of Zn in *E. crypticus*.

Discussion

Risk assessment procedures focus on total substrate concentrations in relation to toxic effects, e.g., through the ratio of PEC and NEC. For example, the PEC_{soil} is thereby derived with environmental chemical data or models, and the NEC_{soil} is often extrapolated from single-species laboratory toxicity experiments with soil invertebrates. Van Wensem et al. [11], Van Straalen [12], and Lanno et al. (Chapter 5) show that the use of body residues reduces uncertainties in risk assessment procedures, which supports the approach presented in this chapter. Accuracy of risk quantification for a contaminated site will thus likely improve when PEC and NEC are based on body concentrations. The PEC_{body}, the predicted (or observed) body concentration of a species, should be predicted for individuals exposed at a particular field site by which the influences of soil factors upon exposure are taken into account. The NEC_{body}, the threshold body concentration for toxic effects of a species, can be derived in laboratory experiments. PEC/NEC ratios calculated in this way for a species collected in the field can be used to specify risks for a specific organism, or PEC/NEC ratios of various species can be averaged to give an estimate of overall contamination risk for a contaminated site.

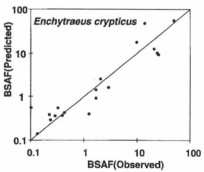

Figure 10-4   Relationship between observed and predicted BSAFs of Zn in *Enchytraeus crypticus* exposed in 20 metal-contaminated field soils Legend: Black dots: cases.

Importance of various soil factors

Soil acidity was identified as the most important solid-phase characteristic for both Zn supply (abiotic partitioning) and demand (uptake in worms) from soil. Uptake of Zn was weakly associated to the Al_{ox} and clay contents of the soil. The similarity of the formulas for K_p and BSAF + BSAR suggests that Zn is taken up from the pore water, a soil compartment in which Zn behavior covaries with the pore water, or a combination of both pools.

The K_p formulas imply that a reasonable approximation of the Zn porewater concentration for Dutch field soils can be made when only total soil Zn concentration (HNO_3 digestion) and soil acidity (pH[$CaCl_2$]) are known. Similarly, Zn uptake in oligochaete worms can reasonably be approximated on the basis of the same parameters. This outcome clearly meets the criterion for simple application in daily practice. However, the formulas also suggest that the BSAFs differ between organisms, since the coefficients for pH($CaCl_2$) were not identical for both worm species (0.39 versus 0.52). Abiotic partitioning and uptake are thus apparently influenced by the same soil factor, but the biological demand has a species-specific component. In laboratory conditions, the causes of the differences between species may be related to allometric differences (large versus small) and differences in food preference and crawling behavior (encounter rate). In field conditions, ecological differences between species and soil heterogeneity may be additional sources of variation in uptake between species. The species-specific aspects of the bioavailability concept should be investigated further in future studies.

Evaluation of the method

Empirical aspects

Empirical formulas do not necessarily represent true equilibration phenomena of chemicals between soil constituents and organisms; at best they reflect these phenomena. Important considerations related to this are as follows:

1) External validation of the formulas is needed, using other soils. Van Gestel et al. [13], e.g., expressed log(K_p) and log(BSAF) for Zn as a function of 3 fixed soil characteristics, i.e., pH(H_2O) and the fractions organic matter and clay, using field soils, pretreated before the assays by drying and rewetting with a nutrient solution. The crucial role of soil acidity for Zn partitioning in soil and for uptake in worms was confirmed for radish and lettuce. Other studies also demonstrated that low pH was associated with increased uptake and effects of metals (e.g., [14–16]), but linkage to partitioning in soils was not made in these studies. Thus, the formulas found for Zn in this study are confirmed by other authors.

2) Direct effects of soil type on organism performance should be addressed, since direct effects of, e.g., pH may bias BSAFs (and, less likely, BSARs), rendering the formulas inaccurate or invalid. Bodyweight reductions were observed in this study. For *E. andrei*, weight loss appeared to be related to soil acidity, and the NEC_{body} was exceeded in some soils [8]. The formulas derived with and without

soils with low pH(CaCl$_2$) were, however, grossly similar, as expected from the high Q^2 value.

3) The soil type for which metal availability is to be predicted should have soil characteristics within the ranges of soil properties encountered in the study dataset. Predictions will be less accurate or even invalid for soils that exceed these ranges, both for solid-phase characteristics and metal concentrations.

4) Further work should substantiate whether the formulas derived for K$_p$ have a systematic relationship to BSAF or BSAR in other organisms. A variety of organisms should be studied, with particular focus on those organisms for which uptake via the pore water is not obvious. Without validation through uptake studies, FIAM studies show only how abiotic processes differ among soils while neglecting the meaning of differences for uptake in organisms and for risks. In the daily practice of risk assessment, grouping of species with similar uptake profiles seems most practical in the future, and averaged formulas for such groups should be developed to take the differences among species into account [17].

5) The formulas should not be biased by nonlinearity of the relationship between soil and body concentrations. The regression procedure implicitly assumes that soil and body concentrations are linearly related on a double-logarithmic scale over the full concentration range. However, in the high concentration range, toxic effects may result in altered uptake, and in the low concentration range, regulation of micronutrients may occur [18]. In the case of *E. crypticus*, data on Zn accumulation in an artificial substrate suggest that the relationships are linear over the full exposure range tested, but for Cu, there was a tendency for regulation in the low-exposure range [19] (Figure 10-5).

Compartment modeling

Dynamic modeling of body residues is preferred over static analyses, irrespective of peculiarities or practical difficulties, since it allows for a better distinction of the principal processes. For *E. crypticus*, a peculiarity was the difference between elimination rate constants in different soils, i.e., between 2 and 86 percent.d^{-1}. The estimated k_2 values may, however, be flawed because the data covered only the initial 14 d of exposure, whereas k_2 estimation should preferably be based upon elimination studies. Nonetheless, the large variability of k_2 is at variance with the idea that the elimination rate constant is an organism-specific constant for a certain chemical. The results reported here suggest that the elimination rate of *E. crypticus* depends on the physical characteristics of the substrate, e.g., Zn elimination could be influenced by soil bulk density through its effect on the rate of soil gut passage. Since the equilibrium body concentration (and thus toxicity) depends on the ratio of k_1 and k_2, this observation suggests that physical soil characteristics may influence toxicity.

One-compartment modeling may not be appropriate for all types of organisms. Higher-order models may be more appropriate if an organism contains various physiological

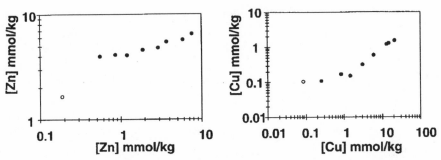

Figure 10-5   Body concentrations of Zn and Cu in *Enchytraeus crypticus* as a function of total metal concentrations in OECD artificial medium Legend: Black marks indicate medium amended with metal salts; the open mark indicates the concentration in animals in control medium.

compartments with different elimination rates. Long-term elimination studies can be applied to investigate this and to improve the accuracy of parameter estimation. Furthermore, uptake may not lead to equilibrium concentrations after prolonged exposure, as found in some species [20,21]. In such cases, only the uptake rate constant is useful to quantify the dynamics of the uptake process.

Kinetic models are empirical, but they can also be based on mechanistic knowledge. In this respect, reference can be made to physiology-based pharmacokinetic (PB-PK) models [22], which have been developed to analyze drug kinetics using organ-specific behavior in vertebrates. At present, PB-PK modeling has not been applied in soil ecotoxicology, although it would form the logical biological counterpart of the mechanistic FIAM approach. In summary, it is obvious that tailor-made toxicodynamic models may be needed to analyze uptake in certain organisms, although presently 1-compartment modeling seems appropriate to analyze Zn uptake in worms.

Mechanistic aspects

The empirical approach could be developed further by deriving the formulas on the basis of mechanistic considerations. Hare and Tessier [23] studied Cd body concentrations in aquatic insect larvae from different lakes, and tried to explain the variation on the basis of mechanistic considerations. Body concentrations could be reasonably predicted using the free ion activity. Extension of the model with the factors "competition with hydrogen at biological uptake sites" and "Cd complexation by natural organic matter" explained a larger part of the observed variation. For soils, such mechanistic-based modeling forms the major challenge for the near future.

Differences in biological demand between organisms

Various physiological strategies for coping with metal exposure have been identified in soil organisms. In some species, a large fraction of metals is almost immediately toxicologically inactivated due to inert binding in the organism tissues, e.g, springtails [24], isopods [25], and plants [26,27]. For the animals, it can be hypothesized that the freely circulating metals cause the toxic effects. In some plants, especially in *Silene cucubalus*, it was shown that the toxicity of Cu was associated with damage to the outer root cells. This suggests that in some species external metal concentrations, rather than internal concentrations, may determine toxic effects. In conclusion, metal physiology may differ between species, causing them to differ with respect to the aspect of biological demand. It is obvious that physiological data may be helpful in designing dynamic investigations into bioavailability.

Physiological studies have also demonstrated that some metals are regulated [18], as observed in this study of *E. crypticus*: the body concentration of Zn in stock culture is maintained at a concentration of 1.38 mmol/kg. In all soils, the estimated equilibrium concentration exceeded this level, but in some soils the increase of the body concentration started only after some days of exposure. The issue of regulation may take 2 forms that should be distinguished before deriving the multivariate regression equations, namely that 1) organisms may show no or delayed uptake when true bioavailability is low due to substrate characteristics (high Zn concentration but low environmental availability) or 2) no uptake may occur when the environmental concentration is too low to exceed the regulation capacity, irrespective of the substrate characteristics. The latter substrates should not be incorporated in regression procedures because in those cases the substrate characteristics are irrelevant for uptake. Observations on body residues in stock cultures and on the regulation capacity of a species when exposed to a range of concentrations in a single substrate (see Figure 10-5) may be helpful for proper decision-making and to avoid circular reasoning.

Outlook

Because different organisms have different uptake mechanisms and explore their heterogeneous habitats in different ways, the composite concept of bioavailability needs specification, both with respect to the characteristics and homogeneity of the soil and to the species (and ecotypes) involved. Species-specific aspects play a role in the uptake processes themselves and in the redistribution processes inside the organism tissues. In this chapter, it was demonstrated that Zn partitioning in soil and uptake in 2 oligochaete worms can reasonably be approximated for Dutch field soils using simple soil characteristics: soil acidity and total soil concentration of Zn. Further attention should be paid to the issues of regulated versus nonregulated chemicals, the toxicokinetics of metals in organisms from other taxonomic groups, and the species that have an influence on chemical sorption in their substrate. True progress depends on further development of the FIAM approach but only when accompanied by validation of the abiotic findings with toxicokinetic analyses for different organisms. This shows that profitable results can be obtained only from a close collaboration between soil chemists and soil biologists.

Acknowledgments - We thank the organizing committee of the 2nd International Workshop on Earthworm Ecotoxicology for their invitation to prepare this paper. We thank R. Baerselman, R.P.T. Janssen, M. Ott, R.P.M Van Veen, and P. Zweers for measurements of soil and animal characteristics and subsequent parts of the data analyses.

We thank our colleagues from the Laboratory of Ecotoxicology for stimulating discussions. The comments on an earlier draft of the paper, made by Tjalling Jager, Joke van Wensem, and 3 anonymous reviewers, have improved the paper.

References

1. Allen HE. 1997. Standards for metals should not be based on total concentrations. *SETAC-Europe News* 8:7–9.

2. Sposito G. 1984. The future of an illusion: ion activities in soil solutions. *Soil Sci Soc Am J* 48:531–536.

3. De Rooij NM, Smits JGC. 1997. Methodology of heavy metal standards for soil. Phase I: Definition study. Report. No. T2004T01.01. Delft, The Netherlands: Delft Hydraulics.

4. Ure AM, Davidson CM. 1995. Chemical speciation in the environment. London UK: Chapman & Hall.

5. Dickson KL, Giesy JP, Wolfe L. 1994. Summary and conclusions. In: Hamelink JL, Landrum PF, Bergman HL, Benson WH, editors. Bioavailability. Physical, chemical and biological interactions. Boca Raton FL: CRC. p 221–230.

6. Van Straalen NM, Van Wensem J. 1986. Heavy metal content of forest litter arthropods as related to body-size and trophic level. *Environ Pollut (Ser A)* 42:209–221.

7. Janssen RPT, Peijnenburg WJGM, Posthuma L, Van Den Hoop MAGT. 1997. Equilibrium partitioning of heavy metals in Dutch field soils. I. Relationships between metal partition coefficients and soil characteristics. *Environ Toxicol Chem* 16:2470–2478.

8. Janssen RPT, Posthuma L, Baerselman R, Den Hollander HA, Van Veen RPM, Peijnenburg WJGM. 1997. Equilibrium partitioning of heavy metals in Dutch field soils. II. Prediction of metal accumulation in earthworms. *Environ Toxicol Chem* 16:2479–2488.

9. Van Gestel CAM, Dirven-Van Breemen EM, Baerselman R. 1993. Accumulation and elimination of cadmium, chromium and Zn and effects on growth and reproduction in *Eisenia andrei* (Oligochaeta, Annelida). *Sci Tot Environ* Suppl. 1993:585–597.

10. Loehr RC, Webster MT. 1996. Behavior of fresh vs. aged chemicals in soil. *J Soil Contam* 5:361–383.

11. Van Wensem J, Vegter JJ, Van Straalen NM. 1994. Soil quality criteria derived from critical body concentrations of metals in soil invertebrates. *Appl Soil Ecol* 1:185–191.

12. Van Straalen NM. 1996. Critical body concentrations: their use in bioindication. In: Van Straalen NM, Krivolutsky DA, editors. Bioindicator systems for soil pollution. Amsterdam, the Netherlands: Kluwer Academic. p 5–16.

13. Van Gestel CAM, Dirven-van Breemen EM, Kamerman JW. 1992. Beoordeling van gereinigde grond. IV.Toepassing van bioassays met planten en regenwormen op referentiegronden. Report no. 216402004 RIVM. Bilthoven, The Netherlands.

14. Ma W-C. 1982. The influence of soil properties and worm-related factors on the concentration of heavy metals in earthworms. *Pedobiologia* 24:109–119.

15. Corp N, Morgan AJ. 1991. Accumulation of heavy metals from polluted soils by the earthworm, *Lumbricus rubellus*: Can laboratory exposure of control worms reduce biomonitoring problems? *Environ Pollut* 74:39–52.

16. Spurgeon DJ, Hopkin SP. 1996. Effects of variations of the organic matter content and pH of soils on the availability and toxicity of Zn to the earthworm *Eisenia fetida*. *Pedobiologia* 40:80–96.

17. Peijnenburg WJGM, Posthuma L, Eijsackers HJP, Allen HE. 1997. A conceptual framework for implementation of bioavailability of metals for environmental management purposes. *Ecotoxicol Environ Saf* 37:163–172.

18. Chapman PM, Allen HE, Godtfredsen K, Z'Graggen MN. 1996. Evaluation of BCFs as measures for classifying and regulating metals. *Environ Science Technol* 30:448A–452A.

19. Posthuma L, Baerselman R, Van Veen RPM, Dirven-Van Breemen EM. In press. Single and joint toxic effects of copper and Zn on reproduction of *Enchytraeus crypticus* in relation to sorption of metals in soils. *Ecotoxicol Environ Saf*.

20. Janssen MPM, Bruins A, De Vries Th, Van Straalen NM. 1991. Comparison of cadmium kinetics in four soil arthropod species. *Arch Environ Contam Toxicol* 20:305–312.

21. Dallinger R. 1993. Strategies of metal detoxification in terrestrial invertebrates. In: Dallinger R, Rainbow PS, editors. Ecotoxicology of metals in invertebrates. Chelsea MI: Lewis.

22. Krishnan K, Andersen ME. 1994. Physiologically based pharmacokinetic modelling in toxicology. In: Hayes AW, editor. Principles and methods of toxicology. 3rd ed. New York: Raven. p 149–188.

23. Hare L, Tessier A. 1996. Predicting animal cadmium concentrations in lakes. *Nature* 380:430–432.

24. Posthuma L, Hogervorst RF, Van Straalen NM. 1992. Adaptation to soil pollution by cadmium excretion in natural populations of *Orchesella cincta* (L.) (Collembola). *Arch Environ Contam Toxicol* 22:146–156.

25. Donker MH. 1992. Energy reserves and distrubution of metals in populations of the isopod *Porcellio scaber* from metal-contaminated sites. *Funct Ecol* 6:445–454.

26. Harmsen H. 1993. Physiology of Zn in *Silene vulgaris* [Ph.D. Thesis]. Amsterdam, the Netherlands: Vrije Universiteit.

27. De Vos CH. 1991. Copper-induced oxidative stress and free radical damage in roots of copper tolerant and sensitive *Silene cucbalus* [Ph.D. Thesis]. Amsterdam, the Netherlands: Vrije Universiteit.

Chapter 11
Effects of additions of organic matter and different ions on partitioning of ^{134}Cs in a sandy soil and uptake by earthworms

M.P.M. Janssen, P. Glastra, and J.F.M.M. Lembrechts

The partitioning and uptake of ^{134}Cs from soil by 2 earthworm species (*Lumbricus rubellus* and *Eisenia andrei*) have been described by using a 3-phase model, which comprises a soil-solid phase, a liquid phase, and an organism. The assumptions that some environmental parameters (e.g., pH) affect mainly the solid-liquid distribution of ^{134}Cs and that others (e.g., temperature) predominantly change the distribution between the liquid phase and the earthworm were verified. Previous studies in a liquid medium showed that the addition of K and stable Cs to a liquid medium significantly decreased ^{134}Cs concentrations in the 2 earthworm species studied. Addition of Ca and NH_4^+ did not result in a significant change, nor did it change in pH. Addition of different elements to potted soils affected the distribution between the solid and liquid phases of the soil significantly, but effects on the ^{134}Cs concentrations in earthworms were limited. K and organic matter were important parameters in reducing the ^{134}Cs concentration in earthworms relative to the ^{134}Cs concentration in soil solution. For a correct interpretation of the availability of Cs, it is necessary to combine site-specific information on partitioning of Cs and uptake of Cs by biota. The results showed that the effects of additions on the distribution between the liquid and solid phases of the soil may be counteracted by effects on uptake and that partitioning may change with time and in the presence of earthworms.

In many studies, toxicity has been directly linked to soil concentrations. Data from such studies are often difficult to interpret, as they ignore kinetics and provide limited insight into the soil characteristics that affect distribution between soil and soil solution and uptake. Differences in toxicity are often explained by differences in bioavailability of the soil, although "bioavailability" is poorly defined in most of these cases.

The effects of soil characteristics on the toxicity of heavy metals for earthworms are not easy to quantify because the dominant soil characteristics seem to be different for each element [1]. Van Gestel [1] proposed that an approach that accounts for partitioning may be useful in understanding the effects of environmental parameters on heavy metal toxicity. Differences in partitioning may explain differences in toxicity of soils with comparable total concentrations of a certain element but with different soil characteristics. However, data were not available at that time to support his assumption. Recently, a limited number of studies that incorporate partitioning to interpret uptake and toxicity have become available [2,3,4].

The present study focuses on soil parameters modulating partitioning and uptake of radiocesium (^{134}Cs, ^{137}Cs) from a sandy soil by earthworms. Although most ecotoxicologists are not familiar with radionuclides, the mechanisms and problems encountered are very similar to those in ecotoxicology. In contrast to heavy metals and other toxic elements, radionuclides offer the possibility of studying exchange and uptake processes at very low concentrations. Using radionuclide techniques allows small changes in concentration to be detected.

Radiocesium was deposited on the upper soil layers in Europe as a result of the Chernobyl accident [5,6]. In certain highland areas (Scotland, Scandinavia), radiocesium concentrations in livestock approach or exceed human-health threshold levels [7]. Wet deposition of radiocesium has been relatively high in these areas, and the soil conditions enable a relatively high uptake by the vegetation [7].

Addition of K, which chemically resembles Cs, has been recommended as a countermeasure to decrease the uptake of radiocesium by vegetation and thus decrease the transfer to livestock. Additions of other cations, e.g., Ca, ammonium, or of organic matter have been studied as well, but these seem to be less effective [8]. In practice, the effectiveness of additions strongly depends on the soil conditions: in certain soils, additions are highly effective in reducing Cs uptake, whereas in others they are not. The effects of additions on the partition of Cs between the solid phase of the soil and soil solution, which is thought to play a key role in the uptake, are not yet fully understood. The relationship between uptake and soil conditions, which have all but been ignored, also requires a better understanding so as to interpret results from countermeasures [9]. Since Cs enters the organism through active transport systems for K or ammonium, competition with these ions at the site of uptake is thus likely to occur [10,11,12,13].

Earthworm data have been used to study the bioavailability of radionuclides in different soils in comparison to that of plants [14]. In contrast to the uptake by plants, uptake by earthworms can be achieved in a limited period of time. Transfer of Cs from soil to earthworms and plants is relatively low compared to the transfer of other elements, e.g., cadmium, Cu, and Zn [7,15]. In studying the effect of additions on the uptake of radiocesium (^{134}Cs) by earthworms, it was assumed that radiocesium was partitioned among 3 compartments: solid soil, soil solution, and the earthworm (Figure 11-1) [1,16]. Uptake was assumed to be directly from the soil solution. Some environmental parameters, such as organic matter or H$^+$, were hypothesized to affect mainly the exchange between solid soil and soil solution, whereas others mainly affect uptake [4,17]. In a previous study, temperature, rather than the exchange between the solid and liquid phases of the soil, was shown to affect uptake [4]. K, Ca, and ammonium were thought to have an

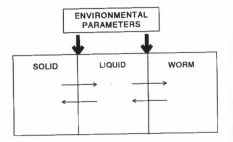

Figure 11-1   Theoretical approach of the effect of environmental parameters on the partitioning of ^{134}Cs between solid and liquid phases of the soil and on ^{134}Cs uptake from the liquid phase.

effect on the exchange between solid soil and soil solution as well as on uptake [9,18]. Surprisingly, limited data are available on the processes of Cs exchange or exchange of other elements between the different compartments.

Consequently, this study will focus on 3 subjects:

1) What is the effect of different additions on the distribution of [134]Cs between the solid and liquid phases of the soil, and what is the time course of these effects?

2) What is the effect of earthworms (*Lumbricus rubellus* and *Eisenia andrei*) on the distribution of [134]Cs?

3) What are the effects of different additions on the uptake of [134]Cs by *L. rubellus* compared to the [134]Cs concentration in the soil and in soil solution?

Materials and methods

Earthworms

The effects of different additions on partitioning and uptake of [134]Cs was studied in the presence of the earthworm species *L. rubellus* and *E. andrei*. *L. rubellus* is a common species in western Europe and has been used in a number of ecotoxicological studies [19], and *E. andrei* is closely related to *E. fetida* [20], which is often used in toxicity tests. Individuals of *L. rubellus* were obtained from a commercial dealer a few days before the experiments were carried out, whereas individuals of *E. andrei* were obtained from a 6-week-old laboratory culture. The earthworms were kept on moist filter paper prior to the experiments in order to empty their guts [4,14].

Soil characteristics and additions

The experiments were carried out in a sandy soil from a pasture soil in Bilthoven, the Netherlands (89% particles > 16 mm, 5.0% organic matter, pH-KCl 4.4). The soil was air dried, sieved, and stored at 4 °C. About 1 month before the actual experiments began, the soil was thoroughly mixed with a [134]Cs solution and, where appropriate, with various additions using a cement mixer to achieve even distribution. Preliminary experiments showed that equilibration of [134]Cs between soil and soil solution generally took less than 1 month. The [134]Cs was added to the soil up to a concentration of 42.4 Bq/g ($= 6.4 \times 10^{-10}$ mmol/g), which is negligible compared to the natural [133]Cs concentration of this soil (7.7×10^{-3} mmol/g). Concentrations of [134]Cs added to the soil were comparable to concentrations used in earlier experiments with liquid media [4].

The experiments were carried out using 18 different treatments: 9 combinations of Ca and ammonium and 9 of K and organic matter (OM) were added to the soil. Ca was added to the soil as $Ca(NO_3)_2$, ammonium as NH_4NO_3, and K as KNO_3. Organic matter was added as dried cow dung (brand: Naturado), which was ground to less than 0.5 mm. The dried dung contained 19.6 g total N kg^{-1}, 20.6 g P_2O_5 kg^{-1}, and 32.7 g K_2O kg^{-1}. The concentrations of each addition used in the different treatments are given in Table 11-1. There were 3 replicates for each combination per species, resulting in a total of 54 pots for each earthworm species.

Table 11-1 Eighteen treatments and sampling occasions

Treatment				Sampling of soil solution (d)		
Ca (mmol·kg⁻¹)	NH₄ (mmol·kg⁻¹)	K (mmol·kg⁻¹)	OM (g·kg⁻¹)	In soil without earthworms[1]	At removal of L. rubellus[2,4]	At removal of E. andrei[3,4]
0	0			0, 7, 18, 33, 82	47	96
0.84	0			47, 89	61	103
8.4	0			0, 7, 18, 54, 96	68	110
0	0.38			54, 96	68	110
0.84	0.38			33, 82	47	96
8.4	0.38			47, 89	61	103
0	3.8			0, 7, 18, 47, 89	61	103
0.84	3.8			54, 96	68	110
8.4	3.8			0, 7, 18, 33, 82	47	96
		0	0	0, 7, 18, 34, 83	48	97
		0.59	0	48, 90	62	104
		5.9	0	0, 7, 18, 55, 97	69	111
		0	10	55, 97	69	111
		0.59	10	34, 83	48	97
		5.9	10	48, 90	62	104
		0	50	0, 7, 18, 48, 90	62	104
		0.59	50	55, 97	69	111
		5.9	50	0, 7, 18, 34, 83	48	97

[1]Number replicates per sampling: 1
[2]Number replicates per sampling: 3
[3]Number replicates per sampling: 3
[4]L. rubellus and E. andrei were exposed one after the other, being exposed for 14 d.

Experimental conditions and set up

The experiments were carried out in climate chambers at 15 °C. Humidity in the climate chambers varied between 50 and 70%. Moisture content of the soil was kept at 18% by adding water every 2 or 3 d. The experiments with *L. rubellus* were conducted in plastic pots containing 360 g of soil dry weight, which was considered to be sufficient enough to keep them for 2 weeks. Cast production of *L. rubellus* in 14 d is about 1/6 of this amount [21]. The experiments with *E. andrei* were conducted in pots containing 150 g of soil dry weight because of their much smaller size. Each pot contained 3 earthworms to generate measurable amounts of radiocesium.

Experiments were first carried out with *L. rubellus* and then with *E. andrei*. To be able to carry out the sampling of soil solution and the earthworms, earthworms were exposed to 3 treatments of Ca/NH$_4$ or K/OM at a time. Pots were randomized within each set of 9 (3 treatments × 3 replicates). The sampling days are summarized in Table 11-1.

Soil-solution sampling

Soil-solution samples were taken on 3 occasions from a number of selected pots during the equilibration period before the earthworms were added. Soil solution samples were also taken from all pots when the earthworms were added or removed (see Table 11-1). Soil solution was extracted by centrifugation of 25 to 30 g soil together with 20 ml chloroform at 16,000 rpm for 1 h [22]. After filtration, the conductivity, pH, and the K, Ca, and [134]Cs concentrations of the soil solution were measured for each pot separately. Ammonium concentration was measured only in soil solution from pots of the ammonium/Ca treatments. These parameters were thought to be important in determining the uptake of [134]Cs or the exchange between the solid and liquid phases [9,18]. K and Ca were not measured in pots from which *E. andrei* were removed.

[134]Cs was measured with a Minaxi gamma counter. K and Ca concentrations were measured with flame atomic absorption. La$_2$O$_3$ was added to prevent interference between Ca and phosphate. The quality of the K and Ca measurements was assured by regular measurement of Standard Reference Material. Ammonium was measured using the ammonium colometry kit from Merck Laboratories [23]. In a few pots, total [134]Cs concentration of the soil was determined at the start and the end of the accumulation period. Measurements of the amount of stable Cs were carried out on a few soil and earthworm samples through instrumental neutron activation analysis [24].

Sampling and measuring [134]Cs in earthworms

After 2 weeks of exposure, 96% of the *L. rubellus* were recovered. For *E. andrei*, the recovery was 98%. After sampling, the earthworms were kept on moist filter paper for 2 d until their guts were voided. Each sample of 3 earthworms was weighed and stored at −18 °C until further analysis could be accomplished. The samples were freeze dried and weighed, and the [134]Cs concentration was measured on a Minaxi gamma counter after the samples were digested in a mixture of HNO$_3$, H$_2$SO$_4$ and HClO$_4$ until dry and the remaining pellet was diluted in 10 or 15 ml demineralized water. The measurements were corrected for radioactive decay to the first day of the experiment. Measurements of the ^{134}Cs concentration of the earthworms were carried out only for *L. rubellus*.

Statistics

The changes in soil-solution characteristics over time in pots without earthworms were tested by regression analysis. The effects of different additions on pH and conductivity and on ^{134}Cs, K, and Ca concentrations in soil solution were tested by 2-way analysis of variance (ANOVA). Two-way ANOVA was also used to investigate the effect of the additions on the dry weight and the ^{134}Cs concentration in *L. rubellus*. Finally, the effect of the ^{134}Cs, Ca, and K concentrations in soil solution on the ^{134}Cs concentration in *L. rubellus* was described by multiple regression. For these elements, actual data on the concentrations in soil solution at the time of removal were available for all pots.

Results

Effects of additions on soil-solution characteristics and time course

Soil-solution samples were taken from soil without earthworms on 5 occasions, up to about 100 d after the start of the experiments (see Table 11-1), and from pots with earthworms at time of removal. The measurements showed significant effects of various additions on the characteristics of the soil solution (Tables 11-2 and 11-3). Although change over time did not always follow a linear pattern, changes were tested by regression analysis. Results are summarized in Table 11-4.

The initial pH was highest (pH 8) in soils with additions of OM. The initial pH in the other treatments varied between 4.5 and 6. K did not affect pH. The pHs were either stable or had decreased with time, the decrease being largest in the pots to which OM was added (Table 11-4; Figures 11-2A, 11-2B).

The treatments in which OM was added showed the highest conductivity of the soil solution, whereas intermediate values were observed in pots with the highest Ca addition. Additions of ammonium and K resulted in a slight elevation (Table 11-2). Conductivity decreased slightly with time in pots with elevated initial conductivity and increased slightly in the other treatments.

Concentrations of ^{134}Cs in soil solution of pots without additions were between 0.5 and 0.8 Bq/ml, whereas additions of Ca, ammonium, and K led to slightly higher values. Highest ^{134}Cs concentrations in soil solution were observed in the pots to which OM or Ca and ammonium had been added (Table 11-2; Figures 11-3A, 11-3B). The partition coefficient (K_d), which was calculated by dividing the average ^{134}Cs concentration in the soil by the ^{134}Cs concentration in soil solution [5,9,25,26], measured 1 week after adding the various additions, varied between 83 and 3.3 ml/g. The decrease in ^{134}Cs concentration in soil solution with time was largest in pots with relatively high initial ^{134}Cs concentrations; however, the decrease in these pots did not reach the concentrations observed in the reference pots containing no additions (Table 11-4; Figures 11-3A, 11-3B).

Table 11-2 Effects of different additions on pH, conductivity, and concentrations of ¹³⁴Cs, K, and Ca in soil solution 1 week after preparing soils[1]

Treatment					Results			
Ca⁺⁺ mmol/kg	NH₄⁺ mmol/kg	K⁺ mmol/kg	OM g/kg	pH	Conduct. (mS·cm⁻¹)	¹³⁴Cs (Bq/ml)	K (mmol/L)	Ca (mmol/L)
0	0			5.63	0.54	0.80	0.64	0.91
8.4	0			4.83	4.96	1.98	1.33	20.7
0	3.8			4.98	1.06	1.39	0.60	1.98
8.4	3.8			4.67	6.63	6.62	1.66	21.6
		0	0	4.69	0.42	0.51	0.42	0.67
		5.9	0	4.79	3.22	1.71	9.16	5.78
		0	50	7.63	10.6	13.0	60.0	5.32
		5.9	50	7.78	11.5	9.88	66.5	5.50

[1]Values are based on 1 observation per treatment but are consistent with other observations over

SETAC Press

Figure 11-2   Time course of the pH in soil solution in potted soils with different additions and the
               effect of earthworms on the pH.

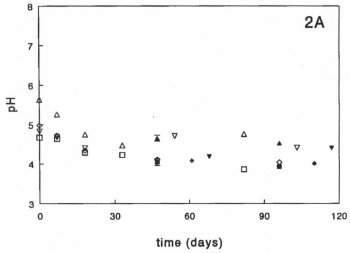

Figure 11-2A $\triangle$, $\blacktriangle$ = 0 mmol/kg Ca, 0 mmol/kg NH$_4$, ∇, $\blacktriangledown$ = 8.4 mmol/kg Ca, 0 mmol/kg NH$_4$, $\Diamond$, $\blacklozenge$
                = 0 mmol/kg Ca, 3.8 mmol/kg NH$_4$, $\square$, $\blacksquare$ = 8.4 mmol/kg Ca, 3.8 mmol/kg NH$_4$.

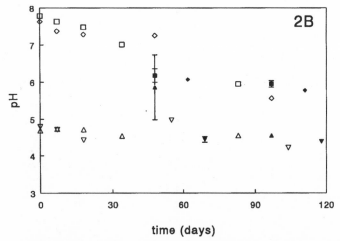

Figure 11-2B   (open symbols = additions; closed symbols = effect of earthworms on pH) $\triangle$, $\blacktriangle$ = 0
                mmol/kg K, 0 g·/g OM, ∇, $\blacktriangledown$ = 5.9 mmol/kg K, 0 g/kg OM, $\Diamond$, $\blacklozenge$ = 0 mmol/kg K, 50 g/
                kg OM, $\square$, $\blacksquare$ = 5.9 mmol/kg K, 50 g/kg^m OM.

The earthworms were kept in the pots for 14 d. Pots with *L. rubellus* were sampled between d 45 and d 70,
E. andrei was sampled between d 95 and d 120 (see also Table 11-1).

The results for pots without earthworms are based on a single observation per occasion; the results from
pots with earthworms are based on 3 observations. Given are mean and standard error.

Table 11-3 Effects of the addition of Ca and ammonium and of K and organic material, respectively, on different soil-solution characteristics at time of removal of earthworms, tested by a 2-way ANOVA

L. rubellus	pH	Conduct.	[134]Cs	K	Ca	NH$_4$
Ca	0.001[a]	0.000	0.005	0.000	0.000	0.031
NH$_4$	0.000	0.000	0.000	0.000	0.000	0.001
Interaction	0.066	0.001	0.469	0.146	0.001	0.023
K	0.418	0.000	0.002	0.000	0.000	
OM	0.001	0.000	0.000	0.000	0.003	
Interaction	0.179	0.000	0.352	0.127	0.000	
E. andrei[b]	pH	Conduct.	[134]Cs	K	Ca	NH$_4$
Ca	0.000	0.000	0.000			0.000
NH$_4$	0.000	0.000	0.000			0.000
Interaction	0.015	0.138	0.646			0.345
K	0.668	0.000	0.021			
OM	0.000	0.000	0.000			
Interaction	0.142	0.047	0.436			

[a]p-values of each test are given, n = 3.
[b]K and Ca were not measured in the E. andrei experiments.

Table 11-4 Effect of time on the change in soil-solution characteristics tested by regression analysis

Treatments				Results[1]				
Ca	NH$_4$	K	OM	pH	Conduct.	[134]Cs	K	Ca
mmol kg^{-1}	mmol kg^{-1}	mmol kg^{-1}	g kg^{-1}					
0	0			0.288	0.136	0.391	0.630	0.033[2]
8.4	0			0.345	0.845	0.272	0.118	0.598
0	3.8			0.071	0.053	0.074	0.231	0.004[2]
8.4	3.8			0.012[2]	0.104	0.006[2]	0.111	0.208
		0	0	0.114	0.002[2]	0.079	0.129	0.081
		5.9	0	0.388	0.012[2]	0.050[2]	0.038[2]	0.280
		0	50	0.020[2]	0.992	0.056	0.457	0.030[2]
		5.9	50	0.000[2]	0.769	0.027[2]	0.103	0.138

[1]Values represent p-values of the regression analysis. Five observation times were used, except for Ca for which 4 were used.
[2]Significant effects

K concentrations in soil solution were low in pots to which no K or OM was added. Highest values were observed in pots to which 50 g of OM were added. Ca concentration was obviously enhanced in pots to which Ca was added. Intermediate values were observed in pots to which either K, OM, or both were added (Table 11-2). The time course of concentrations for K and Ca depended on the addition. K and Ca concentration in soil solution both decreased in the pots to which OM had been added, whereas K concentration in soil solution increased in pots to which K or ammonium had been added. Ca concentration in soil solution decreased in pots without added Ca, whereas in the remaining treatments, it stayed the same or increased slightly. Ammonium concentration in soil solution was measured only in pots to which ammonium and/or Ca had been added and was highest in pots to which ammonium had been added. In all treatments, ammonium decreased with time.

Changes in soil solution due to adding the earthworms

The effect on pH of adding earthworms depended on both the treatment and the earthworm species. Adding *L. rubellus* led to lower pH in soil solution for most treatments, except for the 2 reference treatments containing no additions. Adding *E. andrei* led to higher pH in 3 of the 8 treatments; no effect was observed in 4 of the treatments, whereas a lower pH was observed in 1 of the reference treatments (Figures 11-2A, 11-2B). The conductivity of the soil solution was stable or slightly increased in the reference treatments after the earthworms were added, whereas in the other treatments, the earthworms had either a slightly negative or no effect on the conductivity.

The [134]Cs concentration increased in soil solution after adding the earthworms in most treatments. The increase varied between 0 and 14H the observations 14 d before. The effect on the [134]Cs concentration was largest in the reference pots and smallest in the pots to which OM had been added. In the latter, initial [134]Cs concentrations in soil solution were high but decreased rapidly with time, thus counteracting the effect of the earthworms (Figures 11-3A, 11-3B).

Addition of *L. rubellus* increased the K concentration in soil solution in most treatments. However, K concentration continued to decrease in the treatments in which initial K was high. Addition of *L. rubellus* resulted in decreasing or stable Ca concentrations in the treatments in which Ca or ammonium was added. In the K/OM treatments, Ca concentration increased after the earthworms were added. Ammonium concentration was measured only in pots for the Ca/ammonium treatments. Ammonium concentrations increased in all treatments after either *L. rubellus* or *E. andrei* were added.

Effect of additions on [134]Cs concentration in *L. rubellus*

Mean dry-weight percentage of *L. rubellus* was 13.6%. No differences in dry weight were observed between the different treatments (Table 11-5). At the start of the experiment, the earthworms did not contain a significant amount of [134]Cs (0.06 ± 0.09 Bq/g dry weight, n=6). After 14 d of exposure, average [134]Cs concentrations varied between 10 and 20 Bq/g dry weight, depending on the treatment (Table 11-6). However, the ANOVA did

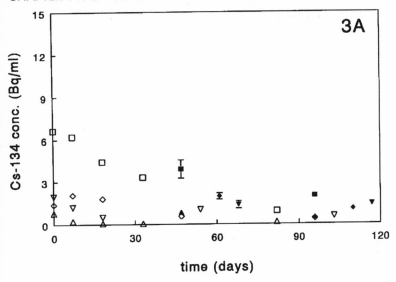

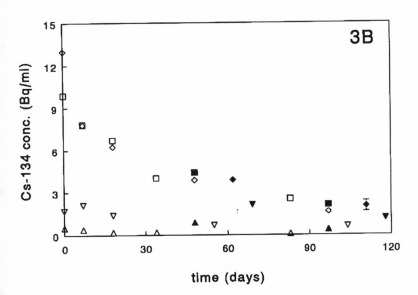

Figure 11-3   Time course of the [134]Cs concentration in soil solution in potted soils with different
                additions and the effect of the earthworm *L. rubellus* on the [134]Cs concentration. Legend:
                Experimental set up and treatments are similar to those in Figures 11-2A and 11-2B.
                (open symbols = additions; closed symbols = effect of earthworms on [134]Cs
                concentration)

not show a significant effect of the additions on the ^{134}Cs concentration of *L. rubellus* (Table 11-5).

Uptake of ^{134}Cs was expected to take place mainly from the soil solution. The results presented in Table 11-2 and Figures 11-3A and 11-3B show considerable differences in ^{134}Cs concentrations in soil solution between the treatments. ^{134}Cs concentrations in the earthworms expressed relative to the ^{134}Cs concentration in soil solution showed considerable differences,

Table 11-5   Effects of the addition of Ca and ammonium and of K and organic material (in g·kg^{-1} soil), respectively, on

Treatment	Results	
	^{134}Cs (Bq·g^{-1} dw)	Dry weight (g)
Ca	0.508[a]	0.197
NH$_4$	0.324	0.983
Interaction	0.763	0.370
K	0.565	0.372
OM	0.129	0.582
Interaction	0.434	0.558

[a]Effects are expressed by the p-values from the 2-way ANOVA.

varying between 2.52 and 17.5 (Table 11-6). The lowest concentration factors were observed in pots to which OM had been added; this indicates that uptake is strongly reduced in these pots. Addition of ammonium seemed to have the lowest effect on uptake.

Table 11-6   Effects of different additions on the concentrations and concentration factors in the earthworm *Lumbricus rubellus*

Treatment				Results	
Ca (Bq·ml^{-1})	NH$_4$ (mmol·kg^{-1})	K (mmol·kg^{-1})	OM (g·kg^{-1})	Mean ^{134}Cs conc. (Bq·g^{-1} dw)	Mean conc. factor (Bq·g^{-1} per Bq·ml^{-1})
0	0			10.9 (0.9)	13.1 (0.8)
0.84	0			13.6 (2.2)	14.9 (3.8)
8.4	0			10.5 (3.3)	7.3 (0.9)
0	0.38			20.1 (5.0)	17.5 (2.9)
0.84	0.38			15.4 (2.2)	12.4 (1.0)
8.4	0.38			12.2 (2.8)	5.8 (0.3)
0	3.8			14.6 (1.9)	7.4 (0.1)
0.84	3.8			15.8 (2.7)	6.2 (0.5)
8.4	3.8			13.8 (2.6)	3.5 (0.1)
		0	0	15.0 (2.3)	17.3 (2.1)
		0.59	0	16.8 (2.0)	13.6 (0.5)
		5.9	0	14.7 (0.5)	7.1 (0.7)
		0	5	11.6 (2.1)	8.0 (0.8)
		0.59	5	11.6 (1.5)	9.5 (0.5)
		5.9	5	14.7 (1.3)	7.0 (0.6)
		0	50	12.1 (0.3)	3.1 (0.1)
		0.59	50	15.1 (1.4)	3.6 (0.1)
		5.9	50	11.1 (0.9)	2.5 (0.1)

[1]Mean values and standard errors (n = 3)

The ^{134}Cs concentrations in the earthworms were described as a function of the actual ^{134}Cs, K, and Ca concentrations in the soil solution. Incorporating actual ^{134}Cs and K concentrations in the regression equation showed the most significant effect. Incorporating only the actual ^{134}Cs concentration, a combination of the ^{134}Cs and Ca concentration, or the K and Ca concentration showed a lower coefficient of determination, whereas an equation with the actual ^{134}Cs, K, and Ca concentrations showed a fit equal to that incorporating the ^{134}Cs and K concentration only (Table 11-7). The ^{134}Cs concentration in *L. rubellus* could thus be described by:

$$^{134}Cs_{worm} = 6.89 \cdot {}^{134}Cs_{soil\ solution} - 0.36 \cdot K_{soil\ solution} \qquad \text{Equation 11-1,}$$

in which [^{134}Cs] is given as Bq/g dry weight and Bq/ml, respectively, and [K] as mmol/L.

Table 11-7   Results of regression analysis of ^{134}Cs concentration in *L. rubellus* and soil-solution characteristics

Variables	Regression equation	r^2(%)
^{134}Cs, K, Ca	$^{134}Cs_{worm} = 7.06[^{134}Cs_{solution}] - 0.38\,[K_{solution}] - 0.03[Ca_{solution}]$	75.6
^{134}Cs, K,	$^{134}Cs_{worm} = 6.89[^{134}Cs_{solution}] - 0.36[K_{solution}]$	75.6
^{134}Cs, Ca	$^{134}Cs_{worm} = 4.04[^{134}Cs_{solution}] - 0.32\,[Ca_{solution}]$	72.1
^{134}Cs,	$^{134}Cs_{worm} = 5.01[^{134}Cs_{solution}]$	69.8
K, Ca	$^{134}Cs_{worm} = 0.31[K_{solution}] - 0.87[Ca_{solution}]$	42.5

[^{134}Cs$_{worm}$] is given as Bq·g^{-1} dry weight, [^{134}Cs$_{solution}$] as Bq·ml^{-1} and [K] and [Ca] as mmol·l^{-1}.

Figure 11-4A shows the relationship between ^{134}Cs concentration in *L. rubellus* and in the soil solution, and Figure 11-4B shows the same relationship, corrected for the K concentration in soil solution, by increasing the ^{134}Cs concentration measured in the earthworms with a value of 0.36·[K$_{solution}$]. For a 100% fit of the regression, this correction should result in a straight line with a slope of 6.89.

Discussion

Effects of additions on concentrations in soil solution

Total concentrations of elements are not directly related to the levels observed in plants and animals [6,25,27]. Thus, it has been recommended to take the concentrations in soil solution into account in predicting uptake and toxicity [1,17]. Previous experiments with liquid media showed ^{134}Cs concentration in earthworms to increase linearly with ^{134}Cs concentration in the medium (unpublished data). However, the amount of trace elements in soil solution depends on how it is in equilibrium with the amount attached to the mineral phase. Environmental factors such as pH and temperature may affect mineral solubility and thus change the amount in soil solution [28].

The effects of fertilizer application on the uptake of cadmium and Zn by crops were studied by Lorenz et al. [29], who found that K and ammonium excess increased the concentrations of K, Ca, Mg, Zn, and cadmium in solution, while the soil pH decreased. Additions of stable Cs, K, and ammonium increased the amount of radiocesium in soil solution, and the effect of stable Cs was the largest [26]. The effect became proportionally smaller as larger amounts of stable Cs were added [26]. Ca and Mg levels and pH

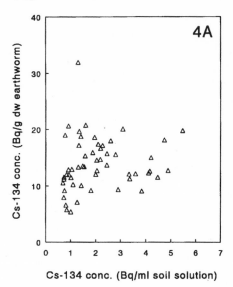

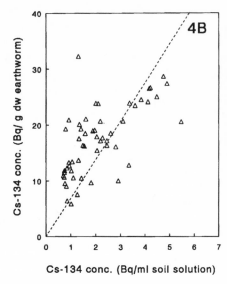

Figure 11-4   Relationship between ^{134}Cs concentration in soil solution (Bq/ml) and ^{134}Cs concentration in the earthworm *Lumbricus rubellus* (Bq/g dry weight) before (4A) and after correction (4B) for K concentration in soil solution Legend: The correction has been carried out by adding $0.36 \cdot K_{soil\ solution}$ to the ^{134}Cs concentration in the earthworm. For a 100% fit of the regression, this correction should result in a straight line with a slope of 6.89. The dotted line represents the equation $^{134}Cs_{earthworm} = 6.89 \cdot {}^{134}Cs_{soil\ solution}$.

slightly affected Cs K_d values in different soils, whereas K and ammonium were highly effective in desorbing Cs [30]. In these experiments, slight effects were also observed for the addition of Ca and ammonium, and the effects of K were smaller (Tables 11-2, 11-3).

High OM content of the soil often results in relatively high radiocesium concentrations in soil solution and thus in low K_d values [5,9,26]. Addition of OM strongly enhanced the amount of radiocesium in soil solution in the experiments. Brown and Bell [31], however, observed a decrease in ammonium-acetate extractable radiocesium after adding grass and apple leaves to a sandy soil. Obviously, the effect of adding OM is much more complex than adding K or ammonium, which only have an effect on the chemical distribution of related chemical elements. Hooda and Alloway [32] have pointed out that OM, applied as sewage sludge, may act both as a source for heavy metals and as a major absorbent, indicating changes in the physicochemical properties of the mixture of sludge and soil to be of paramount importance in determining bioavailability [32].

Concentrations in soil solution with time
This study showed that concentrations of different cations in soil solution change up to 100 d after the addition of K, Ca, ammonium, or OM. Treatments that caused an initial enhancement showed considerable changes with time (Figures 11-2A, 11-2B, 11-3A, 11-3B). Similar changes have been observed in other studies; however, in most studies, single measurements are reported rather than time series.

Cs concentration in soil solution decreased with time in most treatments; the decrease was largest where there were high initial Cs concentrations. These were the treatments in which the highest amount of Ca and ammonium, or the highest amount of K or OM, were added. (Table 11-3; Figures 11-3A, 11-3B). Brown and Bell [31] found a decreasing amount of ammonium-acetate extractable radiocesium with time and attributed this to Cs locking into the soil matrix. The decrease in radiocesium concentrations in earthworms in Norway in the period after Chernobyl was also explained by an increased binding of the deposited radiocesium, as total radiocesium concentration in the soil was relatively constant [33].

In most treatments with an enhanced pH, pH decreased about 1 unit (Figures 11-2A, 11-2B). Hooda and Alloway [32] observed a similar pattern after the addition of sewage sludge to sandy or clay-based soil. They state that decreases of pH with time in soils amended with OM have been reported widely. Lorenz et al. [29] found a highly variable pattern in pH and in K, Ca, Zn, and cadmium concentrations in soil solution after adding fertilizer. The observed patterns could be explained partly by ion exchange mechanisms but also by nitrification and uptake by plants.

Observations indicate that concentration factors based on single measurements may lead to wrong conclusions. During the course of 3 months, ^{134}Cs concentration in soil solution dropped by a factor of up to 6 in some treatments. In such cases, uptake should be integrated over time rather than using a single concentration to estimate concentration factors, especially when accumulation periods are relatively long. Jurinak and Tanji [28] point out that equilibria are generally dynamic instead of static but that, in many cases, traditional soil chemistry equilibrium models are applied.

Effect of earthworms on pH and cation concentrations in soil solution

The effects of earthworms on the solubility of different elements have received limited attention. Earthworms may affect their environment by burrowing, changing the humidity of the soil, increasing aeration (allowing oxidation of chemical compounds and soil OM), and changing the availability as food passes the gut and by excretion products [21].

In previous experiments in sandy soils, both *L. rubellus* and *E. andrei* increased the amount of radiocesium in soil solution [4]. In the results reported here, similar changes were observed. Brown and Bell [31] found no measurable effect of *Aporrectodea longa* on the available fraction of radiocesium after reworking soil and soil-organic mixtures for 3 months. This may be the consequence of snapshot measurements being taken.

Although the effects of the earthworms on Cs concentration in soil solution were largest, significant increases in the soil solution concentrations of other elements were also observed. Ammonium concentration showed a relatively large increase, which is in agreement with the results of Curry et al. [21]. Earthworms may contribute to the nitrogen turnover through the excretion of waste products and through the addition of intestinal and epidermal mucus to the soil [21]. Ammonia and urea are the main components in the excretion products of both *L. rubellus* and *E. fetida* [34].

A relatively small increase in the K concentration in soil solutions was observed after earthworms were added. Increased levels of K in soil solution in the presence of *E. andrei* and *L. rubellus* were observed in previous experiments at 20 °C, but not at 10 °C or 15 °C [4]. Species-specific differences in exchangeable K levels in worm casts were observed by Basker et al. [35]. The effects were more obvious for *L. rubellus* than for *A. caliginosa* and depended on soil type. Basker [35] suggested that earthworms do alter the K distribution between exchangeable and nonexchangeable phases.

The results of the present study indicate that species-dependent changes in pH may occur in the vicinity of earthworms. Similar results have been found in earlier experiments in liquid media and in soil [4,17,18, 36]. The pH change of the immediate environment of the earthworms may have been caused by the excretion of mucus or other excretory products [37]. Schrader [37] showed that the pH of the mucus to be species-specific, with *L. rubellus* excreting mucus with a pH of about 6.1. This species specificity may result in differences in uptake of certain elements.

Uptake of radiocesium and effect of additions on concentrations in earthworms
Recommendations for countermeasures have been partly based on [134]Cs:K ratios in soil solution, assuming that reduction of this ratio will lead to a reduction in [134]Cs uptake [8]. The approach in which ratios are being used to draw conclusions about the effects of countermeasures has been criticized by some authors [6,38]. Ratios assume a linear relationship between the presence of

1 element and the uptake of another. Linear relationships have not been observed consistently between plant and soil concentrations in nature [6,14].

In the experiments reported in this chapter, no effects of the different additions on the [134]Cs concentrations in *L. rubellus* were observed (Table 11-5), although concentrations relative to soil solutions differed. Statistical analysis showed K in soil solution to reduce the uptake of [134]Cs. The relationship between the [134]Cs concentration in soil solution and earthworms could be described by multiple linear regression:

$L\ rubellus$: $^{134}Cs = 6.89 \cdot {}^{134}Cs_{soil\ solution} - 0.36 \cdot K_{soil\ solution}$,   Equation 11-2

where $r^2 = 75.6\%$.

In earlier experiments conducted in liquid media [18], the relationship between the [134]Cs concentration in *L. rubellus* and in liquid media was described by the equation:

$L\ rubellus$: $^{134}Cs = 5.8 \cdot {}^{134}Cs_{medium} - 24.6 \cdot \ln(K_{medium}) - 8.19 \cdot \ln(Cs_{medium})$,   Equation 11-3

where $r^2 = 49.6\%$] [18].

In both equations, [134]Cs concentrations in *L. rubellus* are given in Bq/g dry weight, [134]Cs concentration in medium and soil solution in Bq/ml, and K and stable Cs concentrations in medium and soil solution in mmol/L. The regression coefficients for [134]Cs are comparable for both studies, whereas those for K are dissimilar. This might be attributed to the fact that K concentrations in the experiments in the liquid media were partly carried out in K-poor circumstances (0 to 10 mmol/L), whereas the K concentrations in soil solution in the experiments described above were 0.42 to 66.5 mmol/L. The largest effects of K in the liquid media were observed below 1 mmol K/L [18].

Fawaris and Johanson [9] observed K⁺ to reduce the uptake of ^{137}Cs by sheep's fescue (*Festuca ovina*), whereas stable Cs⁺ and ammonium enhanced the uptake of ^{137}Cs relative to soil. They observed a slight increase of radiocesium transfer factors with increasing OM content [9].

Change in pH may affect the net uptake of various elements. Helmke et al. [15] suggested that pH would have little effect on the uptake of trace elements by earthworms, as soils with different pH yielded similar concentrations in the earthworms. In contrast, Schrader [37] indicated that pH may have a significant effect on bioavailability. Similar remarks have been made by Bernal et al. [39] concerning the effect of change in plant rhizosphere pH on the availability of metals in soils. Kiewiet and Ma [19] observed that bioaccumulation of Pb depended on the pH of the medium, whereas for Cd it did not. Kiewiet and Ma [19] attributed this to the predominant speciation at different pHs and the uptake efficiency of these chemical species. Although the pH in the experiments conducted for this chapter varied between 4 and 8, it can be assumed to have only a small effect on Cs uptake. The pH slightly affects distribution of Cs between solid and liquid phases [30]. Earlier experiments did not show pH to affect uptake of Cs from liquid media significantly, whereas K additions resulted in a large reduction [18].

Mechanisms

The facts that Cs concentration in earthworms is higher than in the surrounding medium, and that K counteracts the uptake of Cs from liquid media suggest an active uptake of Cs [17,18].

Soil OM does not play a role in the specific retention of Cs by soils, and the process of specific binding is exclusively associated with the mineral soil component [5,40]. The mineral soil possesses binding sites with different affinities for Cs. It is assumed that Cs can easily be removed from the regular exchange or planar sites by competitors such as Ca, K, and ammonium, whereas from the frayed edge sites, removal is more specific [11,30,40]. The ability to remove Cs from the clay minerals is ion-specific.

Competition between different ions suggests that both K and ammonium transporters play a key role in uptake of Cs [12,13]. It has also been suggested that the plasma membrane possesses specific transport molecules with different affinities for either K⁺ or ammonium [41]. An approach in which knowledge of the affinities of different binding sites for Cs in soil and in membrane transport proteins is combined with kinetics could be a fruitful exercise. Results from such studies may also enable the correct interpretation of countermeasures. Integrated investigations of uptake kinetics and membrane behavior have also been recommended for understanding uranium uptake [14].

Conclusions

1) The addition of OM showed the highest impact on most soil-solution parameters observed: pH, conductivity, ^{134}Cs, K, and Ca all increased relative to the reference treatments.

2) The other additions (K, Ca, or ammonium) showed minor effects on the soil-solution parameters.

3) In reference pots (without addition of K, Ca, ammonium, or OM) ^{134}Cs concentrations in soil solution were lower than 1 Bq/ml; in pots in which OM had been added, initial concentrations were between 10 and 13 Bq/ml.

4) The largest changes in ^{134}Cs, K, and Ca concentration with time were observed in pots with enhanced initial concentrations (OM treatment).

5) ^{134}Cs concentration in soil solution decreased with time in all treatments.

6) Addition of earthworms (*L. rubellus* and *E. andrei*) resulted in a species-specific effect on the pH in soil solution and an increase in ^{134}Cs and NH_4^+ concentration in soil solution in all treatments.

7) When these initial concentrations were high, the decrease of the K and Ca concentrations in soil solution with time annulled the effect of the earthworms, which increased the K and Ca concentrations in soil solution.

8) The additions had significant effects on the ^{134}Cs concentration in soil solution, but not on the amount in the *L. rubellus*.

Acknowledgments - This study was supported by the CEC Radiation Protection Programme contract F13PCT920010. We would like to thank Leo Posthuma and Rob Baerselman for kindly providing *E. andrei* from their stock and Rob Alkemade for statistical advice. We acknowledge the Interfaculty Reactor Institute of the Technical University Delft for carrying out the neutron activation analysis and the Laboratory of Soil and Crop Research at Oosterbeek for the soil characterization. Three anonymous referees are thanked for their useful remarks in improving the quality of the manuscript.

References

1. Van Gestel CAM. 1992. The influence of soil characteristics on the toxicity of chemicals in earthworms: A review. In: Greig-Smith PW, Becker H, Edwards PJ, Heimbach F, editors. Ecotoxicology of earthworms. Hants, UK: Intercept. p 44–54.

2. Crommentuijn T, Doodeman CJAM, Doornekamp A, Van der Pol JJC, Bedaux JJM, Van Gestel CAM. 1994. Lethal body concentrations and accumulation patterns determine time-dependent toxicity of cadmium in soil arthropods. *Environ Toxicol Chem* 13:1781–1789.

3. Spurgeon DW, Hopkin SP. 1996. Effects of variations of the organic matter content and pH of soils on the availability and toxicity of Zn to the earthworm *Eisenia fetida*. *Pedobiologia* 40:80–96.

4. Janssen MPM, Glastra P, Lembrechts JFMM. 1996b. Uptake of ^{134}Cs from a sandy soil by two earthworm species: The effects of temperature. *Arch Environ Contam Toxicol* 31:184–191.

5. Bunzl K, Schimmack W. 1989. Associations between the fluctuations of the distribution coefficients of Cs, Zn, Sr, Co, Cd, Ce, Ru, Tc, and I in the upper two horizons of a podzol forest soil. *Chemosphere* 18:2109–2120.

6. McGee EJ, Johanson KJ, Keatinge MJ, Synnott HJ, Colgan PA. 1995. An evaluation of ratio systems in radioecological studies. *Health Phys* 70:215–221.

7. Rosén K, Andersson I, Lönsjö H. 1995. Transfer of radiocesium from soil to vegetation and to grazing lambs in a mountain area in Northern Sweden. *J Environ Radioactivity* 26:237–257.

8. Nisbet AF, Mocanu N, Shaw S. 1994. Laboratory investigation into the potential effectiveness of soil-based countermeasures for soil contaminated with radiocesium and radiostrontium. *Sci Tot Environ* 149:145–154.

9. Fawaris BH, Johanson KJ. 1994. Uptake of ^{137}Cs from coniferous forest soil by sheep's fescue in pot experiment. In: Gerzabek MH, editor. Soil-plant relationships. Proceedings 14th Annual ESNA/IUR Meeting; 1994 Sep 12–14; Varna, Bulgaria. Seibersdorf, Austria: FZS. p 186–196.

10. Gadd GM. 1996. Roles of micro-organisms in environmental fate of radionuclides. In: Health impacts of large releases of radionuclides. Ciba Foundation Symposium 203. Chichester UK: Wiley.

11. De Brouwer S, Thiry Y, Myttenaere C. 1994. Availability and fixation of radiocaesium in a forest brown acid soil. *Sci Total Environ* 143:183–191.

12. André B. 1995. An overview of membrane transport proteins in *Saccharomyces cerevisae*. *Yeast* 11:1575–1611.

13. Martinelle K, Häggström L. 1993. Mechanisms of ammonia and ammonium ion toxicity in animal cells: Transport across cell membranes. *J Biotechn* 30:339–350.

14. Sheppard SC, Evenden WG. 1992. Bioavailability indices for uranium: Effect of concentration in eleven soils. *Arch Environ Contam Toxicol* 23:117–124.

15. Helmke PA, Robarge WP, Korotev RL, Schomberg PJ. 1979. Effects of soil-applied sewage sludge on concentrations of elements in earthworms. *J Environ Qual* 8:322–327.

16. Connell DW, Markwell RD. 1990. Bioaccumulation in the soil to earthworm system. *Chemosphere* 20:91–100.

17. Janssen MPM, Glastra P, Lembrechts JFMM. 1996. Uptake of Cs-134 by the earthworm species *Eisenia foetida* and *Lumbricus rubellus*. *Environ Toxicol Chem* 15:873–877.

18. Janssen MPM, Glastra P, Lembrechts JFMM. 1997. The effects of soil chemical characteristics on the ^{134}Cs concentration in earthworms. Uptake from liquid media. *J Environ Radioactivity* 35:313–330.

19. Kiewiet AT, Ma WC. 1991. Effect of pH and Ca on lead and cadmium uptake by earthworms in water. *Ecotoxicol Environ Saf* 21:32–37.

20. Sims RW, Gerard B. 1985. Earthworms. In: Kermack DM, Barnes RSK, editors. Synopsis of the British fauna. London UK: EJ Brill/ W Backhuys.

21. Curry JP, Byrne D, Boyle KE. 1995. The earthworm population of a winter cereal field and its effects on soil and nitrogen turnover. *Biol Fertil Soils* 19:166–172.

22. Mubarak A, Olsen RA. 1976. Immiscible displacement of the soil solution by centrifugation. *Soil Sci Soc Am J* 40:329–331.

23. Anonymous. n.n. Ammonium colometry of blue 2.2'-isopropyl-5.5-methyl-indophenol (Berthelot's reaction). Darmstadt, Germany: E. Merck.

24. Debertin K, Helmer RG. 1988. Gamma- and X-ray spectrometry with semiconductor detectors. Amsterdam, the Netherlands: North-Holland.

25. Sheppard MI. 1985. Radionuclide partitioning coefficients in soils and plants and their correlation. *Health Phys* 49:106–111.

26. Shenber MA, Eriksson Å. 1993. Sorption behaviour of caesium in various soils. *J Environ Radioact* 19:41–51.

27. Zach R, Hawkins JL, Mayoh KR. 1989. Transfer of fallout Cs-137 and natural K-40 in a boreal environment. *J Environ Radioact* 10:19–45.

28. Jurinak JJ, Tanji KK. 1993. Geochemical factors affecting trace element mobility. *J Irrigation Drainage Engin* 119:848–867.

29. Lorenz SE, Hamon RE, McGrath SP, Holm PE, Christensen TH. 1994. Applications of fertilizer cations affect cadmium and Zn concentrations in soil solutions and uptake by plants. *Eur J Soil Sci* 45:159–165.

30. Wauters J, Sweeck L, Valcke E, Elsen A, Cremers A. 1994. Availability of radiocesium in soils: A new methodology. *Sci Total Environ* 157:239–248.

31. Brown SL, Bell JNB. 1995. Earthworms and radionuclides, with experimental investigations on the uptake and exchangeability of radiocaesium. *Environ Pollut* 88:27–35.

32. Hooda PS, Alloway BJ. 1993. Effects of time and temperature on the bioavailability of Cd and Pb from sludge amended soils. *J Soil Sci* 44:97–110.

33. Kålås JA, Bretten S, Byrkjedal I, Njåstad O. 1994. Radiocesium (^{137}Cs) from the Chernobyl reactor in eurasian woodcock and earthworms in Norway. *J Wildl Manage* 58:141–147.

34. Laverack MS. 1963. The physiology of earthworms. London UK: Pergamon.

35. Basker A, Kirkman JH, Macgregor AN. 1994. Changes in K availability and other soil properties due to soil ingestion by earthworms. *Biol Fert Soils* 17:154–158.

36. Kaplan DL, Hartenstein R, Neuhauser EF, Malecki MR. 1980. Physicochemical requirements in the environment of the earthworm *Eisenia foetida*. *Soil Biol Biochem* 12:347–352.

37. Schrader S. 1994. Influence of earthworms on the pH conditions of their environment by cutaneous mucus secretion. *Zool Anz* 233:211–219.

38. Janssen MPM, Glastra P, Lembrechts JFMM. 1997. The effects of environmental parameters on the availability of Cs for earthworms. In: Amiro B, Avadhanula R, Johansson G, Larsson CM, Lüning M, editors. Intern. Symposium on Ionising Radiation. Protection of the Natural Environment; 20–24 May 1996; Stockholm, Sweden. p 224–229.

39. Bernal MP, McGrath SP, Miller AJ, Baker AJM. 1994. Comparison of the chemical changes in the rhizosphere of the nickel hyperaccumulator *Alyssum murale* with the non-accumulator *Raphanus sativus*. *Plant Soil* 164:251–259.

40. Cremers A, Elsen A, De Preter P, Maes A. 1988. Quantitative analysis of radiocaesium retention in soils. *Nature* 335:247–249.

41. Perkins J, Gadd GMG. 1993. Caesium toxicity, accumulation and intracellular localization in yeasts. *Mycol Res* 97:717–724.

Chapter 12

Depuration and uptake kinetics of I, Cs, Mn, Zn, and Cd by the earthworm (*Lumbricus terrestris*) in radiotracer-spiked litter*

Steve C. Sheppard, William G. Evenden, and Teresa C. Cornwell

The relative depuration and uptake kinetics of contaminants should be known in order to interpret appropriately the use of organisms such as earthworms in environmental bioassays and monitoring. For example, 14-d earthworm bioassays should be interpreted with the knowledge that some contaminants will continue to accumulate in tissues for months. The radiotracers ^{125}I, [134]Cs, [54]Mn, [65]Zn, and [109]Cd were applied to deciduous litter, and specimens of *Lumbricus terrestris* were exposed either to litter alone or to litter on the top of soil columns. Depuration was monitored for 120 d and uptake, in a separate experiment, for 20 d. Both depuration and uptake were described using 2-phase, first-order statistical models. Gut clearance had a mean half-time of 1.4 d. The mean half-time for physiological depuration decreased from I (210 d) > Cd (150 d) > Zn (69 d) > Mn (40 d) > Cs (24 d). Both the depuration and the uptake experiments were necessary to resolve even partially the multiphase processes. Earthworm/soil dry weight concentration ratios decreased from Cd > Zn > I ≥ Cs ≥ Mn. The very slow kinetics indicate that tissue concentrations will increase continuously for a long time, with important implications for subsequent food-chain transfers.

Earthworms have been used extensively to monitor contaminated soils [1] and to assess potential impacts of contaminants in controlled laboratory studies [2]. Often, the use is made without regard to the rate at which earthworm tissue concentrations respond to substrate concentration. This has 2 practical implications, 1 for the use of earthworms in monitoring and the other for their use in bioassays.

In field monitoring use, where earthworms are collected as a means to measure contamination of a site, the kinetics of uptake and depuration partially determine the integration area [3,4]. For example, if tissue concentrations respond very rapidly to soil concentrations, then the concentrations observed in the earthworm clearly reflect the immediate area from which it was collected. If the kinetics are slow, the observed concentrations may reflect a soil at some distance from the collection site. This may not be an issue where many earthworms are collected, as the average concentration will be reliable.

In the laboratory, there are other concerns. Ecotoxicological testing of soils with earthworms typically records the effects on the health of earthworms after a relatively short exposure time of several days [5]. However, for elements such as Cd, there is good evidence that the kinetics are very slow and that accumulation of the element, resulting in progressively higher tissue concentrations, may continue for the normal life span of the

---

*This chapter is reprinted from *Environmental Toxicology and Chemistry* 16:2106–2112.

earthworm [6–9]. Obviously, higher tissue concentrations are not necessarily related to health, especially in earthworms that can sequester metals and avoid physiological impact [10–15]. However, it must be remembered that many toxicity data are not obtained at or near steady state and that steady-state tissue concentrations can be much higher than those in a short-term bioassay. In addition, it is important to know when steady state may occur. Steady state is a much preferred condition for standardization of tests: prior to steady state, concentrations are changing rapidly and so may the observable effects on health of the earthworms.

The purpose of the present study was to examine depuration and uptake kinetics with particular emphasis on elements important to the nuclear industry. Iodine is important because of the long-lived isotope [129]I, which is often the most critical nuclide in long-term human impact assessments. Sheppard and Evenden [2] showed that, because of its very long half-life, the chemical toxicity of [129]I may equal or exceed in importance the radiological toxicity for soil animals. Thus, information on the effects of I on soil biota is useful. Sheppard and Evenden [2] included earthworms (*Lumbricus terrestris*) in their toxicity assays. However, there was no effect on survival at concentrations 2 orders of magnitude above where other biota responded. In part, the present study was undertaken to determine if this may be because of slow uptake of I by earthworms.

Cesium is also important to the nuclear industry, and earthworms have significant connections with Cs investigations because of the recycling of Cs from Chernobyl in the litter of contaminated forests [16–18], and because Cs in earthworms is a potential monitoring indicator, especially for buried Cs (Sheppard, unpublished data).

The other elements included in this study, Mn, Zn, and Cd, are important to other industries. Manganese compounds are used in gasoline to replace Pb as an anti-knock compound [19]. These compounds, when emitted to the atmosphere, are rapidly oxidized and inorganic Mn released. However, this Mn is deposited on vegetation and litter without any mediation by soil, so earthworms consuming litter will be exposed to Mn that was not plant-incorporated. The other metals, Zn and Cd, have been extensively investigated in connection with many industries and were included here primarily to allow comparison to the literature.

The experiment used litter spiked with radiotracers of the elements. The radiotracers allowed nondestructive sequential measurement in individual earthworms [20]. Separate experiments were used to measure depuration and uptake. The emphasis was on earthworms exposed exclusively to spiked litter, but earthworms were also exposed to spiked litter on the top of soil columns to represent a more realistic situation.

Materials and methods

Earthworms and containers
The earthworms were obtained from an Ontario commercial bait supplier. The entire supply was determined to be *L. terrestris*, and this was confirmed by an independent spe-

cialist. All earthworms used in the experiments were sexually mature and selected for relatively uniform size and weight.

Two types of containers were used in both the depuration and uptake experiments. The first contained only litter and were 500-ml plastic food containers with small holes in the lids for ventilation. The second were opaque acrylic columns, 5-cm inside diameter by 37 cm deep, filled with a commercial potting soil. These had screen bottoms to allow for free passage of air and water and 500-ml plastic containers fixed to the top to hold the spiked litter. All earthworms were acclimatized in their respective containers and fed unspiked litter for 6 d prior to the experiments. There was 1 earthworm in each litter container and 10 earthworms in each soil column. The experiments and stock of earthworms were kept in a temperature-controlled environment at 18 ± 2 °C with 12 h of light daily.

Litter preparation
Litter was collected in the spring in a mixed forest region in southern Manitoba, Canada. Stems and large debris particles were removed, leaving a well-mixed composite primarily of birch (*Betula papyrifera*) and aspen (*Populus tremuloides*) leaves with some coniferous needles. This was stored moist until used.

A 1-kg wet-weight (69% moisture content) aliquot of litter was spiked, using multiple additions of 10 ml of the spike solutions. The radiotracers were added as 2 different solutions, so that pH could be adjusted to optimize the transfer of the radiotracers. Iodine-125 was added in a 0.2-mol/L solution of NaOH, and ^{134}Cs, ^{54}Mn, ^{65}Zn, and ^{109}Cd were in a 0.1-mol/L solution of HCl. The ^{125}I solution was added first, with thorough mixing of the litter material between each 10-ml addition. This was followed by addition of the other radiotracers in the same manner. A separate aliquot of spiked litter was prepared for the uptake study, but no ^{65}Zn was included because of difficulties in obtaining the radiotracer. The leaf litter used for the experiments had the measured radiotracer concentrations shown in Table 12-1.

Depuration study
The depuration study began with 50 earthworms placed in individual containers with 10 g of the spiked litter. At the same time, the unspiked litter in 15 soil columns was replaced with spiked litter. After 20 d, 20 earthworms were selected from the litter containers based on apparent health (1 died during the study). Only 13 earthworms could be retrieved from the surface of the soil columns without disruption of the columns. All selected worms were rinsed with tap water, weighed, and placed

Table 12-1 Measured concentrations of the radiotracers (Bq/g dry weight) in the litter, showing means and standard deviations (n = 10)[a]

Radiotracer	Depuration experiment	Uptake experiment
I	5700 ± 1600	8400 ± 2400
Cs	98 ± 31	300 ± 130
Mn	180 ± 96	610 ± 270
Zn	8.9 ± 4.2	-
Cd	370 ± 200	1200 ± 500

[a]Analysis was of 2.5-g samples.

alive in 22-ml liquid scintillation counting (LSC) vials for counting. Gamma activity was measured, in 20-min counts, on an Ortec high-purity germanium coaxial photon detector (GMX series with a 25% relative efficiency). After the first count, the earthworms from both the litter and the soil columns were divided into 2 groups. Subsequent counts were done on alternate groups to reduce the counting frequency and stress for individual earthworms. Counts were done 1, 2, 7, 9, 16, 23, 27, 50, 76, and 120 d after removal from spiked litter (only [109]Cd was detectable at 120 d). After each count, the earthworms were placed individually in clean 500-ml containers with fresh, unspiked litter. Earthworms from the soil columns were also given 250 g of clean soil to burrow through.

Uptake study
Forty-five litter containers and 16 soil columns were switched to spiked litter. On 1, 3, 6, 10, 14, and 20 d after the addition of the spiked litter, 6 earthworms from the containers and up to 6 (depending on capture success) from the soil columns were collected. They were placed in new containers and fed moistened cellulose until gut clearance was complete. This normally required about 2 d. The earthworms were then placed into LSC vials and the gamma activity determined. Because of the gut clearance procedure, these earthworms could not be returned to spiked litter for further analyses of uptake, and so all datapoints from this experiment are of different earthworms.

Results and discussion

Depuration
The depuration study included depuration by gut clearance, and in the first few days this process dominated the loss from the earthworms of several of the radiotracers. After this, the slower rate of depuration from the body tissues varied among the radiotracers, because of the different physiological behaviors of the elements. Although the rate of gut clearance was similar for all the radiotracers, the fraction of the radiotracer in the gut versus the tissue varied. For example, an element that was not very bioavailable and was not absorbed to a large extent from the gut would have a rapid early depuration rate because gut clearance would be the dominant process.

The statistical model used to characterize these processes was a 2-phase first-order model:

$$C = A(e^{-\lambda g t}) + (1 - A)(e^{-\lambda p t})$$ Equation 12-1.

The value of A represents the fraction of the radiotracer left in the gut and subject to a gut-clearance depuration rate constant, λg. The remaining fraction, $1 - A$, is subject to a slower, physiologically mediated depuration rate constant, λp. The value of λg was defined and restricted to be the same for all the radiotracers, whereas A and λp were different for each radiotracer. Equation 12-1 was iteratively fit to the data for each earthworm separately, and the resulting coefficients averaged (Table 12-2). The nonlinear (NLIN) procedure of the Statistical Analysis System (SAS) was used; it iteratively selects coefficient values to minimize the sums of squares. The statistical model applied Equation 12-1 for 3 to 5 radiotracers simultaneously. A coefficient was considered not different from 0 when the coefficient divided by its asymptotic standard error of estimate gave a quotient smaller than the corresponding t value.

The data for Zn and Cd were more variable than the other radiotracers, and this caused some numerical convergence difficulties. To remedy this, the value of λg for each earth-

Table 12-2   Coefficients for Equation 12-1 describing the depuration process for worms in litter[a]

Parameter	I	Cs	Mn	Zn	Cd
A mean	0.51	0.38	0.87	0.15	0.27
SD	0.080	0.091	0.088	0.15	0.20
λg mean (per d)	0.51	0.51	0.51	0.51	0.51
SD (per d)	0.15	0.15	0.15	0.15	0.15
Half time (d)	1.4	1.4	1.4	1.4	1.4
λp mean (per d)	0.0033	0.029	0.017	0.010	0.0045
SD (per d)	0.0033	0.0081	0.016	0.0065	0.0069
Half time (d)	210	24	40	69	150

[a]Coefficients were derived from each worm ($n = 19$), and means and standard deviations are given here. All means were significantly greater than 0 ($P < 0.05$). The half times shown are calculated from the means as 0.693/mean. The values λg apply to all 5 elements.

worm was determined with data for 3 radiotracers, Cs, I, and Mn, only, and then this value was used for λg when equations to describe the data for all the radiotracers were determined. The data for each earthworm and the depuration curve based on the average values of the fitted equation coefficients are shown in Figure 12-1.

The values of A indicate relative bioavailability: a large value of A means that the radiotracer was poorly bioavailable compared to the others, and not readily accumulated from the substrate into earthworm tissue. Among the radiotracers, Mn was the least bioavailable (Table 12-2). The mean value shown indicates that when the earthworms were first analyzed, 87% of the ^{54}Mn associated with the earthworm was in the gut contents and not in the earthworm tissue. In contrast, very little of the Zn or Cd was lost by gut clearance, implying that much of these radiotracers was in the tissues.

The mean half-time for gut clearance was 1.4 d, and the range among all the earthworms was 0.94 to 2.7 d. Obviously, this would be dependent on the activity level of the earthworms. In the field, this would vary even more, in response to variables such as temperature [21], light level, food availability, and physiological state.

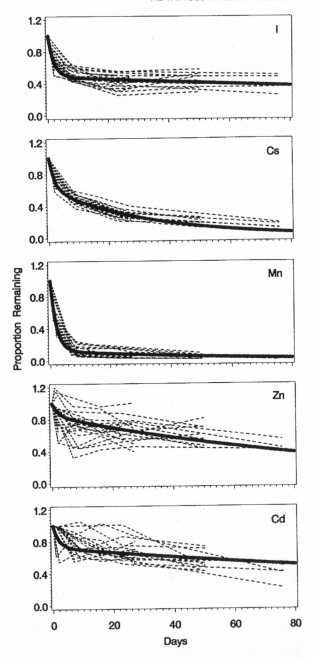

Figure 12-1   Depuration curves showing fraction remaining with time of the total amount of
               radiotracer in earthworms when they were removed from spiked litter Legend: The fine
               lines show the observed results for individual earthworms, and the bold line is drawn
               based on the means of the coefficients determined for Equation 9-1 fit for each
               individual.

The mean half-time for physiological clearance of the tissues was relatively long; 24 d for Cs was the shortest. This has a number of implications. For one, the depuration rate influences the time required for the concentrations in the tissue to reach steady state. Clearly, steady state would not be achieved in an exposure time of 14 d, a common exposure time for the 50% lethal concentration (LC50) and related ecotoxicological endpoints. Secondly, these long depuration times suggest that earthworms collected at a heterogeneous site may integrate the effect of soil concentrations over the area they have visited in the past several weeks or even months. This is advantageous, in that it is a biological integration over space, as opposed to a chemical–analytical/mathematical integration. Also, fewer earthworms need to be sampled to represent a large area. It is a disadvantage if the precise location of an isolated zone of contamination is sought.

The long physiological half-time for I may be the result of both the physiological need for I and the relatively low amount of I available in leaf litter. The I cycle in the earthworm might be very conservative. The radiotracers with the next longest physiological half-times were Zn and Cd. These radiotracers share many of the same biogeochemical attributes, and so half-times were expected to be similar. There is general agreement in the literature [22], and supported here, that Cd remains longer and accumulates to higher relative concentrations than does Zn.

The physiological half-time for Mn is interesting, because it happened to coincide with a very low bioavailability. Only a small portion of the Mn in the litter was bioavailable (high A value, Table 12-2), and what was absorbed was depurated rapidly, with a half-time of 40 d. There are few data in the literature on Mn uptake by earthworms. Ireland [23] speculated that Mn uptake was regulated, largely based on low-observed-concentration ratios ([CRs] concentration in the earthworm on a dry weight basis divided by that in the substrate also on a dry weight basis) in 3 species. Andersen and Laursen [10] give much stronger evidence for regulation in *L. terrestris*, based on very high observed Mn concentrations (up to 1.6 g/kg) in excretory glands. Helmke et al. [24] observed CRs for Mn in *Aporrectodea tuberculata* ranging from 0.01 to 0.09, and Hendriks et al. [25] in *Lumbricus rubellus* ranging from 0.08 to 0.15. The only kinetic data found were by Honda et al. [6], who related tissue concentrations to body length as an index of age and found no increase in Mn concentrations with length. The range of lengths was 5 to 20 cm, and in the species studied, *Pheretima hilgendorfi*, this may represent a time range of 1 to 10 months. For steady state at 1 month and given the confidence bounds of the data, the net half-time would be < 30 d.

The physiological half-time of Cs was the shortest of the radiotracers measured. The values obtained are consistent with Brown and Bell [16], who observed physiological clearance half-times for *Aporrectodea longa* of 15 to 54 d. Crossley et al. [26] measured Cs depuration kinetics in several earthworm species, and for *L. terrestris* the gut clearance half-time was 0.5 d and the physiological clearance half-time was 2 d. However, Crossley et al. [26] only exposed the earthworms to spiked litter for 2 d, and the results reported here and those of Brown and Bell indicate this was not long enough to reach steady-state tissue concentrations. Crossley et al. [26] also observed that earthworms (*Octolasium lacteum*) absorbed Cs only from spiked litter and not from spiked soil, whereas Brown and Bell [16] observed uptake from both media. The behavior of Cs is often related to the abundance of K in the environment because the 2 elements have similar properties and can substitute for each other in some processes. The relatively rapid kinetics for Cs may reflect that of K, when K is regulated to some extent as an osmotic agent.

This hypothesis is supported by the results for depuration of the radiotracers by earthworms from the soil columns (Table 12-3), where they burrowed in soil but still had access to the spiked litter. These earthworms had much lower initial concentrations of the radiotracers than those exposed only to litter. Presumably this is because they were able to utilize soil organic matter as an additional source of elements, and they burrowed in the soil and so were less exposed to the litter. The low initial concentrations meant the depuration curves for Cs, Mn, and Zn were not characterized as well as for the litter-only earthworms because the detection limit was reached in as little as 10 to 20 d. One other notable difference between the litter and soil column earthworms is the values of *A*. For each radiotracer, *A* was somewhat to substantially larger for the earthworms from the soil columns. This indicates lower bioavailability of the radiotracers and may be related to the fact that the earthworms had alternate sources for these elements and that soil in the gut may have attenuated the radiotracers and reduced absorption.

Table 12-3 Coefficients for Equation 12-1 describing the depuration process for worms from soil columns[a]

Parameter	I	Cs	Mn	Zn	Cd
A mean	0.61	0.75	0.81	0.57	0.32
SD	0.14	0.20	0.17	0.17	0.14
λg mean (per d)	0.54	0.54	0.54	0.54	0.54
SD (per d)	0.32	0.32	0.32	0.32	0.32
Half time (d)	1.3	1.3	1.3	1.3	1.3
λp mean (per d)	0.0041	NS[b]	NS	NS	0.012
SD (per d)	0.0041				0.019
Half time (d)	170				58

[a]Coefficients were derived from each worm (*n* = 13), and means and standard deviations are given here. All means except l *p* for Cs, Mn, and Zn were significantly greater than 0 (*P* = 0.05). The half times shown are calculated from the means as 0.693/mean. The values for l *g* apply to all 5 elements.
[b]NS = not significant

This effect was most distinct for Cs and Zn. As discussed previously, Cs is an analog element to K, and the soil was rich in K. Thus, the higher *A* values for Cs, indicating decreased bioavailability, likely reflect the presence of K in the soils. The same may be true for Zn.

Although the depuration study provided estimates of both slow and rapid phases of depuration, the most compelling information is about the long-term retention. It was very clear that I and Cd, in particular, were retained in the tissues. These radiotracers remained at 20 to 50% of their initial tissue concentrations after 80 d, and Cd was at 25% of its initial concentration after 120 d.

Uptake

The methodology for the uptake study differed most from the depuration study in that each earthworm was measured only once. Because their guts were allowed to clear for 2 d prior to analysis, worms measured and then replaced in the litter would not meet the criteria of steady-state exposure needed to measure uptake repeatedly. As a result, Equation 12-2 was applied to all earthworms measured, not to individuals, and each point on the curves is independent.

$$C = B[A(1 - e^{-\lambda rt}) + (1 - A)(1 - e^{\lambda pt})]$$ Equation 12-2.

The coefficient B was iteratively fitted, has concentration units, and the values of B are specific to this experiment. The coefficient A requires a different interpretation here: a high value of A indicates that rapid processes dominated the uptake. The coefficient λp is as defined for Equation 12-1. The rapid elimination rate coefficient, λr, is different from λg of Equation 12-1 because gut clearance here was done prior to the measurements. The processes reflected by λr are assumed to be physiological, in that the radiotracers are not in the gut, but they are more rapid than those determining λp. Because the experiment lasted only 20 d, it did not provide a good basis for estimation of λp. Instead, the mean values from the depuration experiment (Table 12-2) were substituted as constants in Equation 12-2. The iteratively determined values of A and $8r$ are given in Table 12-4. Only the data for earthworms in litter are shown: the concentrations in earthworms exposed in the soil columns were highly variable and showed no consistent trends with time. From these, it can only be said that uptake of I, Cs, Zn, and Cd was rapid enough that concentrations were relatively uniform after only 1 d.

Table 12-4   Coefficients for Equation 12-2 describing the uptake of elements by worms in litter[a]

	I	Cs	Mn	Cd
A	0.19	1.0	0.64	0.0002
λr (per d)	0.85	0.15	2.2	0.00002
Half time (d)	0.82	4.6	0.32	> 100

[a]Half times shown are calculated from the mean as $0.69/\lambda r$.
Values for λp from Table 12-2 were used in Equation 12-2.

The very low value of A for Cd indicates that the uptake was dominated by long-term processes as characterized in the depuration study. The value of λr for Cd in Table 12-4 has little meaning because A is so small, except that it emphasizes the slow processes involved. The plots of the data (Figure 12-2) show almost linear increase in tissue concentration with time. The low value of A for I also indicates that the uptake was dominated by long-term processes. However, the plot (Figure 12-2) shows there was an important rapid early uptake phase for I.

In contrast to Cd and I, the A value for Cs was at the maximum possible value (unity), suggesting that uptake of Cs was dominated by rapid processes. This is consistent with the results of the depuration study, in that Cs had the shortest half-time of the 5 radiotracers in that experiment. Clearly, Cs is rapidly accumulated compared to the other elements studied here. Janssen et al. [21] also showed rapid uptake of Cs, with a half-time of 2.2 to 5.0 d.

The results for Mn are different again. For it, 64% was absorbed very rapidly, more rapidly than the methods employed in this study could accurately characterize, and the remaining 36% was accumulated slowly. Certainly, without the evidence of some long-term retention found in the depuration study (Figure 12-1), the uptake of Mn would be assumed to be at steady state almost immediately (Figure 12-2).

The uptake study was most suited to detect rapid uptake, whereas the depuration study provided good evidence of long-term retention. Neither alone provided complete information. However, the variation among the radiotracers is quite apparent, reflecting the complexity of the underlying processes, and Equations 12-1 and 12-2 must be interpreted as only empirical representations of those processes.

Concentration ratios

Concentration ratios can be computed in a number of ways from this study. In all cases, the substrate was the spiked litter, so that these data cannot be interpreted in the same way as those in the literature computed with soil as the substrate. Using an average moisture content for the earthworms of 85% (±1.6%, $n = 10$), the observed concentrations at the start of the depuration study give CR values in the range of 2.0 for Mn to 12.6 for Cd (Table 12-5). These data include gut contents and are appropriate for food-chain interpretations where a predator would consume the earthworm whole. The derived depuration curves can be used to adjust these values for gut clearance, and this decreases the CR values by a large amount for Mn but less for the other radiotracers (Table 12-5).

The CR values observed at the start of the depuration study for earthworms from the soil columns are almost an

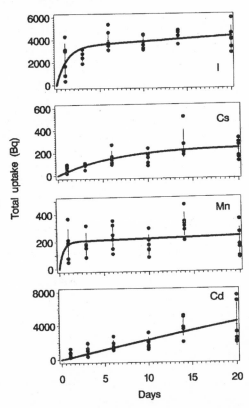

Figure 12-2   Uptake curves for worms on spiked litter only, showing the total amount of radiotracer (Bq per earthworm) with time Legend: The points are individual observations, the vertical bars are 1 standard deviation above and below the mean (the mean is the center of the vertical bar), and the bold line is the result of fitting Equation 12-2.

order of magnitude lower than those exposed solely to litter because the substrate for both is assumed in the calculation to be the spiked litter. As mentioned previously, this probably reflects the supplementary uptake of elements from the soil, the role of soil in the gut attenuating the absorption of the radiotracers, the decreased exposure to the litter as a result of burrowing, as well as the fact that the gut contents will include some unspiked soil.

The coefficient B from Equation 12-2 is an estimate of the concentration in the earthworm tissue (excluding gut contents) at steady state (Table 12-5). These CR values may best represent field conditions. The value most distinctly different from those obtained in the depuration study are for Cd. As is well established in the literature, Cd has the potential to accumulate to a very high degree in earthworms [27–31]. The fitted coefficient of Equation 12-2 can be used to compute tissue concentrations and then CR values corresponding to a 20-d exposure to the spiked litter (Table 12-5). These are most comparable to the results of the start of the depuration study, and at 20 d the estimated CR values for Cd correspond quite well to the observed values.

Table 12-5   Concentration ratios computed as worm/litter ratio of dry-weight concentrations[a]

Element	Observed at 20 d		Estimate at 20 d	Estimate at steady state
	Litter	Soil		
I	2.1 (1.3)	0.60 (2.0)	0.73	3.0
Cs	3.6 (1.6)	0.37 (2.1)	1.1	1.2
Mn	2.0 (1.5)	0.41 (2.2)	0.55	0.74
Zn	4.9 (1.8)	1.8 (1.4)	-	-
Cd	12.6 (2.0)	1.4 (2.2)	5.3	62

[a]Observations at 20 d were for worms prior to the beginning of the depuration experiment and included gut contents. Estimates at 20 d and at steady state were based on equations fit to describe the data of the uptake experiment and do not include gut contents.

The present data agree well with the literature. Few other studies rank the elements studied here. Geometric mean CRs computed from Ma [32] ranked Cd > Zn > Mn, with values of 30, 10, and 0.14, respectively. Voigt et al. [33] ranked transfer to milk as I > Cs > Mn, as observed for the estimated CRs (Table 12-5).

Some caution must be expressed with regard to the use in this study of carrier-free radiotracers. Radiotracers will reflect the behavior of stable isotopes of equal bioavailability, until concentration-dependent processes become involved. For example, the uptake rates reported here for Cd reflect a situation of low elemental Cd concentration. In a contaminated setting, the elevated Cd concentrations may induce a lower uptake rate, a greater depuration rate, or an altered steady-state tissue concentration because of concentration-induced effects. However, the observations reported in this study agree reasonably well with other studies that used elevated metal concentrations and are appropriate for many contamination settings.

Conclusions

Rapid depuration of the radiotracers was the result of gut clearance, with a mean half-time for earthworms exclusively in litter of 1.4 d. For Mn, 87% of the radiotracer in the earthworm when it was extracted from the spiked litter was lost as a result of gut clearance. This fraction was much smaller for the other radiotracers, especially for Zn and Cd where less than 30% of the radiotracer in the earthworm was lost by gut clearance. For all the radiotracers, some portion was retained in the tissues, with loss half-times of 24 to 210 d. The longest loss half-times were for I (210 d) and Cd (150 d), and for these it is apparent that concentrations of the radiotracer in the tissues will continue to increase for much of the normal life span of the earthworms. This has important implications for food-chain transfers. The shortest half-time associated with loss from tissues was for Cs (24 d), but even here 20% of the initial concentration was still present in the tissues after 80 d. The depuration study was most useful to illustrate the phases of the elements in the earthworms that responded to slow processes.

The uptake study for earthworms exclusively in litter was most useful to illustrate the processes that responded rapidly, apart from gut clearance. For I and Mn in particular, it was evident that the initial uptake was quite rapid. However, for all the elements, the longer-term processes likely controlled the steady-state tissue concentrations.

The earthworms exposed to spiked litter on the top of a soil column showed similar kinetics to those exclusively in litter, with some exceptions. For one, the tissue concentrations were consistently lower for earthworms that had access to soil. This may be because there was an additional supply of the stable elements from the soil and because the earthworms were less exposed to the litter. This explanation may also hold for Cs, because K is a close chemical analog of Cs. The provision of extra K by the soil may have been what increased the depuration rate of Cs and caused lower steady-state concentrations.

The combination of depuration and uptake studies was necessary to illustrate the various rate processes involved. Each experimental method was better suited to measure specific ranges of rates.

Acknowledgments - This project contributed to programs of the Deep River Science Academy (DRSA). The students involved, J. Miklos and J. Thompson, both won awards associated with the project. The authors acknowledge financial support from Atomic Energy of Canada Limited, Ontario Hydro under the auspices of the CANDU Owners Group, and the DRSA. The support of A. Tomlin and staff was most appreciated. The manuscript was improved by the thoughtful comments provided by G.A. Bird and M. Stephenson.

References

1. Callahan CA, Menzie CA, Burmaster DE, Wilborn DC, Ernst T. 1991. On site methods for assessing chemical impact on the soil environment using earthworms: A case study at the Baird and McGuire Superfund site, Holbrook, Massachusetts. *Environ Toxicol Chem* 10:817–826.

2. Sheppard SC, Evenden WG. 1995. Toxicity of soil iodine to terrestrial biota, with implications for ^{129}I. *J Environ Radioact* 27:99–116.

3. Marinussen MPJC, van der Zee SEATM, deHaan FAM, Bouwman LM, Hefting MM. 1997. Heavy metal (copper, lead, and zinc) accumulation and excretion by the earthworm, *Dendrobaena veneta*. *J Environ Qual* 26:278–284.

4. Morgan JE, Morgan AJ.1993. Seasonal changes in the tissue metal (Cd, Zn, and Pb) concentrations in two ecophysiologically dissimilar earthworm species: pollution monitoring implications. *Environ Pollut* 82:1–7.

5. [OECD] Organization for Economic Cooperation and Development. 1984. Organization for Economic Cooperation and Development guidelines for testing of chemicals: earthworm acute toxicity test. OECD Guideline No. 207. Paris, France: OECD.

6. Honda K, Nasu T, Tatsukawa R. 1984. Metal distribution in the earthworm, *Pheretima hilgendorfi*, and their variations with growth. *Arch Environ Contam Toxicol* 13:427–432.

7. Neuhauser EF, Cukic ZV, Malecki MR, Loehr RC, Durkin PR.1995. Bioconcentration and biokinetics of heavy metals in the earthworm. *Environ Pollut* 89:293–301

8. Perämäki P, Itämies J, Karttunen V, Lajunen LHJ, Pulliainen E. 1992. Influence of pH on the accumulation of cadmium and lead in earthworms (*Aporrectodea caliginosa*) under controlled conditions. *Ann Zool Fenn* 29:105–111.

9. van Gestel CAM, Dirven-van Breemen EM, Baerselman R. 1993. Accumulation and elimination of cadmium, chromium and zinc and effects on growth and reproduction in *Eisenia andrei* (Oligochaeta, Annelida). *Sci Total Environ* Suppl:585–597.

10. Andersen C, Laursen J. 1982. Distribution of heavy metals in *Lumbricus terrestris*, *Aporrectodea longa* and *A. rosea* measured by atomic absorption and X-ray fluorescence spectrometry. *Pedobiologia* 24:347–356.

11. Honeycutt ME, Roberts BL, Roane DS. 1995. Cadmium disposition in the earthworm *Eisenia fetida*. *Ecotoxicol Environ Saf* 30:143–150.

12. Ireland MP, Richards KS. 1981. Metal content, after exposure to cadmium, of two species of earthworms of known differing calcium metabolic activity. *Environ Pollut* 26:69–78.

13. Morgan JE, Morgan AJ. 1990. The distribution of cadmium, copper, lead, zinc and calcium in the tissues of the earthworm *Lumbricus rubellus* sampled from one uncontaminated and 4 polluted soils. *Oecologia* 84:559–566.

14. Morgan JE, Morgan AJ. 1989. Zinc sequestration by earthworm (Annelida: Oligochaeta) chloragocytes. An in vivo investigation using fully quantitative electron probe X-ray micro-analysis. *Histochemistry* 90:405–411.

15. Suzuki KT, Yamamura M, Mori T. 1980. Cadmium-binding proteins induced in the earthworm. *Arch Environ Contam Toxicol* 9:415–424.

16. Brown SL, Bell JNB. 1995. Earthworms and radionuclides, with experimental investigations on the uptake and exchangeability of radiocaesium. *Environ Pollut* 88:27–39.

17. Kålås JA, Bretten S, Byrkjedal I, Njåstad O. 1994. Radiocesium (^{137}Cs) from the Chernobyl reactor in Eurasian Woodcock and earthworms in Norway. *J Wildl Manage* 58:141–147.

18. Müller-Lemans H, van Dorp F. 1996. Bioturbation as a mechanism for radionuclide transport in soil: relevance of earthworms. *J Environ Radioact* 13:7–20.

19. Loranger S, Zayed J. 1994. Manganese and lead concentrations in ambient air and emission rates from unleaded and leaded gasoline between 1981 and 1992 in Canada: a comparative study. *Atmos Environ* 28:1645–1651.

20. Crossley Jr DA, Blood ER, Hendrix PF, Seastedt TR. 1995. Turnover of cobalt-60 by earthworms (*Eisenia foetida*) (Lumbricidae, Oligochaeta). *Appl Soil Ecol* 2:71–75.

21. Janssen MPM, Glasstra P, Lembrechts JFMM. 1996. Uptake of cesium-134 by the earthworm species *Eisenia foetida* and *Lumbricus rubellus*. *Environ Toxicol Chem* 15:873–877.

22. Pizl V, Josens G. 1995. The influence of traffic pollution on earthworms and their heavy metal contents in an urban ecosystem. *Pedobiologia* 39:442–453.

23. Ireland MP. 1979. Metal accumulation by the earthworms *Lumbricus rubellus, Dendrobaena veneta* and *Eiseniella tetraedra* living in heavy metal polluted sites. *Environ Pollut* 79:201–206.

24. Helmke PA, Robarge WP, Korotev RL, Schomberg PJ. 1979. Effects of soil-applied sewage sludge on concentrations of elements in earthworms. *J Environ Qual* 8:322–327.

25. Hendriks AJ, Ma W-C, Brouns JJ, de Ruiter-Dijkman EM, Gast R. 1995. Modeling and monitoring organochlorine and heavy metal accumulation in soils, earthworms, and shrews in Rhine-Delta floodplains. *Arch Environ Contam Toxicol* 29:115–127.

26. Crossley DA, Reichle DE, Edwards CA. 1971. Intake and turnover of radioactive cesium by earthworms (Lumbricidae). *Pedobiologia* 11:71–76.

27. Andersen C. 1979. Cadmium, lead and calcium content, number and biomass, in earthworms (Lumbricidae) from sewage sludge-treated soil. *Pedobiologia* 19:309–319.

28. Ash CPJ, Lee DL. 1980. Lead, cadmium, copper and iron in earthworms from roadside sites. *Environ Pollut* 22:59–67.

29. Brewer SR, Barrett GW. 1995. Heavy metal concentrations in earthworms following long-term nutrient enrichment. *Bull Environ Contam Toxicol* 54:120–127.

30. Gonzalez MJ, Ramos L, Hernandez LM. 1994. Distribution of trace metals in sediments and the relationship with their accumulation in earthworms. *Int J Environ Anal Chem* 57:135–150.

31. Wright MA, Stringer A. 1980. Lead, zinc and cadmium content of earthworms from pasture in the vicinity of an industrial smelting complex. *Environ Pollut* 23:313–321.

32. Ma W-C. 1982. The influence of soil properties and worm-related factors on the concentration of heavy metals in earthworms. *Pedobiologia* 24:109–119.

33. Voigt G, Henrichs K, Pröhl G, Paretzke HG. 1988. Measurement of transfer coefficients for ^{137}Cs, ^{60}Co, ^{54}Mn, ^{22}Na, ^{131}I and ^{95m}Tc from feed into milk and beef. *Radiat Environ Biophys* 27:143–152.

Chapter 13

Toxicity of transuranic elements and earthworm radiosensitivity

D.A. Krivolutsky, A.G. Viktorov, V.Z. Martjushov

The biological effects of transuranic elements (^{239}Pu, ^{241}Am, ^{239}Np, and ^{244}Cm) were studied in field and laboratory experiments using earthworms. The maximum concentrations of ^{239}Pu and ^{241}Am in earthworms were recorded after 7 d and for ^{239}Np after 3 d. The dynamics of ^{239}Pu, ^{241}Am, and ^{239}Np accumulation by earthworms over 65 d and the doses of irradiation from the incorporated radionuclides and gut-soil radionuclides were determined. Unlike the other radionuclides tested, ^{244}Cm caused mortality of earthworms within 12 to 24 h of exposure to relatively low concentrations of this element; this could not be accounted for by the radioactivity dose, but was a function of the intrinsic chemical toxicity of this element.

During recent years, integrated studies on the baseline monitoring of the impact of radioactive substances on terrestrial biota, as a result of the Chernobyl accident, have been undertaken [1]. About 1×10^{15} Bq (^{239}Pu + ^{240}Pu), 30×10^8 Bq ^{239}Np, 5×10^{13} Bq ^{241}Am, and 2×10^{14} Bq ^{244}Cm escaped into the environment during the accident. Soil radioactive pollution in the 30-km zone was as high as 90 kBq/kg (^{239}Pu + ^{240}Pu) and 70 kBq/kg ^{241}Am. High radiological and chemical toxicity continue, and with the long half-lives of ^{239}Pu, ^{241}Am, and ^{244}Cm, these radionuclides may have long-term adverse effects on biota [1,2]. In this chapter, the impact of various transuranic elements on earthworm populations both in the field and in laboratory tests is assessed.

Materials and methods

Laboratory experiments

Earthworms used in the present experiments were sexually mature *Aporrectodea caliginosa* Savigny (1826). They were placed in plant pots of 0.008 m^3 that were filled to a depth of 18 cm with 3 kg of dry soil. The air temperature ranged from 13 to 18 °C, and soil in the pots was moistened periodically. To study the accumulation of ^{239}Pu and ^{241}Am, 150 earthworms were placed in each of 3 pots with the soil containing 3.5 MBq/kg of ^{239}Pu and 1.7 MBq/kg of ^{241}Am artificially applied as Pu and Am nitrates. To study the accumulation of ^{239}Np and ^{244}Cm, 100 earthworms were placed in each of 3 pots with soil containing 1.3 MBq/kg ^{239}Np and 0.22 MBq/kg ^{244}Cm artificially applied as Np and Cm nitrates.

In the course of the experiments with ^{239}Pu and ^{241}Am, earthworms in batches of 10 and casts were sampled after 1, 2, 4, 6, 10, 20, 30, and 65 d. The captured earthworms were placed in a glass dish on moist filter paper and allowed to remain there until their gut contents were fully evacuated. Subsequently, all samples were washed with distilled water to remove soil remnants, and the samples were dried.

For the ^{239}Np studies, the earthworms were sampled in batches of 10 after 6 h and 1, 2, 3, 4, 6, 10, and 20 d. These worms were washed with distilled water and fixed in formaldehyde; the gastrointestinal tract with its contents was dissected and dried.

In the course of the experiments with ^{244}Cm, earthworms were sampled in batches of 10 after 12 h.

The ^{239}Np, ^{244}Cm, and ^{239}Np contents were measured in the earthworms using an ion exchanger and the alpha-radiometer ARS-1. The data were statistically analyzed using standard techniques.

Field experiments

An experiment with ^{241}Am was conducted under natural conditions in a sheep fescue meadow in the Southern Urals. The annual dry weight productivity of the meadow was 910 g m^{-2}, and the weight of dead plant on the soil surface 847 g m^{-2}. The main soil characteristics are given in Table 13-1.

Table 13-1   Physico-chemical characteristics of the soil of the experimental plot

pH (H$_2$O)	OM (%)	Ca	Mg (mmol/100g^{-1})	K	Na	Sum	Cation exchange capacity (mmol/100g)
0.0	5.9	31.2	7.1	1.5	0.5	40.3	50.0

The ^{241}Am was applied during the first days of August on the surface of the meadow at a rate of 1.5 kBq/m^2, as Am nitrate, sprayed at a rate of 0.8 L/m^2. Distribution of dose power of ^{241}Am gamma-rays in the soil profile in September is shown in Figure 13-1.

Each sample plot was 10 m^2, and the experiment was performed in 3 replicates. Ten $25 \times 25 \times 30$ cm soil samples were taken from each of the experimental and control plots in September. The worms present were sampled by hand sorting. The earthworms were represented by 1 species, *Dendrobaena octaedra* Savigny (1826). The data were statistically analyzed using standard techniques.

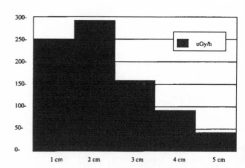

Figure 13-1   Distribution of dose power of Am-241 gamma-rays in soil profile

Results and discussion

Accumulation of ^{239}Pu and ^{241}Am: laboratory experiments

The radionuclides ^{239}Pu and ^{241}Am are very important among artificial radionuclides entering the biosphere because they have long half-lives and high radiotoxicity. As can be seen from Table 13-2, the intensive accumulation of ^{238}Pu into earthworm body tissues continued for 7 d. During that time, the earthworms accumulated the peak concentration of ^{239}Pu (12 kBq/kg). During the last days of the experiment, some decrease in the concentration of the radionuclide to 2.9 kBq/kg dry weight of earthworms was recorded. Concurrently, ^{239}Pu accumulated in the casts, where concentrations 10- to 12-fold higher than the body concentrations of the radionuclide were measured.

The ^{241}Am began to accumulate in earthworm bodies only on the second day after the animals were placed in the radionuclide-polluted soil. The concentration of ^{241}Am in the earthworm reached its maximum by day 7 and was 6.5 kBq/kg, about twice as low as the concentration of ^{239}Pu at the same time. There was twice as much ^{239}Pu as ^{241}Am applied to the soils. After day 7, concentration of ^{241}Am decreased a little during the following days of the experiment. Why ^{239}Pu was detected before ^{241}Am in the body is not yet understood, but there is some evidence that ^{241}Am is more strongly bound to inorganic soil particles and hence is absorbed by the gastrointestinal tract at a slower rate than ^{239}Pu.

It was found that the ^{241}Am concentration during the first 21 d of the experiment was only 1 to 7 × greater in the casts than in the body tissue. Comparison of the behavior of the 2 elements reveals similarity of accumulation of ^{239}Pu and ^{241}Am by earthworms living in chernozem soil. At the same time, some differences in the accumulation of the radionuclides were observed in casts, where the concentration of ^{239}Pu exceeds that of ^{241}Am several times.

Based on the experimental data, the concentration ratios of ^{239}Pu and ^{241}Am by earthworms ranged within 0.0003 to 0.0004: these elements are not readily incorporated into biological tissues. A gradual increase in the concentration ratio of ^{239}Pu and ^{241}Am was observed on day 10 of the exposure.

Accumulation of ^{239}Np: laboratory experiments

It took 3 d for ^{239}Np to reach the peak concentration in the earthworm's body (Table 13-3). During that period, the earthworms accumulated 510 Bq ^{239}Np/g dry weight. In the following days of the experiment, some decline of radionuclide concentration to 310 Bq/kg was recorded. Concurrently, an increase in the content of ^{239}Np took place in the intestinal tract, where the radionuclide concentration on day 20 was 3 × as high as that in the body. The latter is associated with a decline of the deposition of ^{239}Np in the tissues.

The observed concentration ratio of ^{239}Np ranged between 0.7 and 0.39, indicating that this element is incorporated in tissues at only half the rate previously observed for ^{90}Sr [3]. For ^{90}Sr, concentration ratios of 0.52 to 0.67 were obtained, according to the soil type. The maximum accumulation of ^{239}Np in earthworms occurred within the first 3 d of exposure. During subsequent days, the accumulation of ^{239}Np by earthworms in a leached chernozem was essentially similar to that of ^{90}Sr [3].

Effects of ^{344}Cm on earthworms: laboratory experiments

After the first 12 h of exposure to the soil contaminated with ^{244}Cm, earthworms became less active. After 24 h, the entire population was eliminated. Concentration of ^{244}Cm in the present experiment was lower, as compared to other radionuclides. Radiation alone could not be responsible for the death of the earthworms. This is supported by the re-

Table 13-2　Accumulation of ^{239}Pu and ^{241}Am in the body of earthworm *A. caliginosa* (kBq/kg dry weight) placed in soil containing 3.5 MBq/kg of ^{239}Pu and 1.7 MBq/kg of ^{241}Am

	Sampling after beginning of experiment (d)							
	1	2	4	7	11	21	31	65
Body	0.7±0.1	3.6±0.5	3.7±0.6	12.0±1.8	9.1±1.4	6.9±1.0	–	2.9±0.4
Casts	0.9±0.1	69.0±10.4	71.0±10.7	10.0±1.5	19.0±2.9	17.0±7.1	19.0±2.9	–
Body	nd	0.9±0.1	0.8±0.1	6.5±0.9	2.3±0.3	2.1±0.3	–	1.1±0.1
Casts	nd	19.0±2.9	22.0±3.3	4.7±0.7	5.7±0.9	14.0±2.1	9.4±1.4	2.4±0.4

Table 13-3   Concentration of ^{239}Np in the body of earthworm *A. caliginosa* (kBq/kg dry matter) exposed to a leached chernozem containing 1.3×10^3 kBq/kg of ^{239}Np

Sampling after beginning of experiment (d)					
	0.25	1	2	3	10	20
Earthworms	200 ± 30	92 ± 13	320 ± 48	510 ± 76	170 ± 25	310 ± 46
GIT[1] content	22 ± 3	200 ± 30	145 ± 21	333 ± 49	590 ± 88	920 ± 138
Body/soil ratio	0.15	0.07	0.25	0.39	0.13	0.24
GIT[1]/soil ratio	0.02	0.20	0.11	0.25	0.45	0.71

[1] GIT – gastrointestinal tract

sults of the experiment in which higher concentrations of ^{239}Pu, ^{241}Am, ^{239}Np were involved [3]. Therefore, in the experiments on ^{244}Cm, a chemical factor must have been important. The ^{244}Cm concentration of 220 kBq/kg dry soil is clearly highly toxic for earthworms.

Effects of soil contamination with ^{241}Am on earthworms: field observations

The application of ^{241}Am at a rate of 1.5 kBq/m^2 caused changes in earthworm numbers and biomass (Table 13-4). There was a sharp decline in earthworm numbers and especially in biomass in the presence of the ^{241}Am. Changes in the structure of the earthworm

Table 13-4   Numbers and biomass of *D. octaedra* in control and ^{241}Am-contaminated plots in a sheep's fescue meadow

^{241}Am treated plot		Control plot	
Numbers (individuals/m^2)	Biomass (g/m^2)	Numbers (ind/m^2)	Biomass (g/m^2)
Soil depth of 0 to 5 cm			
11.2 ± 3.4	0.98 ± 0.40	16.0 ± 7.2	2.6 ± 1.2
Soil depth of 0 to 30 cm			
12.8 ± 4.9	1.0 ± 0.4	18.2 ± 7.4	2.7 ± 1.1

population in ^{241}Am-contaminated plots were also notable. About 64% of the earthworm population were sexually mature in the control plot, while no mature earthworm cocoons were found in the experimental plot. No earthworms were found below the 10-cm depth in either plot.

Based on the pattern of the radionuclide distribution in the soil profile, the dose of gamma-radiation absorbed from ^{241}Am was calculated. Earthworms in the upper 5 cm of soil were exposed to the dose of 0.003 Gy over 30 d. However, the main dose is expected from the alpha-irradiation of ^{241}Am, which has stronger biological effects than gamma-radiation (quality quotients of 20 and 1, respectively). Based on observed concentration factors and assuming a regular distribution of ^{241}Am within the animal, the absorbed dose was estimated at less than 0.3 Gy/month. Thus, when ^{241}Am was applied to the surface of the meadow grass-sheep fescue meadow at 1.5 kBq/m^2, the maximum total absorbed irradiation dose to earthworms was calculated as about 0.6 Gy/month. The values of LD50/30 are much higher (Table 13-5).

Conclusion

The dynamics of the ^{239}Pu and ^{241}Am accumulation by earthworms has been first estimated in experiments with earthworms dwelling under conditions of soil radioactive pollution, and concentration ratios of ^{239}Pu and ^{241}Am were calculated for those elements. The concentration of ^{239}Pu in earthworm casts during the entire observation period was 2 to 7 × higher than the body radioactivity, and the concentration of ^{239}Pu in casts was considerably higher than the concentration of ^{241}Am. During the entire period of observation, the concentrations of ^{239}Pu, ^{241}Am, and ^{239}Np in earthworm bodies and casts did not exceed their concentrations in the soil. This is probably due to the fact that ^{239}Pu, ^{241}Am, and ^{239}Np are strongly bound to soil complexes and do not readily pass into a soluble state.

Table 13-5 Radiosensitivity of different earthworms species in relation to their cytotype based on references cited

Species	Cytotype	LD50/30 kGy
Aporrectodea caliginosa		
Kursk region	2x=36	1.12 ± 0.03 [4]
Ural region	2x=36	1.34 ± 0.06 [4]
A. rosea	3x=40	0.57 ± 0.06 [4]
Eisenia nordenskioldi	6x=108	1.06 ± 0.10 [4]
E. fetida	2x=22	0.6 [5]
Dendrobaena octaedr (cocoons)	6x=108	0.02 [4]
Lumbricus terrestris	2x=36	0.678 [6]
Lumbricus terrestris	2x=36	1.0 [7]

References

1. Sokolov VE, Krivolutzkii DA, Ryabov IN, Taskaev AI, Shevchenko VA. 1989. Bioindication of after-effects of the Chernobyl atomic power station accident in 1986–1987. *Biology International* 18:6–11.

2. Krivolutzkii DA, Pokarzhevsky AD, Viktorov AG. 1992a. Earthworm populations in soils contaminated by the Chernobyl atomic power station accident, 1986–1988. *Soil Biol Biochem* 224:1729–1731.

3. Krivolutzky DA, Kozhevnikova TL, Martjushov VZ, Antonenko GI. 1992b. Effects of transuranic (^{239}Pu, ^{239}Np, ^{241}Am) on soil fauna. *Biol Fertil Soils* 213:79–84.

4. Viktorov AG. 1995. Differential reactions of lumbricids to radioactive pollution in various biotopes of Chernobyl region. In: van don Brink WJ, Bosman R, Arendt F, editors. Contaminated soil '95. Amsterdam, the Netherlands: Kluwer Academic. p 265–266.

5. Suzuki J, Egami N. 1984. Radiation-induced damage and recovery from it in germ cells in the earthworm, *Eisenia foetida*. The University of Tokyo *J Faculty Sci*, Sec. IV, 215:329–342.

6. Reichle DE, Witherspoon JP, Mitchell MJ, Styron CL. 1972. Effects of beta-gamma radiation of earthworms under simulated-fallout conditions. USACC Symposium Series Survival of Food Crops, and Livestock in the Event of Nuclear War, Conf-700909; Jan 1972. p 527–534.

7. Hancock RL. 1962. Lethal doses of irradiation for Lumbricids. *Life Sciences* 21:625–628.

Earthworm biomarkers: advances relevant to earthworm ecotoxicology

Chapter 14

Review of selected biomarkers in earthworms

Janeck J. Scott-Fordsmand and Jason M. Weeks

This chapter reviews 5 selected biomarkers measured in earthworms: DNA alterations, metal-binding proteins (MBPs), esterase activity, lysosomal membrane stability, and immunological responses. The biomarkers are reviewed with special regard to their possible use in risk assessment procedures, evaluating key features such as appearance in various earthworm species, chemicals known to induce the biomarkers, possible dose–response relationship between a chemical and the biomarker response, links between the biomarker response and effects on the individual or population level, and the natural variability of the biomarker expression. The selected biomarkers were identified in most species investigated; most studies were performed on various *Lumbricus* and *Eisenia* species, and far fewer were performed with other species. A narrow range of chemicals were tested, except for esterase activity, where among others, many organophosphorous and carbamate compounds were tested. Only a few of the biomarkers showed a dose–response relationship between the chemical concentration and the biomarker response measured, i.e., the cholinesterase activity, the lysosomal membrane stability, and the immunological responses showed some indication of an increasing response with increasing contamination levels. The lysosomal-membrane stability and the immunological responses showed limited success linking the biomarker response to effects at the population level. Very little is known regarding the persistence of the various biomarker responses and their natural variability.

In recent years, the use of biological responses other than reproduction, growth, and mortality to estimate either exposure or resultant effects of chemicals has received increased attention [1–7]. Such responses, typically biochemical changes, are mostly referred to as "biomarker responses" [2]. Recently, "biomarker" was defined "as a biological response that can be related to an exposure to, or toxic effect of, an environmental chemical or chemicals" [8]. In this chapter, this definition is adopted, and selected biological responses to pollutants at the molecular, subcellular, and cellular levels in earthworms will be discussed in relation to their utility as biomarkers in the process of risk assessment.

One major reason for the current interest in biomarkers is the limitation of the classical approach to environmental toxicology that involves measuring the amount of the chemical present in an animal or plant and then relating this to adverse effects on mortality, reproduction, and growth. First of all, in the classical approach, the bioavailability and the toxicity of a compound may differ in laboratory tests compared to those observed in the field. Second, under field conditions, multiple toxicants may be present simultaneously, producing an even more complex situation. Finally, only a few (if any) of the conventional endpoints from the classical approach can be assessed as in situ experiments or surveys. Thus, results from classical laboratory ecotoxicology are often difficult to project to a field situation and are indeed difficult, if not impossible, to validate under field conditions, although progress is being made in this area (Holmstrup et al., Chapter 27).

The major strength of the biomarker approach lies in its potential to circumvent the serious limitations of the classical approach to environmental toxicology [6]. Biomarkers deal with the question of bioavailability of chemicals by reacting only to the biologically available fraction of a pollutant, this is especially relevant in soil where the bioavailability of the chemical is known to vary with different soil characteristics. Biomarkers also have the advantages that they can exhibit the effect caused by many toxic compounds present at the same time (complex pollution), and they are applicable under both laboratory and field conditions.

In order for a biomarker, or a battery of biomarkers, to be useful to ecological risk assessment, a basic understanding of the characteristics of the markers is required. Five important considerations or key features should be considered:

1) Species: The marker must be identified in the species of concern, and preferably the species used should be of ecological importance (i.e., key species) in the ecosystem investigated. In this respect, earthworms are ideal organisms, as they are ubiquitous, and they are important for maintaining soil fertility and in the breakdown of organic matter (OM) in most terrestrial ecosystems [9].

2) Chemicals: Knowledge is required on the range of hazards (e.g., toxic compounds) that elicit a biomarker response. A chemically nonspecific biomarker might be used in monitoring or screening studies, whereas chemical-specific biomarkers might be preferred in cases of known pollutants.

3) Dose response: To estimate the magnitude of the problem, a dose–response relationship between the measured biomarker response and the concentration of the pollutant is desirable.

4) Linking to higher levels: If a biomarker is to be used as more than a measure of exposure, then a correlation between the observed responses and deleterious effects to the individuals or populations/communities (e.g., reproduction and mortality) should be established. If a biomarker links with effects at other levels, a subcellular biomarker may, for example, act as an early warning of effects at the population level.

5) Natural variability and persistence in time: For a biomarker to be useful in the field, any response should ideally have a low inherent variability with a low, or at least known, dependence on physiological and physicochemical conditions, i.e., baseline data must be available. The induction time and the persistence of the biomarker response should be known in order to estimate the likelihood and significance of detecting a response in field samples.

Finally, for practical and economic considerations, the reproducibility, costs, and ease of use of a potential biomarker procedure need to be considered. These considerations will not be discussed herein.

This chapter will focus on 5 types of relevant biomarkers that have a proven record in earthworm research. For a selected suite of biomarkers that have potential to be used in risk assessment, the key features described above are summarized.

DNA alterations

In studies on mammals, in particular on humans, many environmental pollutants have been found to have genotoxic properties. Many chemical compounds, or their metabolites, also react with the DNA of earthworms. Such reactions may lead to 1) covalent bindings of the chemical, or its metabolites, to DNA (this process is termed "adduct formation"); 2) strand breakage; 3) base exchange; or, 4) increased unscheduled DNA synthesis [10–12]. These changes may be investigated by a number of different techniques; however, few of these techniques have been applied to earthworms. Two types of DNA changes have been investigated in earthworms, e.g., strand breaks and DNA adducts. The basic principle of the strand breakage is the disconnection in 1 strand of the DNA from the double helix. A single-strand breakage may result from chemical exposure, but the detection of a strand breakage does not necessarily provide any information on the cause for that particular breakage. The basic principle for the formation of DNA adducts is a covalent binding of a compound to the DNA. The identification of DNA adducts will thus provide information on the chemical causing a DNA adduct.

Species

These DNA changes have been tested in only 3 earthworm species: *Lumbricus terrestris* (for both techniques) [13–15], *L. castaneus* (for strand breakage) [16], and *Eisenia fetida* (for the strand breaks with only a few experimental results reported) [15,17].

Chemicals

Stressors such as a mixture of dioxins (polychlorinated dibenzodioxins/polychlorinated dibenzofurans [PCDDs/PCDFs]), X-rays, mitocin C, and soil contaminated with a range of organic compounds, e.g., coke, benzene, and aniline, have been tested for strand breakage using a single cell electrophoresis assay [15–17]. For these, an increased number of strand breakages were found with increased degree of contamination. Earthworms exposed to industrially contaminated soils (including polycyclic aromatic hydrocarbons [PAHs] and other organic compounds) have been tested for DNA adducts using the ^{32}P-postlabelling technique [13–14] with an increased number of adducts found for worms exposed to contaminated soil.

Dose response

For strand breaks, there is an indication of a dose–response relationship for the dioxins, although comparatively few concentrations have been tested. A dose–response relationship was not shown, however, for DNA adduct formation, but there appeared to be a time-dependent increase in the level of DNA adduct formation following exposure to PAHs and to soil contaminated with coke [14]. These results agree with studies on beluga whales where benzo[a]pyrene (BaP) DNA adducts were measured following exposure of the whales to BaP [18].

Linking to higher levels

No attempt has been made to link DNA alterations to endpoints (such as a change in earthworm fecundity) that will indicate effects on population size or community structure. Such a possible relationship between DNA alterations and effects on population size or community structure is complicated by the fact that DNA alterations caused by chemical exposure might be repaired or located at an inactive site of the DNA; thus the health of the organism is not adversely affected. However, if the damage is not repaired and is located at an active site, this may cause an effect on the health of the individual or the subsequent generations. For example, damage to the oocytes and sperm may lead to decreased numbers of viable progeny or heritable mutations [19].

Natural variability

Little is known about the natural variation of strand breakages and DNA adducts formation in earthworms. However, studies on other organisms have shown that strand breakages and DNA adducts occur under normal conditions; it is therefore very important for researchers to have appropriate reference material when studying such effects [12,18]. Appropriate reference material should identify the natural occurrence of DNA changes caused by factors other than the presence of toxic compounds in the environment.

Persistence in time

Little is known about the persistence of these markers. It has been shown that DNA adducts formed in earthworms following exposure to PAHs may persist for a considerable time [14]. Strand breakage may be repaired rather more rapidly [10].

Metallothionein and other metal-binding proteins

A large number of studies have described the ability of earthworms to accumulate certain metals, especially Cd, Zn, and Cu [20]. Heavy metals entering earthworms may be bound in various ways within the organism. One way of binding and detoxifying heavy metals is by binding to metallothionein (MT) and/or other MBPs. Metallothionein and other MBPs are well known and occur throughout the animal kingdom as well as in several species of plants and microorganisms (eukaryote), but they are little studied in terrestrial invertebrates [21–22]. The role of MTs and other MBPs is by no means fully understood, but in general these proteins are thought to be involved in the intracellular regulation of essential and nonessential metal levels in tissues. Metallothioneins have be divided into 3 molecular classes (Classes I to III) [23] on the basis of their decreasing similarity and homology with horse MT. In earthworms, various MBPs have been identified, some of which resemble MT found in other invertebrates, although in general, they contain fewer cysteines. Others are quite distinct from MTs and contain among other entities large quantities of aromatic amino acids [22, 24–27]. In this chapter, "other metal-binding proteins" are proteins, characterized or noncharacterized, that have been observed to bind metals and cannot be grouped as MTs.

Species
Metallothionein-like proteins and MBPs have been identified in a number of earthworm species such as *Eisenia fetida, Dendrodrillus rubidus, Aporectodea (Allobophora) calignosa,* and various *Lumbricus* species (i.e., *terrestris, rubellus, variegatus,* and *castaneus*) [26–31].

Chemicals
Although a number of metals and some organic compounds are known to induce MT and other MBPs [22,32], only Cd has been investigated in earthworms. Some of the first to report an MBP in earthworms were Suzuki et al. [28]. By exposing *Eisenia fetida* to Cd nitrate at concentrations of 1 to 500 mg Cd/kg composted sewage sludge, they observed a dose–response induction of 3 Cd-binding proteins with a concurrent increase in the Cd concentration in the worms. One of the proteins was characterized as an MT-like protein [24]. Other studies [24–31] also reported induction of MBPs in earthworms after exposure to Cd, but no studies have reported an induction of these proteins by other metals. Morgan et al. [27] reported Zn but not Cu to be present in the Cd-binding proteins. For organisms such as rats and snails, Cu, Zn, and Hg are known to bind to MT-like or other MBPs [22,32]. Other factors, e.g., hormones and certain organic compounds, are known to induce MTs and MBPs [21].

Dose response
A number of studies have shown dose–response induction of Cd MT-like proteins and other Cd-binding proteins [24,28]. However, most emphasis has been put on the problem of identifying the structure and type of the MBPs found.

Linking to higher levels
No studies hitherto have reported linkages between the MBP induced in earthworms and higher-level effects such as reduced reproduction. Given the facts that 1) the functional role of MTs and MBPs is not well understood and 2) that they may be related to internal metal storage, it is unlikely that such linkages will be established in the near future. It seems that these proteins may in the first instance be considered as markers of exposure to chemical stressors, notably metals [21].

Natural variability
No studies were found dealing with the influence of natural stressors on MBPs in earthworms. However, in other organisms, it has been shown that induction of MT may be dependent on a variety of factors such as temperature, nutritional status, hormone levels, and indirect induction by organic compounds [22,32,33]. Furthermore, it should also be considered that the structure of the Cd-binding protein is dependent upon which species is measured and the ability of other factors to influence the MT induction may vary between species.

Persistence in time

The persistence of MT-like and other MBPs has not been studied in earthworms. However, it is known from studies with both vertebrates and aquatic invertebrates that MBPs can be degraded in specific tissues with organelles such as lysosomes [34]. The rate at which the MBP is metabolized may depend on, among other factors, the type of metal bound to the protein. For example, Cd-binding proteins in rats are normally more stable than Zn- and Cu-binding proteins [21,35,36].

Cholinesterase activity

Cholinesterases (ChEs) are used by earthworms for the transmission of nerve signals just as they are used in vertebrates [37,38]. Thus, at the biochemical level, toxins may be expressed as a change in ChE activity. In vertebrates, 2 classes of chemical compounds, notably carbamate and organophosphorous pesticides (OP), can cause a depression in total ChE activity [39]. The efficiency, the degradability, and the low costs of these pesticides have led to their wide agricultural usage [40]. Although OP and carbamates inhibit the ChEs in target organisms, it was soon realized that many other organisms, including mammals, birds, and insects, were also affected by such compounds [40].

Species

The inhibition of ChE in earthworms by chemical compounds has been established in 3 *Eisenia* species (*fetida*, *andrei*, and *veneta* [also named *Dendrobaena veneta*]), in 2 *Lumbricus* species (*terrestris* and *rubellus*), and in 1 of each of the following genera: *Aporectodea* (*calignosa*), *Allobophora* (*chlorotica*), and *Pheretima* (*posthuma*).

Chemicals

A wide variety of pesticides has been shown to cause a decrease in the ChE activity of earthworms (Table 14-1). These compounds include OP [41–46], carbamate [41–48], organochlorine [45] and benzimidazole [45,49] compounds. These compounds have been tested either in in vitro or in vivo studies in which the earthworm has been exposed through food, soil, or direct injection of the compound. Organophosphorous compounds caused the most severe reduction in ChE activity, although a rapid recovery nor-

Table 14-1   Pesticides tested for their effect on cholinesterase activity in earthworms

Pesticide group	Chemical
Organophosphorous (OP)	Azinphos, Dasanit, Malathion, Methidation, N-2596, Parathion, Parathion-methyl, Parathion-ethyl, Paraoxon, Phorate, Triorthotolyl phosphate
Carbamate	Aldicarb, Azak, Carbaryl, Carbofuran, Methomyl, Oxamyl, Prostigmine, Propoxur
Benzimidazole	Benomyl, benzimidazol-2-yl-3-butylurea, Carbenzim, Fuberidazole, Thiabendazole, Thiophanat-methyl
Organochlorine	Bromophosoxon, Copperoxychloride, Endosulfan, Gamma-HCH, Potassium N-hydroxymethyl-N-methyl (dithiocarbamate), Trichloronate

mally took place, with less overall reduction in ChE activity reported for carbamates [41]. Organochlorine, benzimidazole, and Cu oxychloride compounds [45,49] were not reported to cause reduced ChE activity in in vivo and in vitro studies, although metabolic compounds of benzimidazole caused inhibition of ChE in *Lumbricus terrestris* [49]. Thus, it is not always the parent compound that causes the toxic effect but rather a metabolite. For example, the measured reduction in ChE activity in earthworms exposed to benomyl has been attributed to its metabolite butyl-isocyanate [49].

Dose response

In most cases where the ChE activity has been investigated at different dose levels, a dose response for ChE activity has been observed [41,48]. Dikshith and Gupta [47] exposed the earthworm *Pheretima posthuma* to soil treated with different concentrations of carbaryl and found that the inhibition of acetylcholinesterase (AChE) activity was dose dependent, both in coelomic fluid and in nerve tissue.

Linking to higher levels

In vertebrates such as birds, severe depression of the AChE activity has often been related to mortality of the organism [50]. In earthworms, however, there is no clear evidence of a relationship between ChE depression and mortality or adverse effects on reproduction. For example, Stenersen et al. [41] found that the mortality of carbofuran-treated worms was 8 × higher than the mortality for OP-treated worms, despite the latter having the most severely depressed ChE activity (i.e., the ChE activity of OP-exposed worms was reduced by more than 99% , while carbamate-exposed earthworms had reductions of only 30 to 80%). Similar results were obtained by Stringer and Wright [49] who observed no in vivo inhibition of ChE in *Lumbricus terrestris* treated with benomyl, although mortality was observed after 6 d. In birds, an AchE depression of more that 80% is likely to result in death [50]. Gupta and Sundaraman [48], however, found a clear inverse relationship between ChE depression and burrowing activity in earthworms, which may affect the ability of the earthworms to avoid unfavorable conditions (e.g., predation) with mortal consequences.

Natural variability

Few studies have investigated natural variability of the ChE activity in earthworms and variation in responses to toxic chemicals. However, as the depression of the ChE activity may depend on the metabolic compounds [43,49] rather than the parent compound, factors that influence the metabolism of these compounds will thus also influence the biomarker response. Differences between tissues and between species are 2 other important sources of variation. For example, Stenersen et al. [46] found different ChE proportions (termed "E1" and "E2") in *Lumbricus terrestris, Eisenia veneta* (also named *Dendrobaena veneta*), and *E. fetida* and thus differences in the degree of depression following exposure to carbaryl and parathion. From studies on birds, fish, and insects, other factors, including season, temperature, nutritional status, and reproductive stage/activity, are also known to influence ChE activity or other enzyme systems [51–53]; thus these factors also may influence ChE activity in earthworms.

Persistence in time
The persistence of the depression of ChE activity in earthworms has been investigated to a limited extent. Stenersen et al. [41] observed that a depression in ChE activity caused by carbamates was less severe than that caused by OP compounds, with recovery of the former to normal levels within 30 to 40 d, relative to > 50 d for the latter.

Lysosomal membrane integrity
At the subcellular level, the lysosomal system has been identified as a particular target for the toxic effects of xenobiotics [54]. Lysosomes are a morphological heterogeneous group of membrane-bound subcellular organelles containing acid hydrolase and ranging in size from 250 Å to 1 μm in diameter. The function of lysosomes in the cell is to catabolize organelles and macromolecules [55]. A change in lysosomal membrane stability is thought to be a general measure of stress [2]. In stable lysosomes, hydrolases are prevented from reacting with substrates by an intact membrane. The membrane stability decreases in response to stress as membrane permeability increases. The mechanisms causing this alteration are not well understood and may vary with the type of stressor. Pathological alterations in lysosomes have been especially useful in the identification of adverse effects on a range of organisms, with much evidence for aquatic organisms [56–59] but rather limited evidence in the terrestrial environment [60]. The technique used in earthworms to investigate lysosomal stability is based on the ability of lysosomes in coelomocytes to retain a dye (neutral red) over time. The longer the lysosome can retain the dye (neutral red retention time [NRR]), the more stable the membrane.

Species
Changes in lysosomal integrity as a result of chemical stress have been tested in a range of earthworm species including 3 *Lumbricus* species (i.e., *terrestris*, *castaneus*, and *rubellus*) and in *Eisenia andrei*.

Chemicals
Few chemicals have been tested; most studies examined the effects of Cu [60–62] and only a few examined the effect of organic contaminants [63]. For example, clear responses were seen in earthworm coelomocytes following exposure in soil contaminated with pyrolyzed plastics after an industrial accident (Svendsen et al., Chapter 18). In aquatic organisms, effects have been observed for organic compounds such as PAHs [56].

Dose response
Exposure to Cu-contaminated soil in both the laboratory and field mesocosm studies showed a clear dose–response relationship for the lysosomal membrane stability with *Lumbricus rubellus* and *Eisenia andrei* [60–62]. Exposure to increasing soil Cu concentra-

tions resulted in a concomitant decrease in NRR time (Figure 14-1). A similar dose–response relationship was obtained for *E. veneta* (also named *Dendrobaena veneta*) exposed to different concentrations of sewage sludge (T.M.E. Olesen, personal communication).

Linking to higher levels

Although few studies have been performed, correlations between NRR time and reproductive output and mortality have been observed for *Lumbricus rubellus* and *Eisenia andrei*, both in laboratory studies and field mesocosm studies [61,62]. In contrast to this, *E. veneta* (also named *Dendrobaena veneta*) exposed to increasing concentrations of sewage sludge, known to contain high concentrations of Cu and other metals, gave a reduced NRR time (i.e., a reduced membrane stability) but also an increased reproductive output (T.M.E. Olesen, personal communication). This may suggest that high amounts of food may prevent the correlation between reduced NRR time and depressed reproductive output.

Natural variability

Few studies have been concerned with the natural variability of the lysosomal membrane stability; however, from studies on aquatic invertebrates, lysosomal responses can be induced by nonchemical stressors such as hypoxia, hyperthermia, osmotic shock, and dietary depletion [56]. For earthworms, Svendsen (personal communication) reported that the NRR time is independent on temperature, when *E. veneta* (also named *Dendrobaena veneta*) was exposed in soil to Cu within a temperature range of 15 to 25 °C.

Persistence in time

Little is known regarding persistence in time of lysosomal membrane stability in earthworms. However, unpublished data (Svendsen and Spurgeon, personal communication) have shown that worms from long-term metal-contaminated field sites (i.e., exposed for up to 300 generations) still exhibited a decreased lysosomal membrane stability (NRR time) when collected from field sites contaminated with elevated metal concentrations. However, other data (Olesen, personal communication) have shown that 6 months after a single application of sewage sludge, known to contain high concentrations of Cu and other metals, to a series of field mesocosms, the NRR time for surviving earthworms had returned to normal levels. The latter observation suggests that for earthworms that have been exposed to a mixture of contaminants (resulting in a depressed NRR time), but where the exposure has ended, the NRR time may under certain circumstances return to normal levels with time. This recovery may be due to the nature of the pollutant, in the former instance a continuous input of toxic metals, in the latter a single application of sewage.

Immunological responses

The immune system is the body's main defense against invasion by foreign material and biological agents [64]. In recent years, it has been shown that a wide range of chemicals, including metals and organic compounds, can affect the immune system [6]. Severe immune depression may quickly result in morbidity and death, but sublethal changes in special compartments of the immune system often occur first and provide an early indication of a toxic effect. Such changes may be used as immunological biomarkers. In earthworms, the immune system is located in the coelom, which consists of coelomic fluid and coelomocytes. The latter, which are similar to mammalian leucocytes, are sensitive to foreign material [65]. In earthworms, a variety of different immunological responses have been measured in both the nonspecific immune system, in the specific cell-mediated immune system, and in the humeral-mediated defense system (Table 14-2) [66–69].

Table 14-2   Immunological response measures in earthworms following exposure to chemicals[1]

Immune parameter	Parameter measure
Nonspecific	Phagocytosis, nitroblue tetrazolium dye reduction, lysozyme, superoxide, chemiluminescene, natural killer cells, wound healing, inflammation
Specific: cell-mediated	Mixed leukocyte response, blast transformation, Graf ejection
Specific: humoral	Lytic/agglutination response
Coelomocytes	Total cell counts, differential cell counts, cell viability

[1]after [66]

Species
The effect of chemicals on the immune system has been studied in only a few species such as *Lumbricus terrestris* and, to a lesser extent, *Eisenia fetida* and *E. hortensis*. By far, most of the work has been carried out on *L. terrestris*, with only a few studies on the 2 *Eisenia* species.

Chemicals
In vertebrates, immunological responses are known for a large range of chemical compounds [70]. In earthworms, however, the effect of only a few chemicals has been described. Most studies involved the effects of polychlorinated biphenyls (PCBs) (Aroclor 1254) [65,66, 68, 71–78], although the effects of metals (Cu), metal-polluted soil, and fly ash have also been investigated [67,69,73,79]. Finally, the influence of a range of pesticides, including carbamates, OP, azoles, and a few others, on the total immune activity in earthworms has been studied [71]. A large number of the reported studies were performed as filter-paper contact tests rather than as soil tests, giving a less clear impression of the likely effects under natural conditions.

Dose response
A number of studies have observed a dose–response relationship between the dose of the chemical and the immunological response, although most studies have reported only immunological responses following exposure to a single concentration of a chemical. Rodriguez-Grau et al. [74] observed a dose–response relationship between PCB exposure of *Lumbricus terrestris* and the humeral response secretory rosette (SR) formation. A similar relationship was found for the nonspecific (phagocytosis) and humeral (lysosome, SR, and erythrocyte rosette [ER]) response of PCB in *Eisenia fetida* [72,76]. For refuse-

derived fuel ash, Chen et al. [79] observed a dose–response curve for the nonspecific immune function measured as the coelomocyte reduction of nitroblue tetrazolium dye (NBT) for different degrees of contamination. Similar results were obtained for the effect of Cu on the lysosomal activity of the coelomic fluid and coelomocytes [69].

Linking to higher levels

Few studies have been concerned with linking the immunological response to effects at the higher organizational levels. However, Fitzpatrick et al. [72] observed a correlation between the SR formation and mortality in *Eisenia fetida* after exposure to PCB. From studies on vertebrates, it is known that the immunoresponse may have considerable plasticity and thus be difficult to link to effects at higher levels [50].

Natural variability

Few studies have investigated the natural variability of immunological endpoints, although the immunological responses are known to depend on many factors [70]. For most immunological endpoints in *Lumbricus terrestris*, Goven et al. [68,69] and Eyambe et al. [80] found little or no variation with season, except for the total extruded coelomocyte counts that were elevated in winter. Fritzpatrick et al. [72] reported species differences for the effect of PCBs on the humeral response of earthworms.

Persistence in time

The immunological system is known for its flexibility and its adaptability [6]. For earthworms, it has been observed that the immunological depression returns to normal levels after removal of the earthworm from sources of exposure. Goven et al. [68], for example, exposed *Lumbricus terrestris* to PCB and then transferred the earthworms to PCB-free soil for 16 weeks. A range of immunological responses (coelomocyte viability, differential cell counts [DCCs], ER, SR, and phagocytosis [RRBC]), all initially depressed by PCB exposure, returned to normal levels, although the internal earthworm body concentration of PCB did not return to previous levels. It is, however, not known whether the response would have remained depressed with continuous PCB exposure.

Discussion

In this chapter, 5 selected biomarkers were discussed in quite specific detail: DNA changes, MBPs, ChE activity, lysosomal membrane stability, and immunological responses. These markers have been discussed individually in relation to factors that are important should they be involved in the risk assessment process.

These markers have mainly been studied in *Lumbricus terrestris* and *Eisenia fetida* but also to some extent in a range of other earthworm species (see Table 14-3). The studies reported have mostly been conducted under laboratory conditions, using direct injection (e.g., [41]), filter-paper exposure (e.g., [68]), or soil-media exposure (e.g., [17]). A few studies, however, have been conducted under both laboratory and field conditions (e.g., [60]), the latter being the case only for lysosomal membrane stability.

The 5 biomarkers selected have, in general, not been rigidly tested for a broad range of chemical compounds (Table 14-3). Moreover, the research in most cases has been centered around only a few compounds or, in the case of ChE activity, around a few classes of compounds. The lysosomal membrane stability (measured as NRR time [60]) and a range of the immunological responses [65] seem to be promising markers of metal exposure, although more information is needed on how other factors can influence these markers. Cholinesterase activity appears to be a promising biomarker in earthworms for OP and carbamate pesticides, but less promising for benzimidazole compounds and not useful for organochlorine and metal compounds [41]. DNA alterations may in the longer term be used in connection with dioxins and PAH pollution and possibly pollution with other organic compounds [14,17]. Finally, immunological responses may also be promising biomarkers for certain organic chemicals, notably PCBs [72].

The chemicals tested by these biomarkers have been shown in only a few cases to give a dose–response relationship; however, the studies conducted indicate that this may be the case for all of the markers selected. However, a range of environmental factors may impair the relationship between the chemical concentration and the biomarker response. For the DNA alterations, exposure to increasing concentrations of dioxin resulted in increased numbers of strand breakage [16], while PAH induced a with-time increase in the level of DNA adducts [14]. For MT-like and other MBPs, increasing concentrations of Cd resulted in a corresponding increase of these binding proteins [28]. Exposure to increasing concentrations of OP and carbamate compounds caused a dose–response reduction of the ChE activity [41]. For metals, the lysosomal membrane was shown to be less stable with increasing metal concentrations [60]. The immunological depression depended on the concentration of PCB or metals to which the earthworm was exposed [74].

In general, there is a lack of information on the linking between these markers and parameters that are relevant at the population level. However, some evidence links ecologically relevant effects with the stability of the lysosomal membrane [61,62] and links immunological responses with earthworm mortality [72]. The linking between these biomarker responses and effects on population levels are, however, not straightforward and may be influenced by other factors, e.g., food availability.

The impact of natural variability and the temporal aspects of the biomarker response have been tested only in very few cases. Preliminary results on lysosomal membrane stability indicated an independence of temperature. For immunological responses, seasonal changes seem to have little influence [68,69]. For both marker types and ChE activity, the responses appear to return to normal levels when the earthworms are no longer exposed to the stressor. For all the markers selected, there is evidence of differences in the species' sensitivities. However, more studies are needed on the natural variability of biomarkers in order to gain information on the specificity of the biomarker response in relation to environmental factors. For example, if the biomarker response is influenced by nutritional status or temperature, such responses may be of limited use under field conditions.

Table 14-3   Summary of 5 selected biomarkers and current knowledge in relation to selected parameters important for risk assessment

	DNA alterations	Biomarkers MT and MBPs	ChE activity	Lysosomal membrane stability	Immunological response
Species[1]	L. terrestris L. castaneus	L. rubellus L. variegatus L. castaneus L. terrestris A. calignosa D. rubidus E. fetida	L. terrestris L. rubellus E. fetida E. andrei E. veneta (De. Veneta) A. calignosa A. chlorotica P. posthuma	L. terrestris L. castaneus L. rubellus E. andrei	L. terrestris E. fetida E. hortensis
Chemicals	PAH Dioxins (PCDD/PCDF) Mitocin C Contaminated soil	Cd	OP Carbamates Benzimidazol Others (See Table 14-1)	Cu Plastics	PCB Copper Polluted soil Fly ash Organophosphorous Carbamates Others
Dose-response relationship	(Yes)	Yes	Yes	Yes	Yes
Linking to higher-level population	?	?	?	Yes	(Yes)
Natural variability	?	?	?	?	(?)
Persistence	?	?	?	?	?

[1] L. = Lumbricus; E. = Eisenia; P. = Pheretima; A. = Allobophora; D. = Dendrodrillus; De. = Dendrobaena.
Yes = parameter observed
(Yes) = parameter observed but very few data
? = not known.

It can be concluded that some biomarkers provide a forewarning of adverse effects result-ing from environmental exposure. More work is needed in the area of understanding the limitations of biomarker use; especially needed is more information on their natural variability and the influence of factors other than the presence of contaminants on re-sponses. For the biomarkers to be used as early warning tools, more effort is needed in the area of linking biomarker responses at the subcellular and cellular levels with effects at the population level, especially under natural conditions.

References

1. McCarthy JF, Shugart LR. 1990. Biomarkers of environmental contamination. Boca Raton FL: Lewis.

2. Hugget RJ, Kirmle RA, Mehrle OM, Bergman HL. 1992. Biomarkers. Biochemical, physiological, and histological markers of anthropogenenic stress. Boca Raton FL: Lewis.

3. Mineau P. 1991. Cholinesterase inhibiting insecticides—their impact on wildlife and the environment. Amsterdam, the Netherlands: Elsevier.

4. Fossi M, Leonsio C. 1994. Non-destructive biomarkers in vertebrates. Boca Raton FL: Lewis.

5. Depledge MH, Fossi MC. 1994. The role of biomarkers in environmental assessment. Invertebrates. *Ecotoxicology* 3:161–172.

6. Peakall DB. 1994. Biomarkers. The way forward in environmental assessment. *Toxicol Ecotoxicol News* 1:55–60.

7. Weeks JM. 1996. The value of biomarkers for ecological risk assessment: academic toys or legislative tools? *Appl Soil Ecol* 2:215–216.

8. Peakall D, Shugart LR. 1993. Biomarkers: research and application in assessment of environmental health. Berlin, Germany: Springer-Verlag.

9. Edwards CA, Bohlen PJ. 1992. The effects of toxic chemicals on earthworms. *Rev Environ Contam Toxicol* 125:23–100.

10. Shugart L, Bickham J, Jackim G, McMahon G, Ridley W, Stein J, Steinert S. 1992. DNA alterations. In: Huggett RJ, Kimerle RA, Merle PM, Bergman HL, editors. Biomarkers biochemical, physiological, and histological markers of anthropogenic stress. Boca Raton FL: Lewis. p 125–154.

11. Fairbairn DW, Olive PL, O'Niel KL. 1995. The comet assay: a comprehensive review. *Mutat Res* 339:37–59.

12. Lloyd-Jones G. 1995. ^{32}P-postlabelling: a valid biomarker for environmental assessment? *Toxicol Ecotoxicol News* 2:100–105.

13. Walsh P, El Adlouni C, Mukhopadhyay MJ, Viel G, Nadeau D, Poirier GG. 1995. ^{32}P-postlabelling determination of DNA adducts in the earthworm *Lumbricus terrestris* exposed to PAH-contaminated soils. *Bull Environ Contam Toxicol* 54:654–661.

14. Van Schooten FJ, Maas LM, Moonen EJC, Kleinjans JCS, van der Oost R. 1995. DNA dosimetry in biological indicator species living on PAH-contaminated soils and sediments. *Ecotoxicol Environ Safety* 30:171–179.

15. Verschaeve L, Gilles J. 1995. Single cell gel electrophoresis assay in the earthworm for the detection of genotoxic compounds in soils. *Bull Environ Contam Toxicol* 54:112–119.

16. Verschaeve L, Gilles J, Schoeters J, van Cleuvenbergen R, de Frj R. 1993. The single cell gel electrophoresis technique or comet test for monitoring dioxin pollution and effects. In: Fiedler H, Frank H, Hutzinger O, Pazzefal W, Riss A, Safe S, editors. Organohalogen compounds II. Austria: Federal Environmental Agency. p 213–216.

17. Salagoviv J, Gilles J, Verschaeve L, Kalina I. 1996. The comet assay for detection of genotoxic damage in earthworms: a promising toll for assessing the biological hazards of polluted sites. *Folia Biol (Prague)* 42:17–21.

18. Shugart L, Theodorakis C. 1994. Environmental genotoxicity: Probing the underlying mechanisms. *Environ Health Perspect* 102 (Suppl. 12):13–17.

19. Anderson SL, Wild GC. 1994. Linking genotoxic responses and reproductive success in ecotoxicology. *Environ Health Perspect* 102 (Suppl. 12):9–12.

20. Hopkin SP. 1989. Ecophysiology of metals in terrestrial invertebrates. London UK: Elsevier.

21. Stegeman JJ, Brouwer M, Di Giulio RT, Forlin L, Fowler BA, Sanders BM, van Veld PA. 1992. Molecular responses to environmental contamination: Enzyme and protein systems as indicators of chemical exposure and effect. In: Huggett RJ, Kimerle RA, Merle PM, Bergman HL, editors. Biomarkers biochemical, physiological, and histological markers of anthropogenic stress. Boca Raton FL: Lewis. p 235–336.

22. Dallinger R. 1996. Metallothionein research in terrestrial invertebrates: Synopsis and perspectives. *Comp Biochem Physiol* 113C:125–133.

23. Kojima Y. 1991. Definition and nomenclature of metallothioneins. *Methods Enzymol* 205:8–10.

24. Yamamura M, Mori T, Suzuki KT. 1981. Metallothionein induced in the earthworm. *Experientia* 37:1187–1189.

25. Bauer-Hilty A, Dallinger R, Berger B. 1989. Isolation and partial characterisation of a cadmium-binding protein from *Lumbricus variegatus* (Oligochaeta, Annelida). *Comp Biochem Physiol* 94C:373–379.

26. Nejmeddine A, Sautiere P, Dhainaut-Courtois N, Baert J-L. 1992. Isolation and characterisation of a Cd-binding protein from *Allolobophora caliginosa* (Annelida, Oligochaeta): Distinction from metallothioneins. *Comp Biochem Physiol* 101C:601–605.

27. Morgan JE, Norey CG, Morgan AJ, Kay J. 1989. A comparison of the cadmiumbinding proteins isolated from the posterior alimentary canal of the earthworms *Dendrodrilus rubidus* and *Lumbricus Rubellus*. *Comp Biochem Physiol* 92C:15–21.

28. Suzuki KT, Yamamura M, Mori T. 1980. Cadmium-binding proteins induced in the earthworm. *Arch Environ Contam Toxicol* 9:415–424.

29. Furst A, Nguyen Q. 1989. Cadmium-induced metallothionein in earthworms (*Lumbricus terrestris*). *Biol Trace Elem Res* 21:81–85.

30. Ramseier S, Deshusses J, Haerdi W. 1990. Cadmium speciation studies in the intestine of *Lumbricus terrestris* by electrophoresis of metal protein complexes. *Mol Cell Biochem* 97:137–144.

31. Dallinger R. 1994. Part B. Detailed report of the contractors. In: Kammenga JE, editor. Progress report 1995 of BIOPRINT. Biochemical fingerprint techniques as versatile tools for the risk assessment of chemicals in terrestrial invertebrates. Third technical report. Report

from a workshop held in Innsbruck, Austria; 9–10 Feb 1996. Silkeborg, Denmark: National Environmental Research Institute. p 21–24.

32. Roesijadi G. 1993. Response of invertebrate metallothioneins and MT genes to metals and implication for environmental toxicology. In: Suzuki KT, Imura N, Kimura M, editors. Metallothionein III. Biological roles and medical implication. Basel, Switzerland: Birkhauser. p 141–158.

33. Min K-S, Itoh N, Okamoto H, Tanaka K. 1993. Indirect induction of metallothionein by organic compounds. In: Suzuki KT, Imura N, Kimura M editors. Metallothionein III. Biological roles and medical implication. Basel, Switzerland: Birkhauser. p 159–174.

34. Klassen CD, Choudhuri S, McKim JM, Lehrman-McKeeman LD, Kershaw WC. 1993. Degradation of metallothionein. In: Suzuki KT, Imura N, Kimura M, editors. Metallothionein III. Biological roles and medical implication. Basel, Switzerland: Birkhauser. p 207–224.

35. Bremner I, Hoekstra WG, Davies NT, Young BW. 1978. Effect of zinc status on rats on the synthesis and degradation of copper-induced metallothioneins. *Biochem J* 174: 883–892.

36. Riddlington JW, Winge DR, Fowler BA. 1981. Long-term turnover of cadmium metallothionein in liver and kidney following a single dose of cadmium in rats. *Biochem Biophys Acta* 673:177–183.

37. Stenersen J. 1980. Esterases of earthworms. Part I: Characterisation of the cholinesterases in *Eisenia foetida* (Savigny) by substrates and inhibitors. *Comp Biochem Physiol* 66C:37–44.

38. Stenersen J. 1980. Esterases of earthworms. Part II: Characterisation of the cholinesterases in *Eisenia foetida* (Savigny) by ion-exchange chromatography and electrophoresis. *Comp Biochem Physiol* 66C:45–51.

39. Edwards CA, Fischer SW. 1991. The use of cholinesterase measurement in assessing the impact of pesticides on terrestrial and aquatic invertebrates. In: Mineau P, editor. Cholinesterase inhibiting insecticides—their impact on wildlife and the environment. Amsterdam, the Nethelands: Elsevier. p 255–275.

40. Thompson HB, Walker CH. 1993. Blood esterases as indicators of exposure to organophosphorous and carbamate insecticides. In: Fossi M, Leonsio C, editors. Non-destructive biomarkers in vertebrates. Boca Raton FL: Lewis. p 39–62.

41. Stenersen J, Gilman A, Vardanis A 1973. Carbofuran: its toxicity to and metabolism by earthworm (*Lumbricus terrestris*). *J Agric Food Chem* 21:166–171.

42. Stenersen J. 1979. Action of pesticides on earthworms. Part I: The toxicity of cholinesterase-inhibiting insecticides to earthworms as evaluated by laboratory tests. *Pestic Sci* 10:66–74.

43. Stenersen J. 1979. Action of pesticides on earthworms. Part II: Elimination of parathion by the earthworm *Eisenia foetida* (Savigny). *Pestic Sci* 10:104–112.

44. Stenersen J. 1979. Action of pesticides on earthworms. Part III: Inhibition and reactivation of cholinesterases in *Eisenia foetida* (Savigny) after treatment with cholinesterase-inhibiting insecticides. *Pestic Sci* 10:113–122.

45. Niklas von J. 1979. Histochemische untersuchungen zur wirkung von pestiziden als cholinesterase-inhibitoren bei *Lumbricus terrestris* L. *Z Angew Zool* 66:359–368.

46. Stenersen J, Brekke E, Engelstad F. 1992. Earthworms for toxicity testing; Species differences in response towards cholinesterase inhibiting insecticides. *Soil Biol Biochem* 24:1761–1764.

47. Dikshith TSS, Gupta SK. 1981.Carbaryl induced biochemical changes in earthworm (*Pheretima posthuma*). *Indian J Biochem Biophys* 18:154.

48. Gupta SK, Sundararaman V. 1991. Correlation between burrowing capability and AChE activity in the earthworm, *Pheretima posthuma*, on exposure to carbaryl. *Bull Environ Contam Toxicol* 46:859–865.

49. Stringer, A. and M.A. Wright. 1976. The toxicity of benomyl and some related 2-substituted benzimidazoles to the earthworm *Lumbricus terrestris*. *Pestic Sci* 7:459–464.

50. Peakall D. 1992. Animal biomarkers as pollution indicators. London UK: Chapman & Hall.

51. Jimenez BD, Burdis LS, Ezell GH, Egan BZ, Lee NE, McCarthy JF, Beauchamp JJ. 1988. Effects of environmental conditions on the mixed function oxidase system in bluegill sunfish (*Lepomis macrochirus*). *Environ Toxicol Chem* 7:623–634.

52. McDonald IC, Krysan JL, Johnson OA. 1990. Studies of electrophoretic variation in diabrotica as influenced by age, sex or diet of adult beetles (Coleptera; Chrysomelidae). *Ann Entomol Soc Am* 83:1192–1202.

53. Rattner BA, Fairbrother A. 1991. Biological variability and the influence of stress on cholinesterase activity. In: Mineau P, editor. Cholinesterase inhibiting insecticides—their impact on wildlife and the environment. Amsterdam: Elsevier. p 90–107.

54. Moore MN. 1990. Lysosomal cytochemistry in marine environmental monitoring. *Histochemistry* 22:187–191.

55. Mayer FL, Versteeg DJ, McKee MJ, Folmar LC, Graney RL, McCume DC, Rattner BA. 1992. Physiological and nonspecific biomarkers. In: Huggett RJ, Kimerle RA, Merle PM, Bergman HL, editors. Biomarkers biochemical, physiological, and histological markers of anthropogenic stress. Boca Raton FL: Lewis. p 5–86.

56. Moore MN. 1985. Cellular responses to pollutants. *Mar Pollut Bull* 16:134–139.

57. Lowe DM, Moore MN, Evans BM. 1992. Contaminant impact on interactions of molecular probes with lysosomes in living hepatocytes from dab *Limanda limanda*. *Mar Ecol Prog Ser* 91:135–140.

58. Lowe DM, Pipe RK. 1994. Contaminant induced lysosomal membrane damage in marine mussel digestive cells: an in vitro study. *Aquat Toxicol* 30:357–365.

59. Svendsen C, Weeks JM. 1995. The use of a lysosome assay for the rapid assessment of cellular stress from copper to the freshwater snail *Viviparus contectus* (Millet). *Mar Pollut Bull* 31:139–142.

60. Weeks JM, Svendsen C. 1996. Neutral red retention by lysosomes from earthworm (*Lumbricus rubellus*) coelomocytes: a simple biomarker of exposure to soil copper. *Environ Toxicol Chem* 15:1801–1805.

61. Svendsen C, Weeks JM. 1997. Relevance and applicability of a simple earthworm biomarker of copper exposure. I. Links to ecological effects in a laboratory study with *Eisenia andrei*. *Ecotoxicol Environ Saf* 36:72–79.

Chapter 15

Cytochrome P450 activity in 3 earthworm species

C.T. Eason, L.H. Booth, S. Brennan, J. Ataria

Apporectodea caliginosa, Lumbricus rubellus, and *Eisenia andrei* are common earthworm species found in New Zealand and Australia. Their suitability for detecting exposure to chemical contamination, initially through measurement of baseline cytochrome P450 activity, has been assessed. There was no detectable enzyme activity with resorufin-containing substrates, but enzyme activity was detected with the substrate 7-ethoxycoumarin. A higher level of activity was detected in *L. rubellus* (0.44 pmols hydroxycoumarin/min/mg gut) than in *A. caliginosa* and *E. andrei*. This chapter focuses on *L. rubellus* in future earthworm biomarker studies.

Environmental protection in New Zealand and Australia requires methods for the detection of early signs of pollution. The aim here is to improve monitoring by providing a validated suite of biochemical markers. Earthworms were chosen as one of a number of sentinel species because of their beneficial role in promoting soil fertility and because they are general representatives of soil fauna. Furthermore, they are easy to handle, suitable for captive breeding, and widespread. Earthworms are a dominant form of animal biomass in the soil. Since they live, feed, and reproduce within the soil [1], they play an important role in soil ecosystems. The highly dispersed nature of their food source, which includes soil organic matter (OM) and its associated micro-organisms and fauna, requires them to ingest large quantities of soil. For example, temperate earthworms ingest 1 to 2 times their body mass daily, while tropical earthworms ingest 20 to 30 times their mass [2]. Their feeding habits make earthworms valuable in situ sentinels for assessing biological risks from hazardous and toxic compounds in soil environments [3].

Regulations for toxicological evaluation of new pesticides in many countries demand data derived from earthworm tests that are carried out according to 1984 OECD Guidelines for Testing Chemicals. Further laboratory techniques and methods for standardized field tests are being developed to extend the range of information available for the registration of chemicals. In addition to the development and ongoing refinement of standardized regulatory toxicology methodology, it was recommended that "encouragement should be given to the use of earthworms for bioassays and to develop sensitive endpoints such as immune function and other sublethal parameters" to improve the understanding of potential adverse effects of xenobiotics [4].

In earthworms, very little research has been conducted on xenobiotic metabolism. In vertebrates, the cytochrome P450 monooxygenase enzyme system has been very extensively studied because of its involvement in the metabolism of foreign compounds. Originally evolved, perhaps as long as 2000 million years ago, to handle naturally occurring toxic compounds [5], these enzymes now play an important role in the detoxification of manufactured chemicals. The amount of enzyme can be increased by exposure to chemicals, thus providing a greater capability to deal with xenobiotics. The induction of cytochrome P450 mono-oxygenase is caused by a wide variety of chemicals. Hence cytochrome P450 activity, in excess of the normal range, is a useful biomarker.

Previous limited attempts to use induction of cytochrome P450 in earthworms as a sensitive bioassay of exposure have proved difficult because of interference from endogenous pigments [6] and the identification of non-inducible forms of cytochrome P450 [7]. This chapter describes a comparison of enzyme activities in *L. rubellus*, *A. caliginosa*, and *E. andrei*, 3 commonly available earthworms in New Zealand.

Methods
Adult earthworms of each species were collected from locations where exposure to pesticides or other xenobiotics was unlikely. Worms were maintained in sieved soil in the laboratory at 18°C and were fed grassmeal.

Identification of a suitable substrate and metabolite stability
Worms were anesthetized in 10% v/v ethanol in water, and the gut was opened and flushed with deionized water to remove soil debris. Whole body, mid-gut and integument, and mid-gut only sections were prepared. Post-mitochondrial fractions were prepared by homogenizing tissue in 0.1 M phosphate buffer containing 1 mM ethylenediaminetetra-acetic acid (EDTA), 0.1 M KCL, 1 mM dithiothreitol and 0.1 M phenanthroline, and centrifuging at 10,000 g to remove cell debris. This approach of developing a method and utilizing the 10,000-g fraction rather than the 100,000-g microsomal pellet was deliberate. The intention was to have small numbers of worms to use in routine biomarker assessments. Cytochrome P450 activity was assayed using the substrates 10 µM 7-ethoxyresorufin [8], 10 µM 7-pentoxyresorufin, 10 µM 7-benzyloxyresorufin [9], and 2 mM 7-ethoxycoumarin [10]. Worm post-mitochondrial fractions from each of 12 *L. rubellus* and *A. caliginosa* were added to samples spiked with 5 µl resorufin to determine if resorufin (which is the product of cytochrome P450 activity on 7-ethoxyresorufin) is degraded by some component in worms. Activities in mid-gut-only sections of *A. caliginosa*, *L. rubellus*, and *E. andrei* were compared.

Optimization of assay conditions
To determine the optimum conditions for metabolism of the substrate 7-ethoxycoumarin, enzyme activity was assessed in a gut preparation from *E. andrei* at 10, 15, 20, and 25EC and at pH 7.0, 7.2, 7.4, 7.6, and 8.0. The whole gut was used to provide sufficient material; earlier experiments had shown that there was no activity in mid-gut and integument samples in *E. andrei*.

Evidence for presence of cytochrome P450
To provide evidence that the enzyme activity measured in *E. andrei* was due to the presence of cytochrome P450, assays were conducted in the presence and absence of an es-

sential cofactor (nicotinamide-adenine dinucleotide phosphate [NADPH], reduced) and with 50 µl of a known specific cytochrome P450 inhibitor, proadifen hydrochloride (20 mM). NADPH is supplied by a NADPH-generating system that contains 12.5 mM NADP, 1000 units of glucose-6-phosphate dehydrogenase.

Results

Identification of suitable substrate

There was no detectable cytochrome P450 activity with the substrates 7-ethoxyresorufin, 7-pentoxyresorufin, or 7-benzyloxyresorufin. There was no evidence of the presence of enzymes capable of degrading resorufin in worm post-mitochondrial fraction. Enzyme activity was detected toward 7-ethoxycoumarin. Higher activity was detected in *L. rubellus* (mid-gut and integument sections) than in *A. caliginosa* and *E. andrei* (mid-gut sections) (Figure 15-1). There was no activity in *A. caliginosa* whole-body samples and in *E. andrei* whole-body and mid-gut sections. A gut-only sample for *A. caliginosa* was not tested.

Optimization of assay conditions

Optimal temperature was shown to be 15°C (Figure 15-2), while the optimal pH was 7.4 (Figure 15-3).

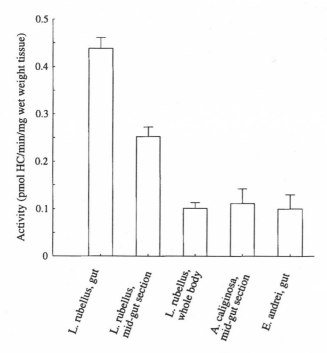

Figure 15-1   Cytochrome P450 enzyme activity (± SE) in *L. rubellus*, *A. caliginosa*, and *E. andrei* expressed as formation of hydroxycoumarin/min/mg wet weight tissue. Legend: Samples were prepared from 10 to 15 worms (N = 3 assays).

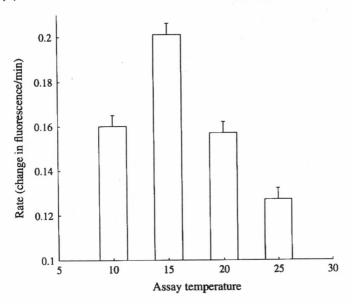

Figure 15-2   Optimization of ethoxycoumarin-o-dealkylase enzyme assay temperature in *E. andrei*
               expressed as change in fluorescence/minute (± SE). Legend: Samples were prepared
               from homogenates of 12 worms. Temperature expressed as °C.

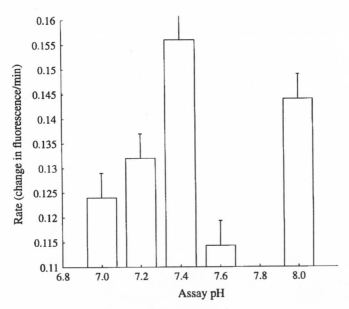

Figure 15-3   Optimization of ethoxycoumarin-o-dealkylase enzyme assay pH in *E. andrei* expressed as
               change in fluorescence/minute (± SE). Legend: Samples were prepared from
               homogenates of 12 worms.

Evidence for presence of cytochrome P450

A complete lack of enzyme activity in the presence of proadifen hydrochloride and the requirement of NADPH for in vitro enzyme activity provided indirect proof that the enzyme system responsible for 7-ethoxycoumarin metabolism in earthworms was cytochrome P450.

Discussion and conclusions

Results with *L. rubellus* would appear to contrast with those of earlier researchers [11] who reported activity in *L. terrestris* with 7-benzyloxyresorufin but no activity with 7-ethoxycoumarin; however, it should be noted that in these studies post-mitochondrial rather than microsomal fractions were used. Nevertheless, the results complement reports of the inability of the gut mono-oxygenase system from *L. terrestris* to dealkylate ethoxyresorufin [6,12] (Table 15-1). *L. rubellus* showed considerably greater activity than that previously reported in *L. terrestris* [6] (440 pmol min/g versus 37 pmol min/g).

The cytochrome P450 activities that were determined in 3 different species of earthworm are extremely low when compared to mammals, fish, and birds [13,14]. This is in agreement with the findings of earlier researchers [6,7]. The low activity level of cytochrome P450 in earthworms might at first seem surprising, since they are widely exposed to xenobiotics

Table 15-1   Cytochrome P450 activity: summary of current and earlier results

Authors	Species	ECOD[1]	EROD[2]	EROD[3]
This chapter	*L. rubellus*	✓	x	x
	A. caliginosa	✓	x	x
	E. andrei	✓	x	x
Berghout et al. [11]	*L. terrestris*	x	x	✓
Stenersen [12]	*L. terrestris*	✓	x	--
Liimatainen and Hanninen [6]	*L. terrestris*	✓	x	--

✓ enzyme activity present
x enzyme activity absent
[1] ethoxycoumarin-o-dealkylase
[2] ethoxyresorufin-o-dealkylase
[3] benzyloxyresorufin-o-dealkylase

in the soil. However, during evolution, earthworms may not have required extensive detoxifying systems, since many naturally occurring plant toxins would be substantially broken down in the soil before earthworms ingested them [15]. Alternatively, there may be other enzyme systems such as glutathione S-transferase (GST) and hydrolases that are important for xenobiotic detoxification in earthworms. There is conflicting data with regard to the inducibility of GST. GST was not induced in *E. andrei* by classic inducers such as methylcholanthrene and benzene exposure [15], but was induced in *Pheretima posthuma* by aldrin, endosulphan, and lindane [16], and in the laboratory induction of GST has been demonstrated with *A. caliginosa* following exposure to chlorpyriphos and diazinon. Ester hydrolysis seems to be a pathway of some importance in earthworms [12]. Basic research on the biotransformation of xenobiotics in earthworms is needed to determine the relative importance of different detoxification enzymes in the elimination process of chemicals.

Despite the low activity level in earthworms, it is not possible to discount cytochrome P450 involvement in detoxification or its potential as a biomarker. Cytochrome P450 has been implicated in the metabolism of carbofuran to 3-hydroxycarbofuran in *L. terrestris* [17] and in one of the few comparative toxicity studies of pesticides to earthworms, *A. caliginosa* (which the preliminary investigation reported here has demonstrated to have lower cytochrome P450 activity than *L. rubellus*) is shown to be much more sensitive than *L. rubellus* to the toxic effects of aldicarb, carbaryl, and carbofuran [18]. It is possible that *L. rubellus* is protected by its inherently greater ability to detoxify these pesticides through the cytochrome P450 mono-oxygenase system. However, high cytochrome P450 activity is not linearly related to a low sensitivity, therefore the comparison of cytochrome P450 activity and pesticide sensitivity of *A. caliginosa* and *L. rubellus* is preliminary.

Good biomarkers are recognized by a substantial difference between normal activity in non-exposed situations and increased activity after exposure to toxicants. Further studies are planned to determine whether cytochrome P450 in earthworms has potential as a biomarker. Cytochrome P450 activity will be assessed with additional substrates. The most important milestone in future projects will be addressed by assessing the inducibility of cytochrome P450 in *L. rubellus* by contaminants and hence the potential of this enzyme as a biomarker of exposure. If induction can be demonstrated, then cytochrome P450 responses in earthworms will be compared with other biomarkers under investigation in earthworms [3,19–23] for a range of different xenobiotics.

Acknowledgments - The New Zealand Foundation for Research, Science and Technology is thanked for funding this research. Dr Graham Sparling, Dr Ravi Gooneratne, and Dr Megan Ogle-Mannering are thanked for helpful comments on this paper.

References

1. Lavelle P. 1988. Earthworm activities and the soil system. *Biol Fert Soils* 6:237–251.
2. Lavelle P, Barvis I, Martin A, Zaidi Z, Schaefer R. 1989. Management of earthworm populations in agroecosystems: a possible way to maintain soil quality? In: Clarholm M, Bergström L, editors. Ecology of arable land: perspectives and challenges. Boston MA: Klewer Academic. p 109–122.
3. Weeks DM, Svendsen C. 1996. Neutral red retention by lysosomes from earthworms (*Lumbricus rubellus*) coelomocytes: a simple biomarker of exposure to soil copper. *Environ Toxicol Chem* 15(10):1801–1805.
4. Greig-Smith PW. 1992. Recommendations of an international workshop on ecotoxicology of earthworms. In: Greig-Smith PW, Becker H, Edwards PJ, Heimbach F, editors. Ecotoxicology of earthworms. Andover UK: Intercept Ltd. p 1.
5. Nebert DW, Gonzalez FJ. 1987. P450 genes: structure, evolution, and regulation. *Animal Rev Biochem* 56:945–993.

6. Liimatainen A, Hanninen O. 1982. Occurrence of cytochrome P450 in the earthworm *Lumbricus terrestris*. In: Hietenen E, Larteinen M, Hanninen O, editors. Cytochrome P450: biochemistry, biophysics, and environmental implication. Amsterdam, the Netherlands: Elsevier. p 255–257.

7. Milligan L, Babish J, Neuhauser F. 1986. Non-inducibility of cytochrome P450 in earthworm *Dendrobaena veneta*. *Compar Biochem Physiol* 85C:85–87.

8. Burke DM, Mayer RT. 1974. Ethoxyresorufin. *Drug Metabolism and Disposition* 2:583–588.

9. Burlie MD, Thompson S, Elcombe CR, Halpert J, Haaparanta T, Mayer RT. 1985. Ethoxy-pentoxy, and benzyloxyphenoxazenes and homologues: a series of substrates to distinguish between different induced cytochromes P450. *Biochem Pharmacol* 34:3337–3345.

10. Ullrich V, Weber P. 1972. The O-dealkylase of 7-ethoxycoumarin by liver microsomes. *Physiolog Chem* 353:1171–1177.

11. Berghout A, Buld J, Wenzel E. 1990. The cytochrome P450-dependent monooxygenase system of the midgut of the earthworm *Lumbricus terrestris*. *European J Pharmacol* 183:1885–1886.

12. Stenersen J. 1984. Detoxification of xenobiotics in earthworms. *Compar Biochem Physiol* 78C:249–252.

13. Moriarty F, Walker CH. 1987. Bioaccumulation in food chains: a rational approach. *Ecotoxicol Environ Safety* 13:208–215.

14. Ronis MJJ, Walker CH. 1989. The microsomal mono-oxygenases of birds. *Rev Biochem Toxicol* 10:310–384.

15. Stokke K, Stenersen J. 1993. Non-inducibility of glutathione transferase of the earthworm *Eisenia andrei*. *Compar Biochem Physiol* 106C(3):753–756.

16. Hans RK, Khan MA, Faroq M, Beg MU. 1993. Glutathione-S-transferase activity in an earthworm (*Pheretima posthuma*) exposed to three insecticides. *Soil Biol Biochem* 25(4):509–511.

17. Stenersen J, Gilman A, Vardanis A. 1973. Carbofuran: its toxicity to and metabolism by earthworms (*Lumbricus terrestris*) *J Agric Food Chem* 21:166–171.

18. Stenersen J. 1979. Action of pesticides on earthworms. Part I: The toxicity of cholinesterase-inhibiting insecticides to earthworms as evaluated by laboratory tests. *Pesticide Sci* 10:66–74.

19. Cikutoric MA, Fitzpatrick LC, Venables BJ, Goven AJ. 1993. Sperm count in earthworms (*Lumbricus terrestris*) as a biomarker for environmental toxicology: effects of cadmium and chlorolase. *Environ Pollut* 81:123–125.

20. Goven AJ, Eyambe GS, Fitzpatrick LC, Venables BJ, Cooper EL. 1993. Cellular biomarkers for measuring toxicity of xenobiotics: effects of polychlorinated biphenyls on earthworm *Lumbricus terrestris* coelomocytes. *Environ Toxicol Chem* 12:863–870.

21. Walsh P, Adlouni C El, Mukhopachay MJ, Viel G, Nadeau D, Poirer GG. 1995. 32P- Post labelling determination of DNA adducts in the earthworm *Lumbricus terrestris* exposed to PAH-contaminated soils. *Bull Environ Toxicol Chem* 54:654–661.

22. Bunn KE, Thompson HM, Tarrant KA. 1996. Effects of agrochemicals on the immune systems of earthworms. *Bull Environ Contam Toxicol* 37(4):632–639.

23. Gibb JOT, Svendsen C, Weeks JM, Nicholson JK. 1997. H NMR spectroscopic investigations of tissue metabolite biomarker response to Cu (II) exposure in terrestrial invertebrates: identification of free histidine as a novel biomarker of exposure to copper in earthworms. *Biomarkers* 2:295–302.

Chapter 16

Immunohistochemical detection of heat shock proteins expression in stressed earthworms

F. Mariño and A.J. Morgan

Heat shock proteins (HSPs) have been exploited as pollution biomarkers in diverse aquatic and terrestrial macroinvertebrates, but not earthworms. An indirect immunoperoxidase histochemical protocol was used on a caudally amputated, stressed earthworm (*Lumbricus rubellus*) model to test the cross-reactivity of 3 commercial (Sigma) monoclonal antibodies. These antibodies, anti-hsp60 (clone LK2), anti-hsp70 (clone BRM-22), and anti-hsp90 (clone AC-16), were cross-reactive and demonstrated that all 3 proteins were upregulated to about 7 d post-amputation. Stress-protein expression subsequently declined, presumably as damaged tissue recovered. A fairly high level of hsp70 and hsp90 constitutive expression was detected in the earthworm chloragocytes. Seminal vesicle regions damaged by the gregarine parasite *Monocystis* strongly expressed all 3 proteins.

Qualitative immunohistochemistry indicated that the monoclonals LK2, BRM-22, and AC-16 may be useful for quantitatively assaying the sublethal responses of *L. rubellus* to numerous environmental stressors, including inorganic and organic residues. Further tests are required to test whether 1 or more of the monoclonals are cross-reactive with HSPs in other earthworm species.

Chemical analyses of pollutant concentrations in soils fail to provide direct information about interactions between residue mixtures and intrinsic soil factors. Nor do such analyses comment on toxicant bioavailability, detoxification, or biological effects at any level of organization. The advent of the biomarker approach in ecotoxicology circumvents some of these limitations.

Biomarkers are a measure of stress expressed at any level of biological organization from the community or population to the cellular or molecular, although they are most frequently associated with lower-level functions. Among the several potential advantages of biomarkers, 2 are outstanding: they provide an integrative assessment of the relationship between toxicant exposure and biological effects [1], and they are predictive in that they provide an early warning of the onset of damage that will eventually be evident at higher ecological levels [2]. Ideally, a good biomarker should yield easily interpretable mechanistic information, have a high sensitivity to either a broad range or a specific class of stressors, and be rapid, cheap, and easy to perform [3].

Standardized earthworm toxicity tests (acute and chronic, lethal and sublethal) are widely used for risk assessment in conjunction with the registration of plant protection chemicals [4–7]. Furthermore, a number of earthworm-based, low organizational level biomarkers have recently been introduced; their primary functions are to detect and monitor functional perturbations caused by complex environmental stressors. Examples include behavioral tests [8], immunological competence tests [9,10], neurophysiological performance tests [11], lysosomal membrane stability tests [12], and molecular genetic tests [13]. The validation of the lysosomal biomarker for the biological assessment of the impact of an industrial fire [14] has provided renewed impetus to the search for new earthworm biomarkers, as recommended by the International Workshop on Earthworm Ecotoxicology [15].

Heat shock proteins are highly conserved proteins belonging to 4 main families whose members have a number of constitutive functions in the normal activities of cells, including maintaining and conferring correct conformation to protein precursors, transmembrane protein trafficking, and assembly and regulation of steroid receptors [16]. Heat shock proteins have several attributes that make them potentially useful ecotoxicological biomarkers [3,17,18]. For example, they are upregulated in response to induction by metals, metalloids, and organic residues whose common biochemical lesion is protein denaturation accompanying faulty molecular folding.

While HSPs have been exploited as biomarkers in a marine bivalve [19], terrestrial isopods and gastropods [20], and a nematode [21], they have not been used in this way in earthworms. Sanders et al. [17] did show, however, that antibodies raised against 2 major stress proteins (hsp70, *chaperonin*60) were more cross-reactive between vertebrate species than between invertebrate species. The practical implication of this finding [17], which included observations on the earthworm *Eisenia fetida*, is that antibodies raised against HSPs from a given species may not be useful for assaying analogous stress proteins in another species.

The present study had 2 main objectives: first, to determine by immunoperoxidase histochemistry (see Ormerod and Imrie [22] for the principles of this labeled-antibody staining of tissue sections technique) whether 3 commercially available antibodies could detect HSP expression (hsp60, hsp70, hsp90) in earthworms stressed by caudal amputation; second, to determine whether constitutive and induced HSP expressions display tissue tropism. These are necessary preliminaries toward the use of HSPs as pollution-monitoring tools.

Materials and methods

Animals and their treatment
Adult *Lumbricus rubellus* were collected by hand sorting from an uncontaminated site at Dinas Powys, South Wales (Ordinance Survey Map Ref. = ST149 723) and transferred in their native soil to the laboratory. After an acclimatization period of about 1 week in a 12-h dark:12-h light constant-temperature room (17 °C), the worms were experimentally stressed. All worms were maintained on moist native soil, with weathered cow manure added as food throughout the experimental period.

The terminal segments (approximately the posterior 1/3 of body length) of each experimental worm were amputated by scalpel. At intervals of 0, 1, 2, 4, 7, or 12 d after amputation, the entire anterior portions of 3 individual worms were fixed in 10% formal-saline solution.

Earthworms have a considerable regenerative capacity, a complex physiological event that entails initial damage, repair, cellular invasion, synthesis, tissue growth, and re-modeling [23,24]. Amputation, therefore, represents a convenient and reproducible form of stress that yields a valid general model for establishing whether the 3 commercial monoclonals were able to detect and locate the earthworm antigens.

Primary antibodies
The 3 antibodies used in this study were raised in mouse hosts and purchased from Sigma (Sigma-Aldrich Company Ltd., Poole, Dorset BH12 4QH, England).

Monoclonal anti-hsp60 (clone no. LK2; code = H3524) was raised against recombinant human hsp60. The Sigma literature indicates that it is cross-reactive with the constitutive and inducible proteins from human, rat, chicken, E. coli, helminth, and spinach. Trials in our laboratory with a dilution series established that optimal immunohistochemical staining on positive controls (human rheumatoid- and osteo-arthritic joints) was ob-tained at 1:50 dilution.

Monoclonal anti-hsp70 (clone no. BRM-22; code = H5147) was raised against hsp70 bovine brain. It is cross-reactive with constitutive (hsp73) and inducible (hsp72) proteins from the brains of human, rat, rabbit, hamster, chicken, and guinea pig, as well as hsp70 from a cell extract in *Drosophila*. (Note that Sanders et al. [17] reported that BRM-22 is also cross-reactive with heat-inducible proteins from the tissues of various Crustacea, Echinodermata, and fish but not from the earthworm *E. fetida*). Optimal dilution for this antibody was 1:100. Monoclonal anti-hsp90 (clone no. AC-16; code = H1775) was raised against hsp90 from the water mold, *Achlya ambisexualis*. It is cross-reactive with consti-tutive and inducible hsp90 from human, rabbit, rat, mouse, chicken, insect, and wheat germ but does not recognize hsp90 from yeast and *E. coli*. Optimal dilution for this anti-body in the laboratory was 1:50.

Immunohistochemistry (indirect immunoperoxidase method)
Fixed specimens were dehydrated in graded alcohols, cleared in xylene, and embedded in paraffin wax. Longitudinal sections nominally 7µm thick were cut on a Bright rotary microtome and mounted on Chance Gold Star glass slides treated with an aqueous sub-bing solution of 1% gelatine in 0.1% chromic potassium sulphate.

Sections were dewaxed with 100% xylene. Endogenous peroxidase was blocked with a mixture of methanol: 30% hydrogen peroxide (377.6 ml: 6.4 ml, v/v) for 30 mins at room temperature. After washing in running tap water for 5 mins, the sections were washed (3 × 3 minutes) in 0.01M phosphate-buffered saline (PBS), pH 7.1. The primary monoclonal antibodies (LK2, BRM-22, AC-16) were applied to separate sections at their optimal dilu-tions in PBS/bovine serum albumin (BSA). The slides were incubated in a moist cham-ber at 4 °C for about 15 h, and then washed (3 × 3 mins) in PBS. Horseradish peroxidase-conjugated rabbit anti-mouse secondary antibody (DAKO Ltd; code = P260) diluted 1:100 with PBS/BSA was applied to each slide for 1 h at room temperature. After 3 × 3-min washes in PBS, the sections were incubated for 5 mins in a chromogen (0.05% di-aminobenzidine dihydrochloride [DAB] in 100 ml PBS) and peroxidase-substrate (4 drops of 30% hydrogen peroxide) solution. After being washed in water, the sections were lightly counterstained with Meyer's haematoxylin, dehydrated, cleared, and mounted in D.P.X Mountant (BDH Chemicals, Ltd., Poole, England). Immunostained sections and controls (primary antibodies omitted) were photographed using an Olympus System Photomicroscope, Model BHC. Positive immunostaining was a distinctive chestnut brown, not to be confused in the black-and-white prints with haematoxylin counterstain (dark blue) and dark brown pigmentation in the body wall.

Results

A systematic study of HSP expression in all tissues along a regeneration gradient, from the posterior zone adjacent to the plane of amputation to the extreme anterior tip of the earthworm, was beyond the scope of the present study. Nevertheless, most of the major organs and tissues of the amputees were qualitatively examined for HSP expression at time intervals up to 12 d post-amputation.

The 3 monoclonal antibodies appeared to be cross-reactive with their corresponding *L. rubellus* HSPs, since positive immunostaining was detected in each case, especially in the chloragocytes enveloping the alimentary canal and major blood vessels (Figures 16-1 through 16-18). Staining with anti-hsp60 (LK2; Figures 16-1 through 16-6) was generally weaker than staining with anti-hsp70 (BRM-22; Figures 16-7 through 16-12) and anti-hsp90 (AC-16; Figures 16-13 through 16-18). The stress-induced expression of the 3 HSPs was tissue tropic; in each case, the distribution pattern was similar.

There was little evidence of constitutive hsp60 staining in "normal" earthworm tissues (0-d controls; Figure 16-1); hsp70 (Figure 16-7) and hsp90 (Figure 16-13) were detected in unstressed controls.

Immunostaining intensity for all 3 shock proteins appeared to increase progressively up to about 7 d post-amputation and then declined by 12 d. The temporal pattern of HSP expression correlated very well with the reported [23] increase in DNA synthesis during the first 4 d, followed by a sharp reduction after about 7 d, in experimentally wounded *E. fetida*. The HSP observations should be viewed with caution, however, because there is concern about the ability to be certain that comparable anatomical areas along the worm's longitudinal axis were examined at each post-amputation time interval in *L. rubellus*.

Seminal vesicles of many lumbricid worms are commonly infected by life stages of the gregarine parasite *Monocystis lumbrici* [25]. It is evident that the sporozoites of the parasite penetrate and destroy clusters of developing earthworm spermatozoa, the sperm morulae (Figures 16-19 through 16-21). The 3 monoclonals yielded positive focal immunostaining in those regions of the seminal vesicles containing parasite-damaged morulae. It was not possible to determine whether the HSPs were expressed in the maturing gregarines or in the damaged host cells. However, the characteristic lozenge-shaped zygocysts of *Monocystis* and uninfected regions of *L. rubellus* seminal vesicles containing apparently undamaged sperm cells did not display significant hsp60, hsp70, or hsp90 immunoreactivity (Figures 16-19 through 16-21).

Immunohistochemical controls, where the primary antibodies were omitted from the protocol, did not produce peroxidase staining (Figures 16-22 through 16-24), thus indicating that the quenching of endogenous peroxidase activity was successful and that the detected tissue-tropic immunoreactivity produced by the unexpurgated protocol reflected antigen distribution.

Discussion

The case for cellular biomarkers in pollution monitoring was succinctly stated by Cajaraville et al. [26]: "Cellular responses to pollutant-induced cell injury anticipate damage at higher levels of biological organization, thus providing highly sensitive indicators of environmental impact . . . [and the ability] to predict responses at organismic level." The so-called HSPs have been suggested in recent years as good candidate biomarkers responsive to a diverse range of environmental and other stress agents [27]. It is encouraging that the induced expression of certain family members of these conserved proteins correlates very well with changes in other cellular and physiological performance indices. For example, Sanders et al. [19] reported a relationship between increased levels of a 60kDa stress protein in Cu-exposed marine mussels and scope-for-growth measurements in the bivalve; and Ryan and Hightower [3], not only detected concentration-dependent upregulation of stress protein in fish cells stressed by either Cu or Cd, but they also found that these sublethal changes coincided with increased cytotoxicity measured by neutral red retention (NRR) time.

The immunohistochemical detection of HSPs in *L. rubellus*, using 3 commercially available monoclonal antibodies, paves the way for the routine quantitative assays that are prerequisite to viable ecotoxicology application. Sanders et al. [17], using polyclonal antibodies in immunoblot tests, detected a chaperonin (*cpn* 60) and an hsp70 in the earthworm *E. fetida*. However, it is prudent to characterize the Sigma monoclonals further before they are exploited in earthworm bioassays, especially in view of the fact that Sanders et al. [17] did not detect by Western blotting the cross-reactivity of 1 of them, BRM-22 (anti-hsp70), with heat-inducible protein from *E. fetida*. This published report does not necessarily contradict measured immunoperoxidase staining with BRM-22 in *L. rubellus* tissues. There is a relatively high interspecies variability in stress protein immunoreactivity, both within and between certain invertebrate taxonomic groups. For example, within the Crustacea, the hsp70 of 1 marine shrimp species, *Crangon crangon*, cross-reacted with BRM-22, but the hsp70 of another shrimp species, *Palaemon adspersus*, did not; within the Annelida, 2 polyclonals that detect heat-inducible proteins in the oligochaete *E. fetida* did not cross-react with corresponding proteins in 2 polychaete species [17]. Furthermore, although the 70 kDa HSP family in the insects *Drosophila* and *Sarcophaga crassipalpis* are immunologically similar, the control of their expression is very difficult [28]. Thus it is possible that *Lumbricus* and *Eisenia* HSPs have differential immunological or expression characteristics.

Heat shock proteins are often considered to be expressed in most tissues in stressed multicellular organisms. The observations of Joplin and Denlinger [28] on HSP responses at different developmental stages of the flesh fly *S. crassipalpis* indicate otherwise, because hsp65 and hsp72 in this insect species show complex developmental and tissue-specific switches of expression. Analogous tissue-specific HSP expression was observed in amputated *L. rubellus*, with the chloragogenous tissue displaying relatively high hsp60, hsp70, and hsp90 immunostaining compared with, for example, body wall muscle. While it remains to be seen whether or not the pattern of earthworm HSP expression is stressor determined, the high level of HSP expression induced by amputation in adjacent chloragocytes is potentially significant for ecotoxicological risk assessment. This tissue plays a central role in the accumulation and detoxificiation of a chemically diverse range of environmental toxicants [29] and may thus be directly exposed to a number of HSP-inducing chemicals in earthworms inhabiting contaminated soils. The present qualitative observations indicate that it may be prudent to exclude the seminal vesicles from earthworm tissue preparations for quantitative HSP assays because of the likely confounding effects engendered by the fairly ubiquitous *Monocystis* infections.

In conclusion, before the encouraging observations that certain commercially available monoclonals detect earthworm HSPs can be exploited quantitatively to measure sublethal toxicant-induced stress, a number of systematic exploratory studies should be undertaken. Some of these have been highlighted in an instructive preliminary report on HSPs in the bacterivorous nematode *Heterocephalobus pauciannulatus* and the slug *Deroceras reticulatum* [21]. First, it would be valuable to know whether anti-HSP monoclonal antibodies from other commercial sources (e.g., Stressgen, Victoria, BC, Canada [18]) are also able to detect the earthworm proteins. Second, the effective monoclonals should be used in quantitative protocols such as SDS-PAGE/Western Blotting [18] or enzyme-linked immunosorbent assay (ELISA) [30] to determine dose responses during exposures to specific toxicants under controlled laboratory conditions. Third, the sensitivity of the HSP biomarker should be compared with the sensitivities of other cellular or biochemical biomarkers (e.g., Ryan and Hightower [3] showed concentration-dependent relationships between increased HSP levels and cytotoxicity as measured by neutral-red uptake assay in cultured fish cells stressed by Cd or Cu) and whole-organism performance indicators (e.g., Kammenga et al. [21] reported a correlation between the dose-dependence of hsp70 levels in slugs exposed to Cd or Zn and growth or reproduction parameters). Finally, attempts should be made to evaluate and validate the use of earthworm HSPs as molecular biomarkers in microcosms and in free-living populations.

Acknowledgments - The authors thank the Xunta de Galicia and the University of Vigo for the financial support that made this collaboration possible.

Labeling code for figures:

a = alimentary epithelium;
b = blood vessel;
c = chloragogenous tissue;
m = circular muscle;
p = *Monocystis* sporozoites;
r = infected sperm morulae;
s = intersegmental septum;
u = undamaged (developing spermatozoa)
w = body wall musculature.

Figure 16-1   Immunoperoxidase staining of hsp60 (antibody = LK2) at 0 d post-amputation. Only faint staining of constitutive protein, confined mainly to the chloragog tissue, was evident. Magnification = 1600×.

Figure 16-2   Immunoperoxidase staining of hsp60 at 1 d. Moderate expression in the chloragogenous tissue and in the blood sinus. No appreciable staining in the body wall musculature. Magnification = 1600×.

Figure 16-3   Immunoperoxidase staining of hsp60 at 2 d. Distribution similar to Figure 16-2. Magnification = 1600×.

Figure 16-4   Immunoperoxidase staining of hsp60 at 4 d. Fairly intense expression in chloragogenous tissue. Magnification = 1600×.

Figure 16-5   Immunoperoxidase staining of hsp60 at 7 d. Intense expression in chloragogenous tissue and within blood vessels. Magnification = 1600×.

Figure 16-6   Immunoperoxidase staining of hsp60 at 12 d. Staining intensity much reduced, although this micrograph may be taken from an anterior region of the alimentary canal, i.e., a considerable distance from the zone of tissue damage and repair. (Note that the calciferous gland secretory epithelium was negative for all 3 antigens at each time interval examined [not illustrated]). Magnification = 1600×.

Figure 16-7   Immunoperoxidase staining of hsp70 (antibody - BRM-22) at 0 d. Moderate constitutive expression in chloragogenous tissue within segments well anterior of the zone of tissue damage. Magnification = 1600×.

Figure 16-8   Immunoperoxidase staining of hsp70 at 1 d. Moderate expression in chloragogenous tissue and in a blood sinus; adjacent musculature has no shock protein expression. Magnification = 1600×.

Figure 16-9   Very intense immunoperoxidase (anti-hsp70) staining at 2 d post-amputation in chloragogenous tissue and basement layer (arrows). Evidence of some expression in the alimentary epithelia (double arrows). Magnification = 1600×.

Figure 16-10   Immunoperoxidase staining of hsp70 at 4 d. Staining distribution and intensity similar to Figure 16-9. Magnification = 1600×.

Figure 16-11   Immunoperoxidase staining of hsp70 at 7 d. Note intense staining in chloragogenous tissue and blood sinuses. Magnification = 1600×.

Figure 16-12   Immunoperoxidase staining of hsp70 at 12 d. Antigen expression (moderate) is restricted to chloragogenous tissue, blood vessels, and basement layer. Alimentary epithelia were only faintly stained; body wall musculature is unstained. Magnification = 800×.

Figure 16-13   Immunoperoxidase staining of hsp90 (antibody = AC-1b) at 0 d. Mild constitutive expression detected in chloragogenous tissue only. Magnification = 1600×.

Figure 16-14   Immunoperoxidase staining of hsp90 at 1 d post-amputation. Note that the chloragogenous tissue and, especially, a blood vessel are intensely stained. Magnification = 1600×.

Figure 16-15   Immunoperoxidase staining of hsp90 at 2 d. Staining intensity is very intense in the chloragogenous tissue and an enclosed blood sinus. The body wall musculature has little or no expression. Magnification = 1600×.

Figure 16-16   Immunoperoxidase staining of hsp90 at 4 d. Staining intensity and distribution similar to Figure 16-15. Note shock protein expression within the inter-segmental septum. Magnification = 800×.

Figure 16-17   Immunoperoxidase staining of hsp90 at 7 d. Moderate to intense staining in chloragogenous tissue. Magnification = 1600×.

Figure 16-18   Immunoperoxidase staining of hsp90 at 12 d in a posterior region near zone of wounding. Note intense staining in blood vessel; moderate staining in chloragogenous tissue, basement layer, and alimentary epithelia. Magnification = 1600×.

Figure 16-19   Immunoperoxidase staining of hsp60 in the seminal vesicles at 4 d post-amputation. Note intense staining in the *Monocystis*-infected, tissue-damaged, sperm morulae. (Note that there was little or no shock-protein expression in the vicinity of the parasitic sporozoites, and no expression in the earthworm's morphologically intact developing spermatozoa [not illustrated]). Magnification = 1600×.

Figure 16-20   Immunoperoxidase staining of hsp70 in the seminal vesicles at 7 d. Antigen distribution similar to Figure 16-19. Magnification = 1600×.

Figure 16-21   Immunoperoxidase staining of hsp90 in the seminal vesicles at 7 d. Antigen distribution similar to Figure 16-19. Magnification = 1600×.

Figure 16-22   Immunoperoxidase control omitting primary antibody. Section taken from an earthworm 7 d post-amputation. Note that there is no peroxidase staining in any tissue. Magnification = 800×.

Figure 16-23   Immunperoxidase control: section of seminal vesicle at 12 d post-amputation. No peroxidase staining. Magnification = 1600×.

Figure 16-24   Immunoperoxidase control: section through alimentary canal and a blood sinus at 12 d. Magnification = 1600×.

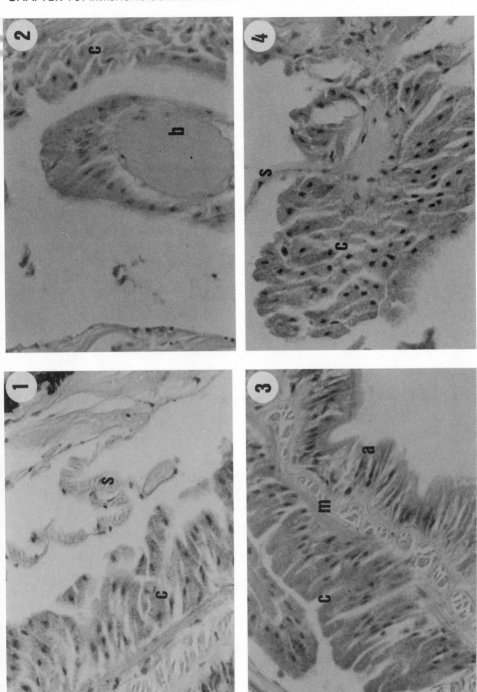

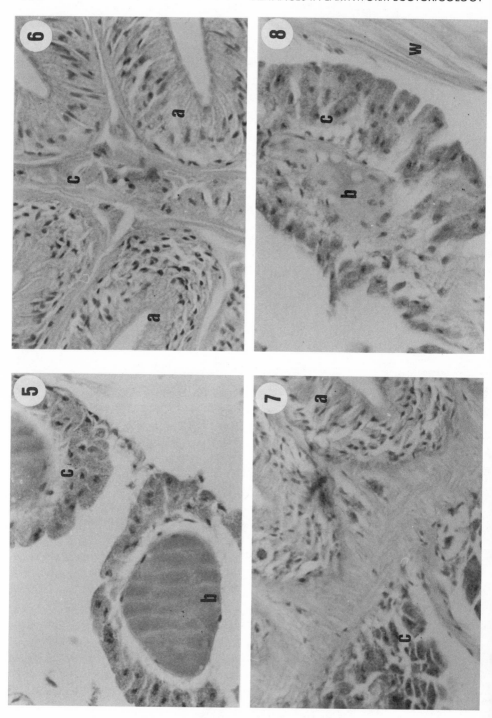

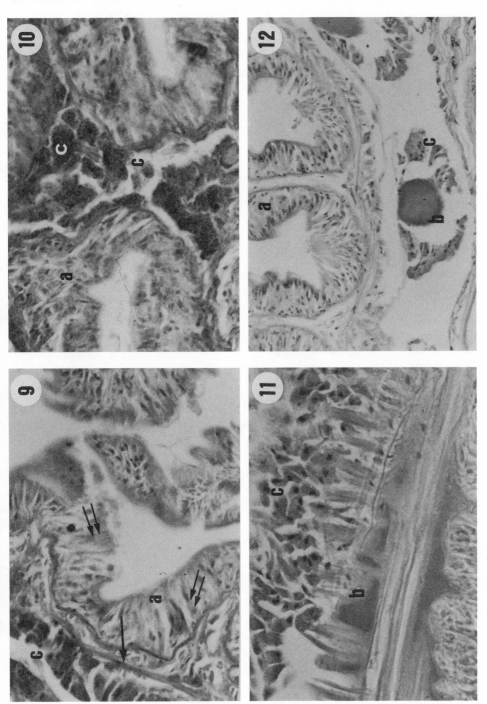

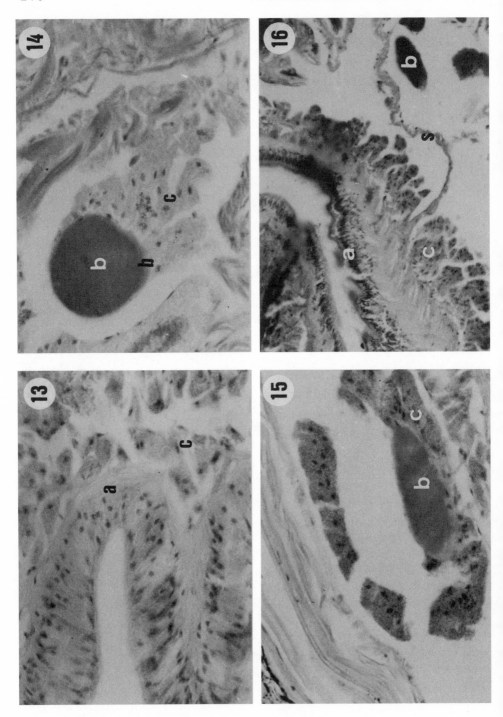

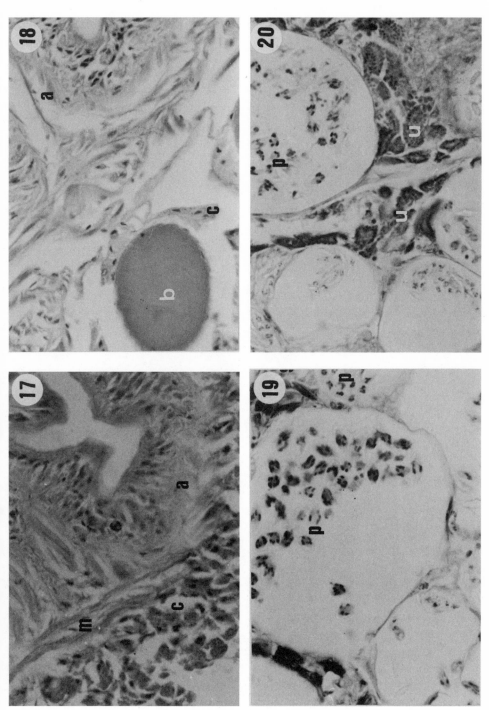

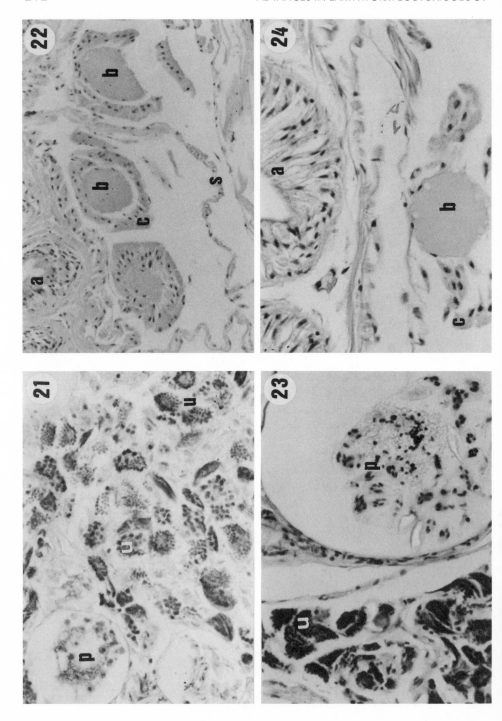

References

1. Cormier SM, Daniel FB. 1994. Biomarkers: taking the science forward. *Environ Toxicol Chem* 13:1011–1012.

2. Moore MN, Köhler A, Lowe DM, Simpson MG. 1994. An integrated approach to cellular biomarkers in fish. In: Fossi MC, Leonzio C, editors. Nondestructive biomarkers in vertebrates. Boca Raton FL: Lewis. p 171–216.

3. Ryan JA, Hightower LE. 1994. Evaluation of heavy-metal ion toxicity in fish cells using a combined stress protein and cytoxicity assay. *Environ Toxicol Chem* 13:1231–1240.

4. Edwards CA, Bohlen PJ. 1992. The effects of toxic chemicals on earthworms. *Rev Environ Contam Toxicol* 125:23–99.

5. Van Gestel CAM. 1992. Validation of earthworm toxicity tests by comparison with field studies: a review of benomyl, carbendazim, carbofuran, and carbaryl. *Ecotoxicol Environ Saf* 23:221–236.

6. Van Gestel CAM, Van Dis WA, Van Breemen EM, Sparenburg PM. 1989. Development of a standardized reproduction toxicity test with the earthworm species *Eisinia fetida andrei* using copper, pentachlorophenol, and 2,4-dichloroaniline. *Ecotoxicol Environ Saf* 18:305–312.

7. Van Gestel CAM, Dirven-Van Breemen EM, Baerselman R, Emans HJB, Janssen JAM, Postuma R, Van Vliet PJM. 1992. Comparison of sublethal and lethal criteria for nine different chemicals in standardized toxicity tests using the earthworm *Eisenia andrei*. *Ecotoxicol Environ Saf* 23:206–220.

8. Yeardley Jr RB, Lazorchak JM, Gast LC. 1996. The potential of an earthworm avoidance test for evaluation of hazardous waste sites. *Environ Toxicol Chem* 15:1532–1537.

9. Chen SC, Fitzpatrick LC, Goven AJ, Venables BJ, Cooper EL. 1992. Nitroblue tetrazolium dye reduction by the earthworm (*Lumbricus terrestris*) coelomocytes: an enzyme assay for nonspecific immunotoxicity of xenobiotics. *Environ Toxicol Chem* 10:1037–1043.

10. Suzuki MM, Cooper EL, Eyambe GS, Goven AJ, Fitzpatrick LC, Venables BJ. 1995. Polychlorinated biphenyls (PCBs) depress allogenic natural cytotoxicity by earthworm coelomocytes. *Environ Toxicol Chem* 14:1697–1700.

11. Drewes CD, Callahan CA. 1988. Electrophysiological detection of sublethal neurotoxic effects in intact earthworms. In: Edwards CA, Neuhauser EF, editors. Earthworms in waste and environmental management. The Hague, The Netherlands: SPB Academic. p 355–366.

12. Weeks JM, Svendsen C. 1996. Neutral-red retention by lysosomes from earthworm (*Lumbricus rubellus*) coelomocytes: A simple biomarker for exposure of soil invertebrates. *Environ Toxicol Chem* 15:1801–1805.

13. Stürzenbaum SR, Morgan AJ, Kille P. 1997. Heavy metal induced molecular responses in the earthworm: genetic fingerprinting by direct differential display. *Appl Soil Ecol* (In press).

14. Svendsen C, Meharg AA, Freestone P, Weeks JM. 1996. Use of an earthworm lysosomal biomarker for the ecological assessment of pollution from an industrial plastics fire. *Appl Soil Ecol* 3:99–107.

15. Greig-Smith PW, Becker H, Edwards PJ, Heimbach F, editors. 1992. Ecotoxicology of earthworms. Recommendations of the International Workshop on Earthworm Ecotoxicology. Andover UK: Intercept Ltd. p 13.

16. Georgopoulos C, Welch WJ. 1993. Role of the major heat shock proteins as molecular chaperones. *Annu Rev Cell Biol* 9:601–634.

17. Sanders BM, Martin LS, Wakagawa PA, Hunter DA, Miller S, Ullrich SJ. 1994. Specific cross-reactivity of antibodies raised against two major stress proteins, stress 70 and chaperonin 60, in diverse species. *Environ Toxicol Chem* 13:1241–1249.

18. Williams JH, Farag AM, Stansbury MA, Young PA, Bergman HL, Petersen NS. 1996. Accumulation of hsp70 in juvenile and adult rainbow trout gill exposed to metal-contaminated water and/or diet. *Environ Toxicol Chem* 15:1324–1328.

19. Sanders BM, Martin LS, Nelson WG, Phelps DK, Welch W. 1991. Relationships between accumulation of a 60 kDa stress protein and scope-for-growth in *Mytilus edulis* exposed to a range of copper concentrations. *Mar Environ Res* 31:8–197.

20. Köhler H-R, Triebskorn R, Stöcker W, Kloetzal P-M, Alberti G. 1992. The 70kD heat shock protein (hsp70) in soil invertebrates: A possible tool for monitoring environmental toxicants. *Arch Environ Toxicol* 22:334–338.

21. Kammenga JE, Köhler H-R, Dallinger R, Simonsen V, Weeks JM, van Gestel CAM. 1995. Progress report 1994 of bioprint. Second technical report. Biochemical fingerprint techniques as versatile tools for the risk assessment of chemicals in terrestrial invertebrates. Silkeborg, Denmark: National Environmental Research Institute. p 32.

22. Ormerod MG, Imrie SF. 1989. Immunohistochemistry. In: Lacey AJ, editor. Light microscopy in biology. A practical approach. Oxford UK: IRL. p 103–136.

23. Burke JM. 1974. Wound healing in *Eisenia foetida* (Oligochaeta). 1. Histology and ^{3}H-thymidine radioautography of the epidermis. *J Exp Zool* 188:49–64.

24. Burke JM. 1974. Wound healing in *Eisenia foetida* (Oligochaeta). 2. A fine structural study of the role of the epidermis. *Cell Tiss Res* 154:61–82.

25. Roberts LS, Janovy Jr J. 1996. Foundations of parasitology. 5th ed. Dubuque IA: Wm. C. Brown. p 114–115.

26. Cajaraville MP, Robledo Y, Etxeberria M, Marigómez I. 1995. Cellular biomarkers as useful tools in the biological monitoring of environmental pollution: molluscan digestive lysosomes. In: Cajaraville MP, editor. Cell biology in environmental toxicology. Del Pais Vasco, Bilbao, Spain: Servicio Editorial Universidad. p 29–55.

27. Sanders BM. 1993. Stress proteins in aquatic organisms: An environmental perspective. *Crit Rev Toxicol* 23:49–75.

28. Joplin KH, Denlinger DL. 1990. Development and tissue specific control of the heat shock induced 70kDa related proteins in the flesh fly, *Sarcophaga crassipalpis*. *J Insect Physiol* 36:239–249.

29. Morgan AJ, Morgan JE, Turner M, Winters C, Yarwood A. 1993. Metal relationships of earthworms. In: Dallinger R, Rainbow PS, editors. Ecotoxicology of metals in invertebrates. Boca Raton FL: Lewis. p 333–358.

30. Yu Z, Magee WE, Spotila JR. 1994. Monoclonal antibody ELISA test indicates that large amounts of constitutive hsp70 are present in salamanders, turtle and fish. *J Therm Biol* 19:41–53.

Chapter 17

Identification of new heavy-metal-responsive biomarkers in the earthworm

S.R. Stürzenbaum, P. Kille, A.J. Morgan

Sensitive molecular genetic tools have facilitated the identification of 10 genes within the earthworm *Lumbricus rubellus* that exhibit elevated expression upon exposure to heavy-metal pollution. The expression profile of 5 of these genes has been characterized in a population of control earthworms, 2 populations of earthworms native to different metal-contaminated habitats, and earthworms either maintained on artificial soil alone or supplemented with acute levels of Cd. Analysis of the heavy-metal-adapted populations indicated that the gene encoding for a membrane-bound lysosomal glycoprotein showed a nonspecific response toward general heavy-metal stress, while carboxypeptidase, the translationally controlled tumor protein (TCTP), and the mitochondrial large ribosomal subunit gave a greater response to Cu pollution. In contrast, phosphoenol pyruvate carboxykinase (guanosine 5'-triphosphate) showed greatest elevation toward the heavy-metal cocktail found in an old Pb/Zn/Cd mine. However, controlled experiments with metal-deficient, artificial soil revealed an elevation in 3 of the 5 genes. It is proposed that this phenomenon is due to the conditions of animal husbandry. Nevertheless, a highly significant induction above laboratory base line was observed in at least 4 genes after acute exposure to Cd. Further experiments are necessary to validate the potential of these genes as candidate biomarkers.

Heavy-metal pollution of terrestrial ecosystems is a widespread phenomenon stemming from the extraction of raw materials from the earth's crust and their potential for industrial consumption. Each contaminated site contains unique characteristics pertaining to the exact heavy-metal cocktail present, the pollutant gradients involved, the potential dispersal route, and the ecology of the surrounding site. Populations of the earthworm *Lumbricus rubellus* (Hoffmeister) have been found throughout a range of metal-polluted sites, presumably adapting to the most extreme of these ecological habitats. Whether the organisms are adapted or acclimated (and the distinction is significant), environmental stress evokes physiological (i.e., phenotypic) responses. Phenotypic changes, of course, reflect the expression of specific genes. The identification of these genes will increase understanding of micro-evolutionary processes and the biochemistry involved with adaptation and acclimation. Furthermore, the specific genes expressed by these populations may act as new genetic indexes (or biomarkers), which may benefit and enhance current testing protocols that evaluate the degree of stress caused by toxic metals under field conditions.

By definition, biomarkers are indicative of pollutant-induced biochemical responses. Ideally, a biomarker should be responsive to either general toxicity (e.g., heavy metals, pesticides), a specific group of toxicants (e.g., heavy metals), or even a single pollutant (e.g., Cu). Three potential cellular or biochemical biomarkers have been identified in the field of earthworm ecotoxicology: the immuno-competence test, an enzyme-based assay assessing toxicity of xenobiotics [1]; the neutral red retention assay, which detects changes in lysosomal membrane stability caused by heavy metals and aromatic hydrocarbons [2]; and cytochrome P450s activity, which is thought to be inducible by certain xenobiotics such as ethoxycoumarin [3].

Rather than assessing cellular, organelle, and enzymatic responses, this study aims to identify and evaluate gene-specific responses to heavy-metal pollution. This chapter describes a study aimed at identifying putative genetic biomarkers using a variety of molecular approaches with preliminary characterization of their expression in populations of worms exposed to heavy metals in native and laboratory environments.

Materials and methods

Earthworms
L. rubellus were sampled from abandoned spoil heaps of an ancient Roman Pb/Zn/Cd mine (> 500 mg Cd/kg dry soil) that was close to the village of Rudry, South Glamorgan, Wales UK (O.S. Grid Reference: ST 200 855). Further specimens were collected from South Caradon Copper Mine, Cornwall UK (O.S. Grid Reference: SX 265 700). Control earthworms were collected from a heavy-metal-deficient site near Dinas Powys, South Glamorgan, Wales UK (O.S. Grid Reference: ST 149 723). Mature (clitellate) earthworms were collected in early spring, all by digging and hand sorting from the upper 20 cm soil layer. All earthworms were transported in their native soil to the laboratory and maintained overnight at a constant temperature of 18 °C. Ribonucleic acid (RNA) isolation was performed the following day. Control earthworms (Dinas Powys) were kept on artificial soil made in accordance with OECD recommendations [4] for an acclimatization period of 2 weeks. Then they were either sacrificed for analysis or transferred to artificial soil supplemented with 250 mg $CdCl_2$/kg dry artificial soil and maintained for an additional 2 weeks.

Isolation of putative biomarker genes
Total RNA was isolated from the posterior alimentary canal and its surrounding tissue using the TRI-Reagent (Sigma Chemical Company, St. Louis, MO). In order to compensate for individual variation, each isolation of total RNA consisted of a pooled sample of 4 earthworms. Subsequent isolation of mRNA was performed with a poly (A^+) RNA isolation kit (Pharmacia, Biotech, Uppsala, Sweden) and qualitatively assessed by means of in vitro translation [5].

A number of molecular genetic techniques have been utilized to identify genes responsive to heavy-metal exposure, thus acting as possible biomarkers (Figure 17-1). Initially, genes expressed in *L. rubellus* native to the Pb/Zn/Cd mine and in a control population of *L. rubellus* maintained for 2 weeks on artificial soil supplemented with 250 mg $CdCl_2$/kg artificial soil were isolated in the form of complimentary deoxyribonucleic acids (cDNAs). In addition, a directed differential display technique was used, comparing banding patterns from worms sampled from control and polluted sites, which identified gene fragments unique to earthworms that inhabit the ancient Roman Pb/Zn/Cd mine. Finally, a subtraction protocol (PCR-Select cDNA, Clontech, Palo Alto CA) was performed, using genetic material from *L. rubellus* inhabiting the control site and the Pb/Zn/Cd mine by enriching gene fragments specific to the metal-induced environment. All isolated fragments were cloned and sequenced as described elsewhere [5].

Cloning and sequencing of polymerase chain reaction products

The amplified cDNA fragments were cloned into pGEM-T Vectors according to the manufacturer's protocol (Promega, Southampton UK). Appropriate plasmid clones were sequenced using the thermosequenase cycle sequencing kit (RPN 2438, Amersham, Buckinghamshire, England) with infrared-labeled universal forward and reverse sequencing primers. Electrophoresis, band detection, and data processing were performed using a Li-cor 4000LS sequencer (Li-cor, Lincoln, NE).

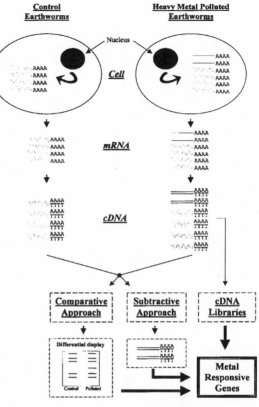

Figure 17-1   Schematic representation describing the experimental strategy employed to identify metal-responsive genes within the earthworm *L. rubellus*

Microvolume rapid polymerase chain reaction

Expression profiles of putative biomarker genes as well as an invariant internal control were determined by quantitative polymerase chain reaction (PCR) using a LightCycler (Idaho Technology Inc., Idaho Falls ID) as described by Wittwer et al. [6]. Product formation was monitored at the end of each 10-sec extension step by measuring the fluorescence emitted ensuing Sybr Green's intercollation into double-stranded DNA molecules. The specificity of each amplified end product was assessed by melting point analysis [7] and agarose gel electrophoresis.

For each target gene, primers (18 to 20 mers) were designed stringently to amplify 100 to 300bp fragments and to minimize mismatching and primer-dimer formation. Amplifications were optimized for MgCl$_2$, Sybr Green (FMC BioProduct, Rockland, ME), and variable annealing temperatures. Amplification reactions consisted of 0.5 μl cDNA template (plasmid-cloned DNA preparation or reverse-transcribed mRNA), 1mM of each primer (synthesized by Gibco BRL, Paisley UK), 0.2 mM dNTPs (New England Biolabs, Beverly, MA), 1:30,000 dilution of Sybr Green and 1 unit of Taq polymerase (Promega), buffered with 50mM Tris, 250 μg/ml bovine serum albumin and supplemented with 2 to 4mM MgCl$_2$ as appropriate (Biogene, Kimbolton, Cambridgeshire UK). The PCR reactions were performed in microvolume capillaries and subjected to 94 °C for < 1 sec, 60 to 70 °C for 3 sec, and 72 °C for 10 sec for 40 cycles.

Quantitative PCR

Stock plasmid preparations containing each of the appropriate target genes were purified by phenol-chloroform extraction followed by ethanol precipitation. Quantification and purity assessments were performed by agarose gel electrophoresis and spectroscopic analysis using diode ray spectrophotometry. Calibration standards were prepared by diluting each purified plasmid to a known concentration in the range of 10 ng to 100 fg. PCR was performed on duplicate samples of each standard. A regression line was generated by plotting the cycle number needed to attain a threshold fluorescence pertaining to the mid logarithmic portion of the amplification against log10 (molecules of target gene) (Figure 17-2).

cDNA was generated as described previously [8] from 0.5 mg of mRNA or 3 mg of total RNA preparations extracted from each population of earthworms investigated. Quantitative PCR was performed on 0.5 μl of a 1:10 dilution of each cDNA as template, except for mitochondrial quantifications, which required a 1:1000 dilution to bring the concentration within the calibrated standard range.

The number of target molecules present in each reverse-transcribed cDNA (derived from different earthworm populations) was quantified by computer-generated regression analysis, extrapolating each unknown over the standard range. All samples quantified were analyzed in triplicate. The resulting absolute values were normalized for differences in reverse-transcription efficiencies by using the expressed level of actin, a housekeeping gene widely accepted to be an invariant.

Results

The application of sensitive molecular genetic tools has led to the isolation of 5 genes putatively affected by heavy-metal pollution. These genes include phosphoenol pyruvate carboxykinase (Genbank Accession no. Y09624) that were isolated from a population of *L. rubellus* inhabiting the Pb/Zn/Cd mine. From the population of *L. rubellus* exposed to Cd in artificial soil, genes were identified as the mitochondrial large ribosomal subunit (Genbank Accession no. Y08157) and the gene encoding for the translationally controlled tumor protein (Genbank Accession no. Y08158). Carboxypeptidase (Genbank Accession no. Y09625) was isolated using a directed differential display technique, and the lysosome-associated glycoprotein (to be submitted) and the mitochondrial large ribosomal subunit (identical to the gene isolated previously) were identified through the subtractive protocol.

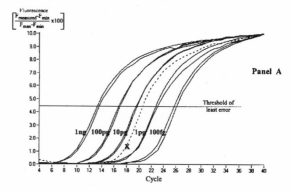

Panel A

The expression profile of each gene within populations of earthworms exposed to different metal regimes was determined by quantitative PCR. Calibration with standards and quantification of unknowns based on regression analysis were performed as described, with a representative example given in Figure 17-2. The absolute values, as given in Table 17-1, were computed after normalizing with actin as an invariable standard.

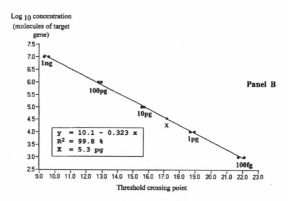

Panel B

Comparing expression profiles of control populations to metal-tolerant populations originating from the Pb/Zn/Cd and Cu mines showed an elevated expression of all the genes investigated with the exception of carboxypeptidase, which showed no significant change in earthworms from the Pb/Zn/Cd mine site. While the lysosomal glycoprotein was equally upregulated in both metal-tolerant populations, the expression profile of other genes was significantly different between tolerant populations, with TCTP, the mitochondrial large ribosomal subunit, and carboxypeptidase

Figure 17-2   Amplification of the earthworm actin gene by quantitative PCR Legend: Panel A depicts the exponential amplification of a dilution series of actin (Genbank Accession no. Y09623) standards (1 ng to 100 fg in solid lines) and 1 sample of cDNA with an unknown actin concentration (X, dashed line). Panel B illustrates the computer-generated regression analysis, extrapolating the actin amplification of unknown concentration over the diluted standard range.

showing significant increases in earthworms sampled at the Cu mine. Phosphoenol pyruvate carboxykinase, on the other hand, was markedly elevated in the population native to the Pb/Zn/Cd mine.

After the 2-week laboratory acclimatization period on artificial soil, the expression profiles of control earthworms did not differ significantly from the profiles of counterparts inhabiting their native soil for 2 genes: the lysosomal glycoprotein and carboxypeptidase.

Table 17-1   Expression profiles of five genes quantified within populations of control (metal deficient) and metal exposed earthworms. Values (± standard error, N = 3) were normalized with actin as an invariable standard and expressed as molecules of target gene per actin molecule.

Gene identification	Control (no metals)	Pb, Zn + Cd mine	Cu mine	Artificial soil	Artificial soil + Cd
Mitochondria (large ribosomal subunit) $\times 10^1$	8.3 (± 0.2)	42.7 (± 2.0)	173.7 (± 8.4)	46.1 (± 8.8)	1431.2 (± 35.1)
Lysosome associated glycoprotein $\times 10^2$	2.0 (± 0.3)	7.6 (± 1.7)	6.2 (± 0.6)	1.6 (± 0.4)	27.1 (± 3.0)
Phosphoenol pyruvate carboxykinase (GTP) $\times 10^{-2}$	2.6 (± 0.2)	191.6 (± 10.0)	133.2 (± 9.5)	19.4 (± 1.1)	29.4 (± 4.6)
Translationally controlled tumor protein (TCTP) $\times 10^{-2}$	7.8 (± 0.1)	33.0 (± 2.6)	2616.1 (± 248.8)	279.1 (± 5.6)	3856.1 (± 99.2)
Carboxypeptidase $\times 10^{-1}$	8.4 (± 0.2)	7.3 (± 0.1)	30.1 (± 1.9)	1.2 (± 0.3)	8.9 (± 0.3)

An increased expression was identified after laboratory acclimatization in the 3 remaining genes. Nevertheless, when control earthworms were exposed to artificial soil supplemented with highly toxic quantities of $CdCl_2$ (unpublished observation), gene expression was increased significantly above control values in the cases of the lysosomal glycoprotein, TCTP, carboxypeptidase, and mitochondrial large ribosomal subunit, but no significant induction of phosphoenol pyruvate carboxykinase could be detected.

Discussion

The overall aim of this work was to address Recommendation 34 formulated at the First International Workshop on Ecotoxicology of Earthworms, Sheffield UK [9], which states that "encouragement should be given to the use of earthworms for bioassays, and to develop new sensitive endpoints such as immune function and other sublethal parameters." The molecular genetic tools described in this chapter have facilitated the identification of potential biomarker genes, which in turn may satisfy the above criteria by creating a new generation of earthworm-based bioassays. The methods described have the advantage of requiring only small amounts of starting material, a distinct benefit when samples are difficult to obtain. In turn, this will consequently reduce the risk of endangering the ecological balance of micro-ecosystems through excessive sampling.

It has been observed throughout biology that adaptation and acclimatization to extreme environments enable organisms to expand into a new niche, thus reducing interspecific competition for survival. The accompanying molecular changes may be general or specific to an environmental insult. The studies described above have identified both general and specific molecular changes, which together begin to reveal how the earthworm can survive severely metal-contaminated habitats.

Ecological theory accepts that there is a metabolic cost entailing enhanced energy expenditure imposed by adaptation [10]. It is therefore not surprising to find mitochondrial genes upregulated within putative adapted populations. In contrast to the nucleus, the mitochondrial genome is relatively small. Nevertheless, representation of mitochondrial messages in total cellular mRNA may be high because of increased copy number within a cell. An organism exposed to an environmental stressor, and in need of an elevated energy supply, can upregulate expression of the mitochondrial genes as well as increase mitochondrial numbers, resulting in a significant increase in the proportion of mitochondrial message within the total cellular RNA. However, the need for energy in order to survive in the toxic metal environment of a mine site may not be the sole reason for increased gene expression. Recent studies have shown that metallothionein, a small, metal-binding protein involved in metal homeostasis and detoxification processes, itself can compromise mitochondrial function [11]. If this occurs within heavy-metal-exposed populations, earthworms will need increased mitochondrial function to compensate for the heavy-metal-mediated mitochondrial dysfunction (Figure 17-3, Panel A).

One of the mechanisms by which heavy metals mediate their toxicity is in the displacement of the essential metal ion Zn from enzymes in which it acts as a specific cofactor.

Foreign metal binding may reduce the activity of the enzyme and therefore have a consequence for cellular function. Carboxypeptidase is one enzyme whose activity is altered significantly if the cofactor Zn is replaced with other heavy metals [12]. In this study, it has been shown that carboxypeptidase is upregulated in worms native to soils from a Cu mine. Is this due to cellular compensation for loss of function due to Cu replacement of Zn, or is it due to some more specific role carboxypeptidase plays within the acclimatization process? The observation that carboxypeptidase is not upregulated in earthworms native to the Pb/Zn/Cd mine may be explained by the probability that most of the Cd will be inactivated by sequestration by metallothionein. This hypothesis is supported by the elevated carboxypeptidase expression found in worms transferred to artificial soil spiked with Cd. In this artificial metal–metal-spiked matrix, it is unlikely that all of the Cd will be bound to metallothionein, mainly because of the quantities involved but also because these control worms are unlikely to be fully acclimated to Cd. Excess Cd might thus be available for metal-substitutional reactions. Furthermore, it has been reported that Cu is a weak inducer of metallothionein [13], possibly

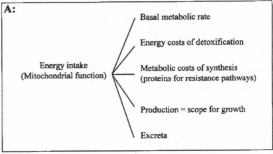

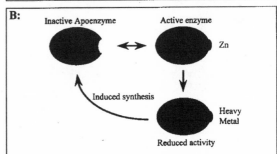

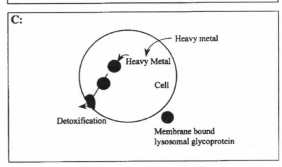

Figure 17-3   Illustrated hypotheses of the underlying mechanistic relationship between a number of identified genes and the effects of heavy metals Legend: Panel A depicts the fundamental functions of mitochondria and its role in acclimatization to toxic heavy metals (adapted from Walker et al., 1996). Panel B shows a schematic representation of a putative feedback response of metal dependent enzymes. Panel C describes the involvement of the lysosomal system in detoxification of heavy metals by expulsion from the cell.

explaining the increased availability of free metals targeting carboxypeptidase in earthworms inhabiting the Cu-contaminated metalliferous soil (Figure 17-3, Panel B).

The increased expression of the membrane-bound lysosomal glycoprotein is most likely due to lysosomal detoxification pathways being switched on. Hence, increased detoxification requires elevated quantities of lysosomal components such as the membrane-bound glycoproteins (Figure 17-3, Panel C). The roles phosphoenol pyruvate carboxykinase and TCTP play in a heavy-metal-mediated response are not fully understood but are most likely to be inflammatory and shock responses, respectively.

Interesting observations were obtained following transfer to artificial soil. After the 2-week acclimatization period, an increased expression was identified in 3 genes (TCTP, phosphoenol pyruvate carboxykinase, and the mitochondrial large ribosomal subunit). It may be argued that the observed upregulation was gene specific and influenced by differences in soil characteristics or animal husbandry. Clearly this phenomenon needs further attention, as it may indicate the limitation of current laboratory-based toxicity protocols with artificial soils. Understandably it is desirable to simplify soil conditions in laboratory situations, but an oversimplification may well result in misleading interpretations of ecotoxicity, especially given that a field situation comprises an array of interacting factors that modulate the bioavailability of stressors.

In conclusion, genetic indices promise to be powerful indicators of heavy-metal stress. Their future potential lies in sensitive detection of deleterious sublethal responses of individual organisms. In addition to evaluating the expression profiles of the other 5 putative biomarker genes, further work has to be undertaken to assess the specificity and dose response of the genes that were the focus of the present study. This assessment should encompass their expression in natural populations inhabiting a wide spectrum of different polluted sites, possibly over a gradient of toxicity, as well as controlled laboratory-based experiments. Through the evaluation of the expression profiles of putative metal-responsive genes, the long-term goal is to pinpoint general, semi-general, and specific biomarkers, thus creating a whole suite of genetic indices to heavy-metal pollution.

Acknowledgments - This study was financially supported by the British Natural Environment Research Council (NERC grants GR9/02083 and GT5/94/ALS) and a Morgan E. Williams Studentship.

References

1. Goven AJ, Chen SC, Fitzpaptrick LC, Venables BJ. 1994. Lysosome activity in earthworm (*Lumbricus terrestris*) coelomic fluid and coelomocytes: Enzyme assay for immunotoxicity of xenobiotics. *Environ Toxicol Chem* 13:607–613.

2. Weeks JM, Svendsen C. 1996. Neutral red retention by lysosomes from earthworm (*Lumbricus rubellus*) coelomocytes—a simple biomarker for exposure of soil invertebrates. *Environ Toxicol Chem* 15:1801–1805.

3. Brennan S, Ataria JA, Booth LH, Eason CT. 1997. Cytochrome P450 as a biomarker in earthworms. [abstract]. Second International Workshop on Earthworm Ecotoxicology; 3–5 Apr 1997; Amsterdam NL. p 16.

4. [OECD] Organization for Economic Cooperation and Development. 1984. Organization for Economic Cooperation and Development guidelines for testing of chemicals: earthworm acute toxicity test. OECD Guideline No. 207. Paris, France: OECD.

5. Stürzenbaum SR, Morgan AJ, Kille P. 1997. Heavy metal induced molecular responses in the earthworm: genetic fingerprinting by directed differential display. *Applied Soil Ecology*. (In press).

6. Wittwer CT, Ririe KM, Andrew RV, David DA, Gundry RA, Balis UJ. 1997. The LightCycler™: a microvolume multisample fluorimeter with rapid temperature control. *Biotechniques* 22:176–181.

7. Ririe KM, Rasmussen RP, Wittwer CT. 1997. Product differentiation by analysis of DNA melting curves during the polymerase chain reaction. *Analytical Biochem* 245:154–160.

8. Kille P, Stephens PE, Kay J. 1991. Elucidation of cDNA sequences for metallothioneins from rainbow trout, stone loach and pike liver using the polymerase chain reaction. *Biochem et Biophys Acta* 1089:407–410.

9. Greig-Smith PW. 1992. Recommendations of an international workshop on ecotoxicology of earthworms. In: Greig-Smith PW, Becker H, Edwards PJ, Heimbach F, editors. Ecotoxicology of earthworms. Andover, Hants UK: Intercept Ltd. p 247–262.

10. Walker CH, Hopkin SP, Sibley RM, Peakall DB. 1996. Physiological effects of pollutants. In: Walker CH, Sibley RM, Peakall DB, editors. Principles of ecotoxicology. London UK: Taylor and Francis Ltd. p 147–154.

11. Simpkins CO, Liao ZH, Gegrehiwot R, Tyndall JA, Torrence CA, Fonong T, Balderman S, Mensah E, Sokolove M. 1996. Opposite effects of metallothionein I and spermine on mitochondrial function. *Life Sci* 58:2091–2099.

12. Hofman T. 1985. Metalloproteinases. In: Harrison P, editor. Metalloproteins, part 2: Metal proteins with non-redox roles-topics in molecular and structural biology 7. London UK: Macmillan p 1–65.

13. Morgan JE, Morgan AJ. 1990. The distribution of Cd, Cu, Pb, Zn and Ca in the tissues of the earthworm *Lumbricus rubellus* sampled from one uncontaminated and four polluted soils. *Oecologia* 84:559–566.

Chapter 18

Lysosomal membrane permeability and earthworm immune-system activity: field-testing on contaminated land

Claus Svendsen, David J Spurgeon, Zvezdelin B Milanov, Jason M Weeks

Two simple and relatively low-cost biomarker assays providing measures of lysosomal membrane permeability and immune activity in coelomocyte cells taken from earthworms (*Lumbricus castaneus*) were applied at the site of an industrial accident to test their applicability in the field. Earthworms and soil samples were collected at the site of a large industrial plastics fire in Thetford UK along three 200-m transects leading from the factory perimeter fence, over a layer of molten, plastic-impregnated soil, and into the surrounding forest. Coelomocyte cells taken from earthworms adjacent to the factory perimeter had the highest lysosomal membrane permeability, measured as the shortest period of neutral red retention (NRR) by the lysosomes (2 min); cells taken from worms further into the surrounding forest had a longer retention time (12 min), while cells taken from worms from a control site showed even greater retention times (25 min). Results from the immune-activity measurements showed that the immune activity was affected only in worms taken immediately adjacent to the factory perimeter; where a significant decrease was measured. This field trial has demonstrated the validity of using both types of in vitro coelomocyte biomarker techniques for use in biological impact assessment along gradients of contamination.

Biomarkers have been used extensively to document and quantify both the exposure to and the effects of environmental pollutants. As monitors of exposure, biomarkers have the advantage of quantifying only biologically available pollutants. There is vast literature on biomarkers in aquatic systems (see reviews [1,2]). However, there is a general paucity of biomarkers for use with terrestrial soil invertebrates. For a review on the current status of earthworm biomarker research, see Scott-Fordsmand and Weeks, Chapter 14. In theory, the types of biological responses that could be considered as biomarkers range from the molecular to effects on the intact organism, the population or community structure, and perhaps also the structure and function of ecosystems. However, if biomarkers are to be used for anything other than stating the obvious (i.e., this species is becoming extinct or a forest area is in decay), they must be applied at the lower end of this continuum, at the cellular, biochemical, and molecular levels, where responses are sensitive and rapid, while being reasonably easy to interpret [2,3].

Biomarkers may in general be divided into nonspecific and specific biomarkers on the basis of their application. Nonspecific biomarkers give a measure of general stress resulting from a broad range of pollutants (e.g., a class of pollutants such as heavy metals), whereas specific biomarkers measure effects that can be linked directly to a specific pollutant (e.g., DNA adducts).

A few carefully chosen nonspecific biomarkers each covering a different class of pollutant could, if used together, serve as an effective first screening during a monitoring program at lower cost and more quickly than traditional chemical analyses. Should this first screening of a land area not give an appropriate all-clear answer, then more sophisticated and specific chemical methods could be employed, thus targeting more effort in the cases where it is most needed.

In order for such a set of biomarkers to be used as a future environmental monitoring tool, it requires validation in the field and calibration against ecologically important parameters [4]. Also, these biomarkers will require calibration against each other to meet the requirements of the first-screen idea. If this between-biomarker calibration was undertaken for a range of different pollutants, different patterns of responses could be expected in the suite of biomarkers for each class of pollutant. Such patterns might point us toward the class of compound causing the effects in each case, thus narrowing the field for any follow-up analytical chemistry.

Scott-Fordsmand and Weeks (Chapter 14) have discussed 2 nonspecific cellular-function-level biomarkers that are reasonably well documented for use in earthworms. The first, lysosomal membrane permeability, has been successfully measured as the ability of the lysosomes in coelomocytes to retain a dye (neutral red) using the technique of Weeks and Svendsen [5]. The second nonspecific biomarker used in earthworms is immune-system activity. Here erythrocyte rosette (ER) formation with, and phagocytosis of, antigenic rabbit red-blood cells are determined after an in vitro incubation of earthworm coelomocytes using the techniques described by Goven et al. [6]. Apart from being well established with regard to dose-response relationships, both of these tests are relatively simple to undertake [7–9].

The present study aims to compare and validate the use and relevance of these 2 coelomocyte-based biomarkers in the field at a well-investigated industrial accident site in the UK [10,11]. Both biomarkers were measured along contamination gradients at the site, and an initial attempt was made to evaluate any effects on the population of the earthworm *Lumbricus castaneus* (Savigny).

Materials and methods

Site
A well-described field site was chosen for this exercise (Figure 18-1). A fire in October 1991 at a plastics recycling factory in Thetford, Norfolk UK involved the pyrolysis of 1000 t of various plastics. The fire blazed for 3 d and severely contaminated a 1200-m^2 area of a neighboring Scots pine and beech stand with molten plastic. This plastic mantle contained a mixture of inorganic (Pb, Cd, and Sb) and organic contaminants (e.g., dioxins) and spread a distance of approximately 50 m into the forest. Chemical analysis of soil and leaf-litter samples taken shortly after the fire showed that the inorganic part of the contamination was restricted to the area covered by the molten plastic [10]. Subsequent to the incident, a shallow soil and leaf-litter layer has developed, overlying the now friable

plastic layer. This new soil layer, although highly polluted, has to a limited extent been recolonized by grass and herb communities, and an impoverished soil invertebrate community has developed. The onsite earthworm community appears to have the same species composition as the surrounding forest area, being dominated by *Lumbricus castaneus* but with generally lower densities (personal observation).

Sampling

Earthworms (*L. castaneus*) and soil cubes were collected from 3 transects running across the plastic mantle and into the surrounding uncontaminated forest (Figure 18-1). As many individuals of *L. castaneus* as possible (up to 6) for the NRR assay were sampled along transect 2 and at a control site 3 km from the factory (see Figure 18-1) on 3 occasions: April 1994, November 1994, and February 1995. Individuals of *L. castaneus* for immune-system assessment were collected from all 3 transects and in the Monks

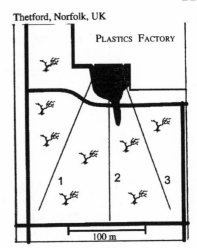

Figure 18-1   Extent of molten plastic mantle at site of fire in Thetford UK, with runoff points of 2 different sources of molten plastic and 3 sample transects

Wood Nature Reserve (Cambridgeshire UK) during April 1996. At the same time, soil cubes for population estimates also were collected; however, these were taken only from transect 2.

Biomarker assays

Small samples of coelomic fluid were taken from individual worms by the careful insertion of a hypodermic needle directly into the coelomic cavity in the region posterior to the clitellum. These samples contained the coelomocytes on which the 2 biomarker assays were measured.

Lysosomal membrane permeability was measured using the technique of Weeks and Svendsen [5] by staining extracted coelomocytes with neutral red and observing the ability of the lysosomes within the coelomocytes to retain the dye in time (after initial uptake). Coelomic fluid samples (~ 20 µl) were placed on microscope slides, and the cells were allowed to adhere to the individual slide's surface for 30 sec prior to the application of 20-µl neutral red working solution (80 mg/ml) and a coverslip. Each slide was scanned continuously for 2-minute intervals under a light microscope (400 ×). Observation was stopped when the ratio of cells with fully stained cytosols reached > 50% of the total number of cells counted. This interval was recorded as the NRR time [5].

The immune-system activity of the earthworms was determined by in vitro incubation of earthworm coelomocytes with stabilized antigenic rabbit red-blood cells for 24 h, using the technique described by Goven et al. [6]. The recording of coelomocytes exhibiting ER formation with, and phagocytosis of, rabbit red-blood cells, was undertaken using a light microscope (400×). A 40-ml sample of the incubated cells was placed in a haemocytometer and stained with Eosin Y to assess the immune status of living coelomocytes. The activity of the immune system was thereafter defined as the percentage of the total number of living cells showing either of the immune-system responses mentioned above.

Population data

Earthworm population data were obtained by gradient wet-sieving (sieve mesh: 5, 2, and 1 mm) 2 replicate soil-cubes, which were cut using a stainless steel box corer (dimensions 250 mm × 250 mm × 160 mm depth, i.e., total volume 10 L) - taken at 0-, 25-, 50-, 100-, 150-, and 200-m distances from the factory perimeter along transect 2 (Figure 18-1). All adult and juvenile earthworms and cocoons (of all species) were handsorted and counted.

Results

Neutral red retention

The results obtained from the NRR assay are shown in Figure 18-2. A 2-way analysis of variance (ANOVA) performed on the data from the April and November 1994 sampling occasions indicated a significant effect on NRR times (P < 0.001) with increasing distance from the factory. Time of sampling, however, was shown to have no effect on the neutral red response. Therefore, the data from all 3 sampling occasions were tested together using a 1-way ANOVA. A significant (P < 0.001) increase in coelomocyte NRR time with increasing distance from the factory perimeter was demonstrated. A Tukey test revealed no significant differences between the short NRR times measured (0 to 2 min) for the earthworms taken from the transect at 0 and 20 m or immediately adjacent to the plastic layer (60 m). The NRR times measured (~ 10 min) in worms from 140 and 200 m were shown to be significantly (P < 0.005) longer than those recorded for worms from the first 60 m of the transect, but there were no significant differences (P > 0.05) in NRR times between these latter 2 distances. Finally, the NRR times measured (~ 25 min) for earthworms from the control site were significantly longer than those measured at any of the locations along transect 2.

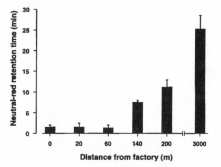

Figure 18-2   Mean (± SEM) neutral red retention times (in mins) for earthworm *Lumbricus castaneus* sampled along transect 2 by distance from factory

Immune-system activity

The results for earthworm immune-system activities (Figure 18-3) showed a significant effect of proximity to the factory on total immune-system activity (ANOVA, $P < 0.05$). However, the only significant decrease in activity occurred in earthworms sampled at 0 m, with activity levels returning to control values beyond this sampling point (Figure 18-3). Thus, decreased immune-system activity was observed only in earthworms collected very near to the old factory perimeter, and background levels of immune-system activity were reached in earthworms still within the zone of the plastic mantle.

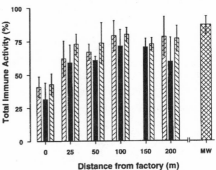

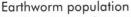

Figure 18-3   Mean (±SD) total immune-system activity for earthworms collected at the respective distances sampled along 3 transects and Monks Wood Legend: transect 1 (/), transect 2 (black), transect 3 (\), and from (x) Monks Wood

Earthworm population

Earthworm population measurements suggested that cocoon production and juvenile density were lowered by the pyrolyzed plastic and the associated contamination. The number of both cocoons and juveniles was reduced (below those at 200 m) up to the 150-m sampling point (Table 18-1). Numbers of both immature and mature earthworms appeared to be unrelated to the distance from the factory. Due to a replicate number of only 2 per interval sampled, the variation within the data was high, and it was difficult to draw reliable conclusions. No statistical analysis was applied to these data.

Table 18-1   Mean (± SD) number of cocoons, juvenile, immature, and mature earthworms collected by wet-sieving 2 replicate 10-L soil cubes taken at 6 distances along transect number 2 (see Figure 18-1)

Distance	Nr. cocoons		Nr. juvenile worms		Nr. immature worms		Nr. mature worms	
(m)	Mean	SD	Mean	SD	Mean	SD	Mean	SD
0	2.5	0.7	6.0	5.7	2.0	0	1.0	1.4
25	6.0	1.4	4.5	0.7	3.0	2.8	1.5	2.1
50	3.5	3.5	4.5	0.7	1.5	2.1	0.5	0.7
100	3.0	0	4.5	0.7	5.0	2.8	1.0	1.4
150	2.5	2.1	2.5	0.7	1.5	2.1	0	0
200	16.0	9.9	10.0	4.2	1.0	0	1.5	0.7

Discussion

The results from the Thetford site, where both the coelomocyte biomarkers were applied, served to validate their use in the field and also to compare their differential sensitivities.

There was a clear decline in the immune-system activity of earthworms collected near the old factory, but background levels of activity were reached within the zone of the plastic mantle formed during the factory fire. For NRR, the sensitivity was much greater and a response gradient persisted over a longer distance, past the edge of the plastic mantel and further into the forest.

The limited population data presented suggest reduced cocoon production and lowered density of juvenile worms to a distance of 150 m from the factory. No such trend with distance was seen in the numbers of immature and mature earthworms. It is well known that due to the heterogeneous distribution of earthworms (both in space and time), large sample numbers are needed to accurately assess their densities in the field. It was hoped that the number of cocoons (due to their immobility) would have provided a more reliable population measure by integrating earthworm presence through time. The data presented here, however limited, may suggest that cocoon numbers could provide a more sensitive measure of differences in the population status of earthworms than do actual worm densities. When effects and differences in the field are addressed within a small spatial scale (200 m in this study), the normal home range of adult worms may obscure any density differences. Before such data become reliable, extensive further investigation is required, preferably in line with and undertaken to an equally high standard as that of the modeling approach developed by Holmstrup et al. (Chapter 27). These authors have demonstrated the ability to backtrack a cocoon's development, from hatching, through a known temperature course, to its time of formation.

The NRR assay has been proven to give clear dose-response relationships after exposure to Cu in both laboratory and semi-field (mesocosm) studies [9,12]. Also, in these more controlled studies, decreased NRR times have shown a clear correlation with the doses at which effects on reproduction, food consumption, and weight gain were observed. Due to the small sample size of the population data used in this study, such a correlation could not be demonstrated.

The NRR data presented here clearly indicated that severe damage to the integrity of the lysosomal membrane had occurred in earthworms taken from or in the immediate vicinity of the plastic layer (0, 20 and 60 m). At the far end of the transect (140 and 200 m), the NRR times were significantly longer, thus indicating more intact (less leaky) lysosomal membranes, although the membrane integrity was still impaired to a greater extent at these positions than at the control site. A downfall of nonspecific biomarkers in general is that they work satisfactorily under controlled laboratory conditions, but when they are applied under field conditions, the results are often influenced by variation in natural parameters, and the applicability of the tests may therefore be further reduced (see review by Niimi [13]). This was not found in this study; the neutral red results did not appear to have been influenced by any natural factors and were independent of the sampling season. The use of this biomarker in the field for identifying and assessing contamination gradients must therefore be considered valid.

Reductions in earthworm immune-system function have been reported following exposure to polychlorinated biphenyls [8], Cu [14], and a range of pesticides of different actions [15]. There is some evidence that a decrease in immune-system activity occurs at exposure concentrations lower than those that may cause significant effects at higher organizational levels (see Scott-Fordsmand and Weeks, Chapter 14). The immune-system assay had, however, never previously been applied under field or semi-field conditions. From this study, it can be seen that it is possible to use the immune-activity assay with organisms collected from the field and obtain consistent results. Although decreased immune activity was demonstrated only at the first station on the very edge of the old factory, there were no significant outliers among the results for all the remaining sample stations on any of the 3 transects or the control site. There is therefore no evidence in this study suggesting that natural variation in the soil physicochemical properties and general habitat affects immune activity in earthworms. However, no statement can be made on the potential influence of seasonal variation. The immune-activity assay should be considered valid for field surveys aiming to compare soil contamination within sites of the same general habitat, but it should be used cautiously for monitoring through time, as further studies on any seasonal effects are needed.

An obvious problem with comparing the coelomocyte biomarker data presented here is the difference in sampling dates for the 2 assays. Any differences in sensitivity between the 2 biomarkers may possibly be explained as a partial recovery of earthworms at the site during the intermediate time. The only way of directly addressing this question of comparability between the datasets would be to measure both assays at the same time, something that was not possible during this study. However, the same difference in sensitivity between the immune-activity and the neutral red assays has been demonstrated in recent laboratory studies with both Cu and 2,4,6-trinitrotoluene (TNT) (C. Svendsen, P-Y. Robidoux, G. I. Sunahara, and J. M. Weeks, unpublished data).

Comparability problems notwithstanding, it is still clear from the data that coelomocyte biomarker assays could be successfully applied in the field without the results being overshadowed by effects from biotic and abiotic factors at the field site.

Acknowledgments - The British Council (UK) is thanked for funding Mr. Z. Milanov during his 3-month stay in the UK.

References

1. McCarthy JF, Shugart LR. 1990. Biomarkers of environmental contamination. Boca Raton FL: Lewis.
2. Huggett RJ, Kimerle RK, Mehrle Jr PM, Bergman HL. 1992. Biomarkers. Biochemical, physiological, and histological markers of anthropogenic stress. Boca Raton FL: Lewis.
3. Peakall DB. 1994. The role of biomarkers in environmental assessment. I. Introduction. *Ecotoxicol* 3:157–160.
4. Weeks JM. 1995. The value of biomarkers for ecological risk assessment: Academic toys or legislative tools. *Appl Soil Ecol* 2:215–216.

5. Weeks JM, Svendsen C. 1996. Neutral red retention by lysosomes from earthworm (*Lumbricus rubellus*) coelomocytes: A simple biomarker for exposure of soil invertebrates. *Environ Toxicol Chem* 15:1801–1805.

6. Goven AJ, Eyambe GS, Fitzpatrick LC, Venables BJ, Cooper EL. 1993. Cellular biomarkers for measuring toxicity of xenobiotics: Effects of polychlorinated biphenyls on earthworm*Lumbricus terrestris* coelomocytes. *Environ Toxicol Chem* 12:863–870.

7. Fitzpatrick LC, Sassani R, Venables BJ, Goven AJ. 1992. Comparative toxicity of polychlorinated biphenyls to earthworms *Eisenia foetida* and *Lumbricus terrestris*. *Environ Pollut* 77:65–69.

8. Rodriguez-Grau J, Venables BJ, Fitzpatrick LC, Goven AJ, Cooper EL. 1989. Suppression of secretory rosette formation by PCBs in *Lumbricus terrestris*: an earthworm assay for humoral immunotoxicity of xenobiotics. *Environ Toxicol Chem* 8:1201–1207.

9. Svendsen C, Weeks JM. 1997. Relevance and applicability of a simple earthworm biomarker of copper exposure. I. Links to ecological effects in a laboratory study with *Eisenia andrei*. *Ecotoxicol Environ Safety* 36:72–79.

10. Svendsen C, Meharg AA, Freestone P, Weeks JM. 1996. Use of an earthworm lysosomal biomarker for the ecological assessment of pollution from an industrial plastics fire. *Appl Soil Ecol* 3:99–107.

11. Meharg AA, Shore RF, Weeks JM, Osborn D, Freestone P, French MC. 1996. Food-chain transfers of cadmium released into a woodland ecosystem by a chemical accident. *Toxicol Environ Chem* 2:1–12.

12. Svendsen C, Weeks JM. 1997. Relevance and applicability of a simple earthworm biomarker of copper exposure. II. Validation and applicability under field conditions in a mesocosm experiment with *Lumbricus rubellus*. *Ecotoxicol Environ Safety*. 36:80–88.

13. Niimi AJ. 1990. Review of biochemical methods and other indicators to assess fish health in aquatic ecosystems containing toxic chemicals. *J Great Lakes Res* 16:529–541.

14. Goven AJ, Fitzpatrick LC, Venables BJ. 1994. Chemical toxicity and host defense inearthworms. *Ann New York Acad Sci* 712:280–300.

15. Bunn KE, Thompson HM, Tarrant KA. 1996. Effects of agrochemicals on the immune systems of earthworms. *Bull Environ Contam Toxicol* 57:632–639.

Experience with field earthworm ecotoxicology

Chapter 19

Comparison of the sensitivities of an earthworm (*Eisenia fetida*) reproduction test and a standardized field test on grassland

Fred Heimbach

For nearly 5 years, a reproduction test has been used as a regulatory Tier II test to investigate the effects of pesticides on earthworms. In this test, earthworms are exposed to a pesticide in a manner that simulates application under practical conditions. The most sensitive endpoint in this test is the number of juveniles per adult. *Eisenia fetida* is used as test species. To investigate the relevance of this method and to compare this to a standardized field test, the results of 78 reproduction studies with 21 pesticides were compared to 28 field studies with 15 pesticides. Ten of these compounds were studied in both tests, and all studies were performed in the same laboratory. For data evaluation, the regression line between the toxicity exposure ratio (TER = $EC50/PEC$; PEC = predicted environmental concentration) and the effects on earthworms was calculated. In this regression, the effects were given as percent deviation from control; in the field studies, they were based on earthworm numbers and the biomass of all species. The results clearly showed that the laboratory reproduction test was at least 5 to 10 × more sensitive than the field test. Hence, the reproduction test fulfills the requirements of a Tier II test for hazard assessment. On the other hand, results indicate that a 5-fold overdose in the reproduction test is too high, since too many compounds are triggered for field testing.

Ecotoxicological studies on earthworms have been performed for the assessment of pesticide toxicity for nearly 20 years. During this period, a number of standardized laboratory tests were developed and described. Three of these have been considered for use in a tiered-test system for registration of crop protection products in different countries, including those of the European Union (EU). These tests include an acute toxicity study (Tier I), a laboratory reproduction study (Tier II), and a field study (Tier III). This tiered procedure can be considered valuable only if laboratory tests allow a prediction of the effects of pesticides on earthworms under field conditions. In a preliminary paper, a good correlation was reported between results from the acute toxicity test (Tier I) and the standardized field test in grassland (Tier III): data from 29 field studies with 21 formulations containing 12 different active ingredients gave a correlation coefficient $r = 0.86$ [1,2]. There has been no comparison of results from the reproduction study (Tier II) and the field study until now. In this chapter, using a variety of pesticides, this comparison is made, and data are presented from a large number of laboratory reproduction studies and standardized field studies.

Materials and methods

Reproduction studies

Since 1992, reproduction studies were performed as described by a Biologische Bundesanstalt Für Land-Und Forstwirtschaft (BBA) guideline [3] and an International Standards Organization (ISO) guideline [4]. The test organism used was *Eisenia fetida* [5]. This earthworm can readily be bred, and its sensitivity to pesticides is comparable to that of *Lumbricus terrestris* [6–8]. The substrate used in this test was an artificial soil composed of peat, sand, and clay mineral. Ten adult worms with individual weights between 250 and 600 mg were added to 16.5 cm × 12 cm × 6 cm boxes containing 0.5 kg dry weight soil (pH = 6.0 ± 0.5). Pesticides were sprayed onto the surface of the soils to simulate pesticide use in agriculture. Four replicates were prepared for each treatment. Earthworms were fed with finely ground cow manure that was distributed on the soil surface several times during the study. In most experiments, the highest recommended application rate of each formulation was used, together with the 2- and/or 5-fold higher application rates in some tests (or even some more rates) (see Table 19-1).

The study duration was 8 weeks. The adult earthworms were exposed for 4 weeks and were fed weekly. Four weeks after the start of the experiment, the surviving adult worms of each box were collected from the substrate and weighed. They were not returned to the test boxes. The number of juveniles produced in each test box was determined after an additional 4 weeks. At the end of each experiment, the numbers of worms in the treated boxes were compared to those in the untreated controls.

The active ingredients used in the laboratory reproduction study were as follows: azinphos-methyl (emulsifiable concentrate [EC] formulation), benomyl (wettable powder [WP]), beta-cyfluthrin and oxydemeton-methyl (EC), carbendazim (suspension concentrate [SC]), cyfluthrin (EC), ethylparathion (EC), fenamiphos (EC), fluthiamide* (water-dispersable granules [WG]), fluthiamide* and metribuzin (WG), imidacloprid (higher rates: soluable concentrate [SL], lower rates: SC), imidacloprid and cyfluthrin (EC), imidacloprid and methamidophos (SL), mercaptodimethur (WP), methylparathion (WP), oxydemeton-methyl (technical active ingredient [tech.]), propoxur (EC), spiroxamide* (tech.), tebuconazole (tech.), tebuconazole and spiroxamide* (emulsion in water [EW]), thiacloprid* (SC), trichlorfon and oxydemeton-methyl (SL). Proposed common names are indicated by "*." This chapter assumes that combinatory effects of several active ingredients higher than additive do not occur. Also, formulation additives of pesticides are considered not to effect the toxicity of active ingredients. All formulations mentioned above were applied as a spray in water (400 to 800 L water/ha).

Field studies

Following the BBA guideline [9] and the ISO guideline [10], standardized field tests have been conducted since 1980 on a perennial grassland with a high abundance (up to 800 individuals per m²) and a high diversity (up to 8 species) of earthworms. Randomized plots of 10 m × 10 m (in a very few cases, 5 m× 5 m in some earlier studies [11]) were

Table 19-1   Comparison of the sensitivities of the two methods: Calculated EC50 values for pesticides tested in the reproduction and the field studies. The quotient (EC50 field / EC50 reproduction) indicates the factor by which the reproduction study is more sensitive than the field study.

Substance	Reproduction studies			Field studies			EC50 field / EC50 reproduction
	PEC (mg/kg)	deviation (% from control)	EC50 (mg/kg)	PEC (mg/kg)	deviation (% from control)	EC50 (mg/kg)	
	0.17	4		4	−41		
A (benomyl)	0.83	−45	0.83	11	−51	6.9	8.3
				16	−73		
	0.8	−57					
B	1.1	−67	0.58			~3.2	~5.5
	4.0	−91					
	3.2	−81		3.2	−59		
	0.3	2					
	0.3	−5					
	0.6	11					
C	0.6	−19	17[*] (> 3.0)	0.6	2	>> 3.0	>> 0.2[*] (> 1.0)
	1.5	−18					
	3.0	−26		3.0	−3		
	3.0	−29					
	0.37	12					
D	0.56	−17	2.7	0.5	9	>> 2.0	>> 0.7
	1.9	−45		2.0	−5		
	2.8	−48					
	0.02	16					
E	0.03	9	>> 0.13	0.05	−11	> 0.20	not calculable
	0.08	12					
	0.13	22		0.20	−25		
	0.17	−51					
F	0.67	−86	0.15	1.0	−10	5.6	37
	3.3	−95		4.0	−40		
	1.3	−73					
G	2.0	−83	0.53	3	−4	13	25
	8.0	−97		12	−45		
	1.3	−81					
H	2.7	−94	< 1.3 (0.24)[*]	2	1	>> 8.0	>> 6.2 (>> 33)[*]
	5.3	−97					
	6.7	−95		8	−6		
	8.0	−87					
I	13.3	−87	< 8.0 (0.35)[*]	13	−14	≥ 53	> 6.6 (≥ 151)[*]
	80	−97		53	−31		
	0.08	−23					
K	0.17	−28		0.25	−2		
	0.33	−41	0.39	0.50	−6	>> 0.50	>> 1.3
	0.83	−75					
	1.70	−77					

*) extrapolated EC50-values which might be slightly vague

treated in duplicate (some in 4 replicates) with the highest recommended application rate of a formulation and in some instances with a higher rate. Following the directions for use, some compounds were applied several times. A toxic standard (benomyl) and a control were included in each study (see [1,2] for more details).

In all but 2 tests, earthworm abundance was determined 4 to 6 weeks after the first application in May, and thereafter in the autumn and again in the spring of the following year. In 2 tests, evaluations were not made 4 to 6 weeks after application; to evaluate data from these tests, results had to be extrapolated. Four or 5 samples were evaluated per plot and sampling interval, i.e., 10 or 16 samples were taken per product, rate, and interval. Sampling methods used were the formalin method [12] or a combined hand-sorting/formalin method. After sampling, the earthworms were identified (see Heimbach [1,2] for more methodological details). Three species categories are used for evaluation of data in this chapter: 1) sum of all earthworm species, 2) sum of all *Lumbricus terrestris*, 3) and sum of all *Aporrectodea caliginosa*. For each of these groups, the number of earthworms and their biomass were determined. The results from treated plots were compared to those from untreated plots (effects in % deviation from control results for numbers and biomass of each sampling; the mean of all 6 results is used as the final result for this data evaluation). Most of the field-study results have already been published by Heimbach [1,2,13].

The active ingredients used in the field were as follows: azinphos-methyl (EC), benomyl (WP), captafol (SC), captan (WP), cyfluthrin (EC), endosulfan (EC), ethiofencarb (EC), fenamiphos (EC), imidacloprid (WP), mercaptodimethur (WP), methamidophos (SL), methylparathion (WP), oxydemeton-methyl (EC), thiacloprid (SC), and propoxur (EC).

Statistical analysis
LC50/EC50s had been calculated by probit analysis. The U test was performed as a procedure to establish significant differences between control and treatments (P = 0.05). For comparison of results from the 2 test procedures, 2 coefficients were calculated: the PEC and the TER.

Predicted environmental concentration
For PEC calculations, an uniform distribution of the applied dosage of a compound in the top 5 cm soil and an average soil density of 1.5 kg/L were assumed. Following these assumptions, an application rate of 1 kg/ha is equivalent to 1.33 mg/kg dry weight soil. To calculate a PEC for pesticides that are applied several times per season, 50% of each further application is added to 100% of the first application. Calculation of the final concentration in the soil as simple multiples of the individual treatments was not used, since it does not take degradation of the compound into consideration. This would ignore a major ecochemical property of modern pesticides.

Calculated PEC values are not assumed to fully represent the actual concentrations of compounds in soils. Nevertheless, they do reflect analyzed mean soil concentrations quite well [1,2]. On the other hand, PEC values were calculated for reproduction and field studies in the same way, and the pesticides had been sprayed on the soil surface in all these studies.

Toxicity exposure ratio
The TER is the quotient of the acute LC50 (in mg/kg dry weight soil) after 14-d test duration and the PEC. Thus, the toxicological hazard of a substance decreases with an increasing TER. The LC50 values given in this chapter were calculated using probit analysis. LC50 values were taken from standardized earthworm acute toxicity tests that were performed in the laboratory according to the OECD test method [14]. Most of the data used for analyses and a description of test methods have been previously published [1,2,15].

Results and discussion
The results from the reproduction studies are summarized in Figure 19-1. To indicate the variability of results, the coefficient of variation was calculated. The coefficient within control replicates is 15% for an average of the studies presented in this chapter. To indicate the range of variation in Figure 19-1, which usually is considered as natural variation, the ± 25% area has been shaded; statistically, this area reflects a no-effect range for the most of the studies. Figure 19-1 shows a good correlation between the TER and the corresponding biological effects: with decreasing TER values (i.e., with increasing toxicity of a substance and/or increasing application rate), the reproduction of earthworms is increasingly inhibited. The variation of results is low and the correlation coefficient of $r = 0.76$ indicates significance (P < 0.1, $n = 78$).

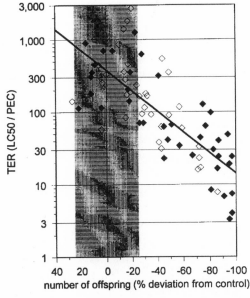

Figure 19-1   Correlation between TER and number of offspring in laboratory reproduction study Legend: ($n = 78$, $r = 0.76$, P < 0.1, ln y = 0.033X + 378; unfilled diamonds give results from pesticides that were not tested in the field)

Figure 19-2 summarizes the results of the field studies. The results indicate a strong correlation between biological effects (reduction in earthworm numbers and biomass) and the TER (toxicity and application rate) of the compounds that were tested. Without exception, all results fit well to this correlation; the correlation coefficient of $r = 0.72$ indicates significance (P < 0.1, $n = 28$).

Endogeic earthworm species such as *Aporrectodea caliginosa* live permanently in soil, while anecic species (*Lumbricus terrestris*) mostly inhabit the soil but crawl on the soil surface during the night to feed on organic matter (OM), e.g., leaves, straw, or small sticks. Accordingly, individuals of this species might be exposed to pesticides applied on the soil surface in a different way than worms that remain below the soil surface. To investigate whether these life strategies result in different toxicities, the results of studies with *L. terrestris* and *A. caliginosa* were evaluated individually, as shown in Figures 19-3 and 19-4. Because of the low abundance of *L. terrestris* and the relatively high biomass of adult individuals of this species, the results show a quite high variability (Figure 19-3). But even with this variation, the correlation (*r* = 0.58, P < 0.1, *n* = 28) is moderately high, indicating no meaningful species-specific sensitivity or exposure of this species. The variation of results of *A. caliginosa* is lower (Figure 19-4) as compared with *L. terrestris*. The variation and correlation (*r* = 0.74, P < 0.1, *n* = 28) are similar to those for the sum of all earthworm species.

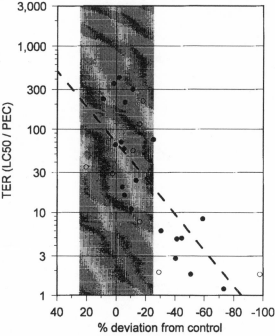

Figure 19-2   Correlation between TER and reduction in earthworm numbers and biomass in field studies for the sum of all earthworm species Legend: (*n* = 28, *r* = 0.72, P < 0.1, ln y = 0.050X + 69; unfilled circles give results from pesticides that were not tested in reproduction studies)

Since the regression and correlation results are similar for *L. terrestris*, *A. caliginosa*, and the sum of all earthworms, it seems justified to use the results of the sum of all earthworm species for all further data evaluations.

To compare results of laboratory and field studies, data from compounds that had been used in both types of studies are summarized in Figure 19-5. In this figure, pesticides are identified by the letters shown in Table 19-1. (Since the regression lines in Figures 19-1 and 19-2 are also based on several pesticides that had not been tested in both studies, there are minor differences between these latter figures and Figure 19-5; this is indicated by the correlation coefficient of *r* = 0.80 [P < 0.1, *n* = 39] for reproduction studies [instead of $r = 0.76$, $n = 78$ in Figure 19-1] and $r = 0.82$ [P < 0.1, $n = 20$] for field studies [instead of $r = 0.72$, $n = 28$ in Figure 19-2]).

The regression lines plotted in Figure 19-5 are not parallel, but they do not cross or overlap in the range of significant biological effects (25 to 100%). These lines, as well as the corresponding individual results, are clearly separated between reproduction and field studies. This demonstrates that, for the compounds tested, the laboratory reproduction study is more sensitive than the field study, although slight differences between sensitivities of earthworm species and exposure scenarios between laboratory and field tests must be considered. This statement is further supported by the individual results from pesticides that had been applied at the similar rates in both studies (i.e., same TER; e.g., substances B, D, F, G, H, and I). Simi-

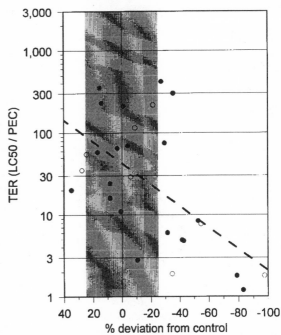

Figure 19-3   Correlation between TER and reduction in earthworm numbers and biomass in field studies for *Lumbricus terrestris* Legend: ($n = 28$, $r = 0.58$, P < 0.1, ln y = 0.030X + 43; unfilled circles give results from pesticides that were not tested in reproduction studies)

larly, in the laboratory study, the same biological effect was found at distinctly lower application rates than in the field (% reduction in offspring numbers in the laboratory and % reduction in earthworm numbers/biomass in the field, respectively (e.g., compounds A, B, F, and G).

The toxicity of benomyl to earthworms is well known and has been described by several authors [1,2,16–19]. It was found to be very toxic to field populations of earthworms at high application rates and moderately toxic at lower rates. The results presented in this chapter support those that have been published thus far. The comparison of biological effects of benomyl in the reproduction and field studies are shown in Figure 19-6. The regression lines of the 2 types of studies show that the sensitivity of the reproduction study is higher than the field test: to get a 50% reduction in earthworm numbers and biomass (EC50) in both tests, the application rate in the field has to be about 8 × higher than that in the reproduction test in the laboratory.

For a quantitative evaluation of the sensitivities of the 2 methods, the results of all these studies were used to calculate EC50 values, i.e., those application rates that caused a 50%

reduction in offspring numbers in the laboratory study and a 50% reduction in earthworm numbers and biomass in the field study. The results are summarized in Table 19-1. In several cases, EC50 values had to be extrapolated because data with clearly more or less than 50% effects were not available. These extrapolated EC50 values are shown in brackets (Table 19-1) and should be interpreted with care. Nevertheless, they can be used to support the conclusions drawn from the other results.

The factor derived by calculating EC50 field/EC50 reproduction (Table 19-1) varies from 5.5 to 37 for results that can be calculated (substances A, B, F, and G) and from >> 0.7 to ≥ 151 (substances C, D, H, I, and K). But even the results of those pesticides providing a factor of >> 0.7 (substance D) and >> 1.3 (substance K) indicate that the reproduction study

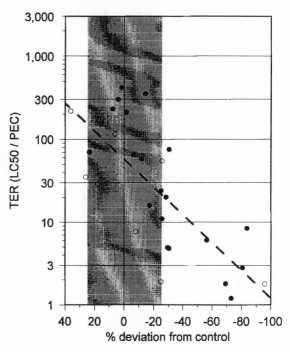

Figure 19-4   Correlation between TER and reduction in earthworm numbers and biomass in field studies for *Aporrectodea caliginosa* Legend: (n = 28, r = 0.74, P < 0.1, ln y = 0.039X + 58; unfilled circles give results from pesticides that were not tested in reproduction studies)

is more sensitive at an application rate that caused about 50% reduction: in the field, these concentrations did not have any effect with either of the pesticides. Substances H and I show the differences in sensitivities of the 2 methods by more than 1 order of magnitude. With 1 single pesticide (substance E), the factor cannot be calculated or estimated because the highest application rates in both the laboratory and field tests fall within the natural variation of results (± 25%).

Conclusions

The results of this analysis clearly show that the reproduction study is more sensitive than the Tier II field study by a factor of at least 5 to 10. Thus, the laboratory reproduction test fulfills the requirements of a test for hazard assessment. Moreover, too many pesticides are triggered for field testing by the 5-fold overdose in the reproduction study, such as that which is currently required by registration authorities in the EU. Even the highest recommended application rate would be conservative.

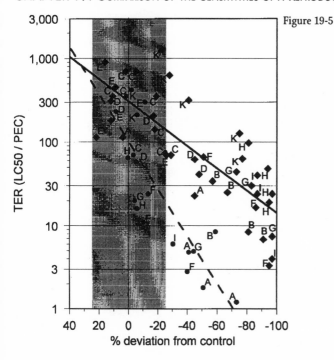

Figure 19-5   Correlation of TER and effects on earthworms in laboratory reproduction studies (♦) and field studies (●). Legend: Results show pesticides that were tested in both types of study. Pesticides are identified by letters (benomyl = A). Reproduction test: $r = 0.80$, $n = 39$, $P < 0.1$, ln y = 0.031X + 307; field test $r = 0.82$, $n = 20$, $P < 0.1$, ln y = 0.065X + 101

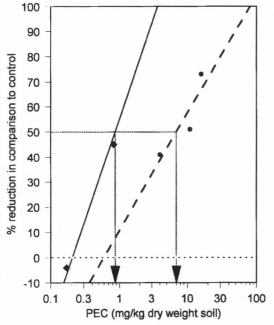

Figure 19-6   Biological effects of benomyl in reproduction studies (♦) and in field studies (●) at different application rates. Legend: The results of field studies are based on 1 study for the lowest application rate, the mean of 5 studies for the interim rate and the mean of 4 studies for the highest rate. The arrows indicate those concentrations (PECs) that caused a 50% reduction in number of offspring (reproduction study), and numbers and biomass (field study).

Acknowledgments - I would like to thank Ms. Gabriele Buettner and Mr. Ronald Mueller for capable technical assistance. I am also grateful for the help provided by Mr. Ralph Vinken in preparing this manuscript.

References

1. Heimbach F. 1992. Effects of pesticides on earthworm populations: comparison of results from laboratory and field test. In: Greig-Smith PW, Becker H, Edwards PJ, Heimbach F, editors. Ecotoxicology of earthworms. Hants UK: Intercept. p 100–106.

2. Heimbach F. 1992. Correlation between data from laboratory and field tests for investigating the toxicity of pesticides to earthworms. *Soil BiolBiochem* 24:1749–1753.

3. [BBA] Federal Biological Research Center for Agriculture and Forestry. 1994. Effects of plant protection products on reproduction and body weight of *Eisenia fetida/Eisenia andrei*. Guidelines for the testing of plant protection products within Registration. Part VI, 2-2. Braunschweig, Germany.

4. [ISO] International Standard Organization. 1995. Guideline DIS 11268-2: Soil quality - effects of pollutants on earthworms (*Eisenia fetida*), Part 2: Determination of effects on reproduction.

5. Bouché MB. 1972. Lombriciens de France, ecologie et systématique. Paris, France: Publ. Institut National de la Recherche Agronomique. p 671.

6. Haque A, Ebing W. 1983. Toxicity determination of pesticides to earthworms in the soil substrate. *J Plant Disease Protect* 90:395-408.

7. Heimbach F. 1985. Comparison of laboratory methods using *Eisenia foetida* and *Lumbricus terrestris* for assessment of the hazard of chemicals to earthworms. *J Plant Disease Protect* 92:186-193.

8. Edwards PJ, Coulson JM. 1992. Choice of earthworm species for laboratory tests. In: Greig-Smith PW, Becker H, Edwards PJ, Heimbach F, editors. Ecotoxicology of earthworms. Hants UK: Intercept. p 36-43.

9. [BBA] Federal Biological Research Center for Agriculture and Forestry. 1994. Effects of plant protection products on earthworms in the field. Guidelines for the testing of plant protection products within registration. Part VI, 2-3. Braunschweig, Germany.

10. [ISO] International Standard Organization. 1996. Draft Guideline CD 11268-3: Soil quality - effects of pollutants on earthworms, Part 3: Field methods.

11. Heimbach F. 1997. Field tests on the side effects of pesticides on earthworms: influence of plot size and cultivation practices. *Soil Biol Biochem* 29:671-676.

12. Raw F. 1959. Estimating earthworm populations by using Formalin. *Nature* 184:1661-1662.

13. Heimbach F. 1993. Use of laboratory toxicity tests for the hazard assessment of chemicals to earthworms representing the soil fauna. In: Eijsackers HJP, Hamers T, editors. Integrated soil and sediment research: a basis for proper protection. Amsterdam, the Netherlands: Kluwer Academic. p 299-302.

14. [OECD] Organization for Economic Cooperation and Development. 1984. Organization for Economic Cooperation and Development guidelines for testing of chemicals: earthworm acute toxicity test. OECD Guideline No. 207. Paris, France: OECD.

15. Heimbach F. 1984. Correlation between three methods for determining the toxicity of chemicals to earthworms. *Pesticide Sci* 15:607-611.

16. Cook ME, Swait AAJ. 1975. Effect of some fungicide treatments on earthworm population and leaf removal in apple orchards. *J Horticult Sci* 50(4):495-499.

17. Edwards PJ, Brown SM. 1982. Use of grassland plots to study the effects of pesticides on earthworms. *Pedobiologia* 24:145-150.

18. Stringer A, Lyons CH. 1977. The effect on earthworm populations of methods of spraying benomyl in an apple orchard. *Pesticide Sci* 8:647-650.

19. Stringer A, Wright MA. 1973. The effect of benomyl and some related compounds on *lumbricus terrestris* and other earthworms. *Pesticide Sci* 4:165-170.

Chapter 20

Comparison of laboratory toxicity tests for pesticides with field effects on earthworm populations: a review

Ainsley Jones and Andrew D.M. Hart

Pesticide risk assessment for earthworms involves comparing the calculated exposure in the field to laboratory toxicity data. A literature review was conducted to assess the correlation between the results of laboratory toxicity tests and field studies and to assess which factors influence this correlation. The results showed that the relationship between laboratory toxicity and field effects is highly variable and that acute toxic effects in the field can occur at much lower concentrations than in laboratory studies. A number of factors were identified that may contribute to the observed differences between laboratory and field effects, of which the most important are susceptibility differences between species. Differences in toxicity between soils and variability in the results of laboratory toxicity measurements are less important factors but may still be significant. Factors for which there was insufficient evidence to assess their significance, but which may contribute to variations in field studies, include the prevalent environmental conditions (e.g., rainfall) and differences in exposure between species. For 1 pesticide, carbofuran, there was some evidence that granular formulations can sometimes cause greater effects than spray formulations. Safety factors to compensate for differences in effects between laboratory and field should be designed to protect against unacceptable field effects. The ability of earthworm populations to recover is an important factor in determining the acceptability of effects, and this was also assessed in the review. The available results suggested that earthworm populations are capable of recovering completely from acute toxic effects of nonpersistent chemicals, i.e., those with a half-life (DT50) in soil of < 50 d, within around 1 y. Recovery was possible from reductions of over 90%. For persistent chemicals (DT50 > 50 d), the ability of earthworm populations to recover from acute toxic effects was reduced. The implications of these results for the assessment of acceptability of effects and design of field trials are discussed.

The importance of earthworms is recognized in ecological risk assessment procedures, and European Council (EC) Directive 91/414/EEC requires that the risk to earthworms be assessed before pesticides can be authorized for use. Assessing the risks of pesticides to earthworms initially involves extrapolating data from laboratory toxicity tests to predict effects under field conditions. The laboratory data are normally generated from a 14-d toxicity test performed according to the Organization for Economic Cooperation and Development (OECD) Guideline 207 [1], using the species *Eisenia fetida*. An acute toxicity exposure ratio (TER) is calculated by comparing the LC50 value to the predicted environmental concentration (PEC) under field conditions at the application rate used. Under the guidelines to EC Directive 91/414/EEC and the European Plant Protection Organization/Council of Europe (EPPO/CoE) guidelines for the environmental risk assessment of pesticides to earthworms [2], the initial PEC is calculated by assuming that all the pesticide is evenly distributed in the top 5 cm of soil. In order to allow for differences between laboratory and field, safety factors are applied to this toxicity/exposure comparison.

Clearly, if laboratory tests are to be used to predict field effects in this way, it is important to assess the correlation between laboratory toxicity data and field effects. This chapter assesses this correlation by comparing published laboratory toxicity data to published results from field experiments.

The safety factors referred to above are designed to protect earthworm populations from unacceptable effects without unnecessarily withholding authorization for pesticides that may be beneficial to agriculture. The ability of earthworm populations to recover is an important factor in assessing the acceptability of pesticide effects. This chapter also reviews available information on the recovery ability of earthworm populations.

Correlation between laboratory tests and field effects

A review of the scientific literature was undertaken to identify field studies in which a quantitative assessment had been made of the effect of a pesticide on earthworm populations. i.e., a reduction compared to a control population and for which there was sufficient information to calculate the PEC. The aim was then to calculate the TER from the PEC and the laboratory toxicity data (if available) and to compare the TER to the field effect.

The review identified 13 field studies covering 14 chemicals in which population effects had been adequately measured against a control and for which the application rate was given. Many chemicals had been studied more than once in field trials so that the total dataset consisted of 71 points. Relatively little data were found on laboratory sublethal toxicity, and this review concentrated on comparing acute LC50 data with field effects. In 1 study [3], some field experiments were repeated 3 y in succession on the same pesticides at the same application rates at the same site. In this case, results have not been included for all 3 y, and only the result corresponding to the first year, when effects were most severe, has been used. Table 20-1 lists the complete set of laboratory and field trials included in this review. In some cases, the effects were measured on total earthworm numbers, while in other cases, effects were measured on populations of particular species: both are included in Table 20-1. Where effects were measured at several times after application, the maximum population reduction was used; the time of maximum effect is shown in Table 20-1. In those cases where more than 1 LC50 value was found in the literature, the lowest LC50 value has been used for calculation of the TER. When the LC50 is reported as a range of values, the median value in that range has been used. For one chemical, methiocarb, data from a 14-d LC50 were not available, so the result from a 28-d toxicity test has been used instead. A 28-d test should give similar results to the 14-d test as is confirmed by Heimbach, who, working with the species *Eisenia andrei* found a 14-d LC50 of 19 mg/kg for benomyl [4] and a 28-d LC50 of 22 mg/kg [5]. LC50 experiments are often performed using the species *E. andrei*, and these data have been used when data for *E. fetida* were not available. This is reasonable, as *E. andrei* is very closely related to *E. fetida* [6]. In all the field trials included, either grass was the only crop present or there was no crop present.

PECs were calculated according to note 2 of the EPPO/CoE guidelines [2], which assumes that all of the pesticide applied reaches the soil unless there is a high degree of plant cover (e.g., cereals in summer). In particular, the guidelines state that for products applied to grassland, the whole amount applied must be taken into consideration. In all the field trials included in Figure 20-1, the products were either applied to grassland or there was no crop present at the time of application. Thus, in all cases, PECs have been calculated by assuming all of the pesticide reached the soil and was evenly distributed in the top 5 cm.

Field effects have been plotted against TER in Figure 20-1. The EPPO/CoE guidelines state that to compensate for differences in organic matter (OM) between the test soil and typical agricultural soils, the LC50 should be reduced by a factor of 2 for chemicals with log $Kow > 2$, and this has been done in Figure 20-1. A similar comparison between laboratory toxicity and field effects was made by Heimbach [7], who performed 21 field studies covering 12 different pesticides on grassland sites in a region of Germany with silty sand or loamy silt soils (Heimbach personal communication). He also determined the laboratory toxicity for each pesticide using a 14-d toxicity test with the earthworm *E. andrei*. It is difficult to make a direct comparison between Heimbach's published results [7] and this review, as in most cases Heimbach used multiple applications in field tests, taking 100% of the PEC for the first application and 50% of the PEC for any further applications. Furthermore Heimbach [7] did not report actual reductions in population but classified effects into broad categories

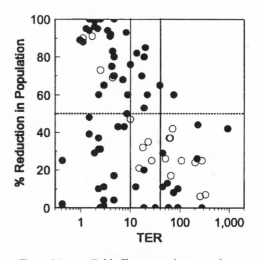

Figure 20-1   Field effect on earthworms after pesticide application versus toxicity/exposure ratio (TER). Legend: Data are maximum percentage reduction in numbers compared to controls. O Data supplied by F. Heimbach (personal communication). l Data from literature by other workers (see Table 20-1).

and integrated the impacts over time by summing the scores for each period to obtain an overall measure of effect. However Heimbach (personal communication) has supplied results giving actual population reductions, on total earthworm numbers at the 3 sampling times used (6 weeks, 6 months, and 1 y after application). These results are also included in Figure 20-1; once again, only the maximum population reduction is plotted. Heimbach's method of calculating PECs has been used for the data in Figure 20-1 but only for those cases where up to 2 applications were used. It should be noted that the pesticides used were not identified, so the LC50 values used are those determined by Heimbach and not the lowest LC50 value found in the literature, as is the case for the other data.

Examination of Figure 20-1 shows that although the relationship between TER and effect on population is not strong, some general trends are apparent. With 1 exception, all reductions of > 50% are associated with TER values < 40. The exception is from a study where a reduction of 60% was observed after a benomyl application with a TER of 150 [3]. It should be noted, however, that the same workers, on the same site 1 y later, also used benomyl at 4-fold this application rate (TER = 38) and observed a reduction of only 53%. It should also be noted that there are 9 cases in which reductions of > 50% are associated with TER values of 10 or greater. These would therefore escape the threshold of 10 for the acute TER given in Annex VI of the EC Directive 91/414/EEC that triggers the requirement for an appropriate risk assessment under field conditions. They would also escape the EPPO/CoE threshold of 10, triggering further testing. However for those cases where the pesticide was considered persistent, a sublethal toxicity test may be required, and the risk assessment under field conditions would still be triggered if the long-term TER is less than 5. Of the 9 cases under consideration, 3 relate to a pesticide (benomyl) that would usually be considered sufficiently toxic and persistent to require a sublethal test. A further 3 relate to a chemical of moderate persistence (carbofuran), which might require a sublethal test.

Heimbach's data in Figure 20-1 are less variable than the other data in this review, and there are no cases in Heimbach's results in which reductions of > 50% on total earthworm populations are associated with TER values of 10 or greater. In contrast, there are 9 such cases for the other data in this review. It is understandable that Heimbach's field studies were less variable than those reviewed in Table 20-1, as the former were all performed using the same methods and in the same locality, so that there would have been less variation due to differences in soil type or earthworm species present.

Factors that may contribute to laboratory–field differences

A number of factors identified from the available literature can contribute to the observed variation between laboratory toxicity and field effects.

Toxicity differences between earthworm species

Five laboratory studies were identified, covering 25 comparisons of 22 chemicals in which inter-species toxicity had been compared using earthworms in the same soil system. These are shown in Table 20-2; it also shows the maximum ratio of toxicity difference between E. fetida and the most sensitive species studied. In general E. fetida tends to be less susceptible than typical agricultural species. In 22 out of 25 cases, the maximum

Table 20-1   Comparison of laboratory toxicity tests for pesticides with field effects on earthworm populations

Chemical	Lab/field	Species	Duration (d)	LC50[1]	App. rate kg/ha	Time[2] (d)	PEC[3]	Species	Tox./exp ratio	Popn. reduct.	Ref.
Benomyl	Lab	E. fetida	14	27							16
	Field				0.125	63	0.18	Total	150	8%	17
	Field				1.25	63	1.8	Total	15	43%	17
	Field				0.56	180	0.8	Total	34	70%	18
	Field				1.12	328	1.6	Total	17	50%	18
	Field				2.24	13	3.2	Total	8.4	75%	18
	Field				0.125	180	0.18	Total	150	60%	3
	Field				0.50	360	0.72	Total	38	53%	3
	Field				2.0	30	2.9	Total	9.3	70%	19
	Field				10	180	14.3	Total	1.9	89%	19
	Field				5.0	30	7.1	L. terrestris	3.8	99%	19
	Field				5.0	30	7.1	L. festivus	3.8	29%	19
	Field				7.8	21	11.1	Total	2.4	95%	20
	Field				12.2	21	17.4	Total	1.55	95%	21
Carbaryl	Lab	E. fetida	14	69							12
	Field				2.5	30	3.6	A. longa	19	80%	22
	Field				2.5	180	3.6	L. terrestris	19	20%	22
	Field				25	30	36	A. longa	1.9	100%	22
	Field				25	30	36	L. terrestris	1.9	96%	22
	Field				2.5	30	3.6	A. longa	19	0%	22
	Field				2.5	30	3.6	L. terrestris	19	0%	22
	Field				25	30	36	A. longa	1.9	100%	22
	Field				25	30	36	L. terrestris	1.9	100%	22
	Field				2.24	21	3.2	Total	22	60%	23
Carben-dazim	Lab	E. andrei	21	5.7							17
	Field				0.3	12	0.43	Total	13	11%	17

Table 20-1 Continued

Chemical	Lab/field	Species	Duration (d)	LC50[1]	App.rate kg/ha	Time[2] (d)	PEC[3]	Species	Tox./exp ratio	Popn. reduct.	Ref.
Carbofuran	Lab	E. fetida	14	28							16
	Field				5.6	21	8.0	Total	3.5	94%	20
	Field				3.4	21	4.9	Total	5.7	43%	20
	Field				2.24	35	3.2	Total	8.8	60%	24
	Field				0.5	42	0.71	Total	39	65%	25
	Field				1.0	42	1.43	Total	20	85%	25
	Field				2.0	42	2.9	Total	10	76%	25
	Field				5.0	42	7.1	Total	3.9	91%	25
	Field				4.5	21	6.4	Total	4.4	83%	23
	Field				2.4	180	3.4	Total	8.2	93%	26
Chlordane	Lab	E. fetida	14	42							4
	Field				5.0	30	7.1	L. terrestris	5.9	4%	22
	Field				5.0	30	7.1	A. longa	5.9	65%	22
	Field				10	180	14.3	L. terrestris	2.95	48%	22
	Field				10	180	14.3	A. longa	2.95	94%	22
	Field				5.0	30	7.1	L. terrestris	5.9	1%	22
	Field				5.0	180	7.1	A. longa	5.9	11%	22
	Field				10	30	14.3	L. terrestris	2.95	39%	22
	Field				10	30	14.3	A. longa	2.95	100%	22
Methiocarb	Lab	E. fetida	28	129							4
	Field				0.5	8	0.71	Total	182	10.1%	27
	Field				1.0	8	1.43	Total	90	29.4%	27
	Field				0.5	8	0.71	Total	182	0%	27
	Field				1.0	8	1.43	Total	90	11%	27
Parathion	Lab	E. fetida	14	277							3
	Field				0.105	30	0.15	Total	1847	42%	3
	Field				0.42	180	0.60	Total	462	44%	3

Table 20-1 Continued

Chemical	Lab/field	Species	Duration (d)	LC50[1]	App. rate kg/ha	Time[2] (d)	PEC[3]	Species	Tox./exp ratio	Popn. reduct.	Ref.
Pentachlo-rophenol	Lab	E. andrei	14	45							22
	Field				12.5	30	17.9	L. terrestris	2.5	31%	22
	Field				12.5	30	17.9	A. longa	2.5	0%	22
	Field				75	30	107	L. terrestris	0.42	25%	22
	Field				75	180	107	A. longa	0.42	2%	22
	Field				12.5	30	17.9	L. terrestris	2.5	0%	22
	Field				12.5	30	17.9	A. longa	2.5	100%	22
	Field				75	30	107	L. terrestris	0.42	0%	22
	Field				75	30	107	A. longa	0.42	0%	*[5]
Phorate	Lab	E. fetida	14	39.6							
	Field				2.0	120	2.9	Allo./App.[4]	14	68%	28
	Field				1.0	70	1.43	Total	27	82%	28
	Field				3.4	21	4.9	Total	8.0	92%	20
Propoxur	Lab	E. fetida	14	58							3
	Field				0.18	30	0.26	Total	223	26%	3
	Field				0.72	360	1.03	Total	56	13%	3
Thiopha-nateme	Lab	E. andrei	14	20.1							22
	Field				3.0	30	4.3	L. terrestris	4.7	17%	22
	Field				3.0	30	4.3	A. longa	4.7	80%	22
	Field				6.0	180	8.6	L. terrestris	2.3	31%	22
	Field				6.0	30	8.6	A. longa	2.3	60%	22
	Field				3.0	30	4.3	L. terrestris	4.7	4%	22
	Field				3.0	30	4.3	A. longa	4.7	100%	22
	Field				6.0	30	8.6	L. terrestris	2.3	10%	22
	Field				6.0	30	8.6	A. longa	2.3	37%	22
	Field				12.2	21	17.5	Total	1.1	88	22

Table 20-1 Continued

Chemical	Lab/field	Species	Duration (d)	LC50[1]	App. rate kg/ha	Time[2] (d)	PEC[3]	Species	Tox./exp ratio	Popn. reduct.	Ref.
Cyfluthrin	Lab	E. fetida	14	15.4							*[5]
	Field				0.16	21	0.23	Total	63	0%	21
Bifenthrin	Lab	E. fetida	14	18.9							*[5]
	Field				0.11	21	0.16	Total	118	0%	21
Cyproco-nazole	Lab	E. fetida	14	335							*[5]
	Field				0.41	21	0.60	Total	558	0%	21

[1]LC50 in mg (a.i.)/kg of soil (dry weight)
[2]Time between application and sampling
[3]PEC calculated by assuming that all of the pesticide reaches the soil, is evenly distributed in the top 5 cm, and that the bulk density of the soil is 1400 kg/m^3
[4]Combined Allolobophora/Aporrectodea genera
[5]Unpublished data supplied by Pesticide Safety Directorate, Ministry of Agriculture, Fisheries and Food, UK

toxicity difference is a factor of 10 or less. However for 1 chemical, propoxur, toxicity differences of much greater than 10 between *E. fetida* and some species have been reported.

Differences in toxicity associated with different soils

The available data on the influence of soil characteristics on earthworm toxicity has been reviewed by van Gestel [8]. For organic chemicals, differences in toxicity between soils can be explained largely by their OM content. This approach is based on the premise that earthworms are exposed to chemicals mainly through the pore water in the soil. However, there are a number of limitations to this approach. For chemicals with Log Kow < 2 toxicity differences between soils cannot be extrapolated on the basis of OM content without taking into account the water content of the soil. The approach also runs into difficulties for highly lipophilic chemicals (e.g., Log Kow > 6) because, for these chemicals, exposure is not solely through the pore water, and oral uptake by ingestion of soil becomes increasingly significant. Another complication is that, for ionizable compounds such as organic acids or bases, soil pH may influence toxicity.

This approach suggests that toxicity differences between soils are relatively small. The OM content of the OECD test soil is 8 to 10%, and the lowest OM content of soil likely to be considered in earthworm risk assessment is 3 to 4%. Thus the maximum toxicity difference, due to soil differences, between laboratory and field conditions is likely to be a factor of 2 to 3. This provides the basis for the application of a factor of 2 to give reduced LC50 values for pesticides with log Kow > 2 in the EPPO/CoE scheme.

Measurement of laboratory toxicity

A certain degree of variability is associated with the toxicity value derived from the 14-d laboratory LC50. This is partly due to the variability inherent in any toxicity trial caused by differences in the preparation of the test medium, age and condition of test animal, etc. Table 20-3 compares results from tests performed apparently identically and gives some idea of the degree of variability associated with laboratory toxicity testing. It appears that the results of laboratory toxicity tests may vary by a factor of 2 or more.

Behavioral differences between species leading to differing exposure

The distribution of earthworms within the soil is obviously important in determining exposure. Some earthworms such as *Allolobophora chlorotica* and *Apporectodea caliginosa* inhabit the top 5 cm or so of soil, except during unfavorable conditions [9]. The risk assessment procedure is likely to be reasonably accurate in estimating exposure to such species. Other species such as *Lumbricus terrestris* live in deep burrows but feed on the soil surface. While in the burrows, they are unlikely to be exposed unless the pesticide is leached down into the burrow. However, while feeding, they may be exposed to higher concentrations of pesticides.

Table 20-2   Toxicity differences among species from laboratory studies with pesticides

Reference	Chemical	LC50 in mg (a.i.)/kg of soil (dry weight)							Max. ratio
		E. fetida	A. caliginosa	A. chlorotica	L. terrestris	L. rubellus	A. longa	A. nocturna	
[29]	aldicarb	16	4	4		4 to 8			4
[29]	carbaryl	>64	<4	4		<4			16
[29]	carbofuran	>64	<4	4		<4			16
[29]	paraoxon	32 to 64	32	32					2
[29]	dithiocarbamate	16 to 32	16 to 32	8		16			4
[30]	parathion	277	116	206	159		60		4.6
[30]	propoxur	58	1.4	10.8	7.2		0.8		72
[31]	benomyl	1 to 10	1 to 10		1 to 10				10
[31]	paraquat	3200	580		1000				5.5
[31]	chloracetamide	75	40	47	54			80	1.9
[5]	benomyl	27			3.5				7.7
[5]	terbufos	6.6			4.6				1.4
[5]	carbofuran	28			4.7				5.9
[5]	propoxur	27			5.2				5.2
[5]	methidathion	4.8			7.6				0.63
[5]	endosulfan	9.4			9.0				1.0
[5]	aldicarb	3.3			26				0.13
[5]	methamidophos	17			110				0.15
[5]	lindane	59			113				0.52
[5]	Dialifos	218			174				1.25
[5]	triazophos	116			210				0.55
[5]	captan	612			237				2.6
[5]	atrazine	131			444				0.3
[5]	Folpet	339			459				0.75
[32]	chlorpyrifos	1077	755		458	129	778		8.3

Environmental conditions

There is obviously no standardization in the temperature or rainfall during the period of the field trials. These can have a significant effect on exposure of earthworms because at some times of the year, when it is very cold or the soil is dry, earthworms may not reside in the top few cm of soil. Instead they may retreat to depths where they are unlikely to be exposed to pesticides in the top few cm [9]. Heavy rainfall may also have an effect on the distribution of the pesticide or the behavior of the worms. The more mobile pesticides may be transported down to greater depths than are normal. This would have the effect of decreasing the duration of exposure to species that live near the surface but may increasing exposure to deep-burrowing species. However, deep-burrowing species may come to the surface to feed after heavy rainfall. This behavior may decrease contact with pesticides leached into burrows but increase contact with pesticides present on the surface, e.g., granular formulations. Unfortunately, little information is given about environmental conditions in most of the other studies in Table 20-1, but it is likely that the range of conditions varied considerably. Some studies may have been conducted in conditions that approach the worst possible case, while in others, conditions may have been far from the worst-case situation.

Table 20-3	Range of reported LC50 laboratory test values	
Chemical	LC50 mg (a.i.)/kg soil (dry weight)	Reference
Carbaryl	69	[18]
	109	[33]
Aldicarb	3.3	[17]
	8.0	[4]
Lindane	136	[17]
	59	[4]

Effect of crop type

Some crops may allow the majority of the pesticide to reach the soil surface, while others would provide a great deal of cover so that only a fraction reaches the soil. In all the studies included in Figure 20-1, either the crop present was grass or there was no crop present, so this is unlikely to be a large source of variation between laboratory and field results in this analysis.

Type of formulation

It has often been suggested that different formulations may cause differences in earthworm exposure. In particular, granules and seed dressings are unlikely to result in a homogeneous distribution in the top 5 cm of soil. There is special concern regarding L. terrestris [10] and granules because this species sometimes moves over the soil surface and may come into direct contact with, or even ingest, the granules. No experiments in Figure 20-1 involve seed dressings, but some of the experiments involving carbofuran did make use of granules. Figure 20-2 compares field effects from granular formulations of carbofuran with those from other formulations. It is difficult to conclude whether there are substantial differences between granular and liquid formulations because data on the latter are limited. It should be noted that effects from granular formulations are likely to

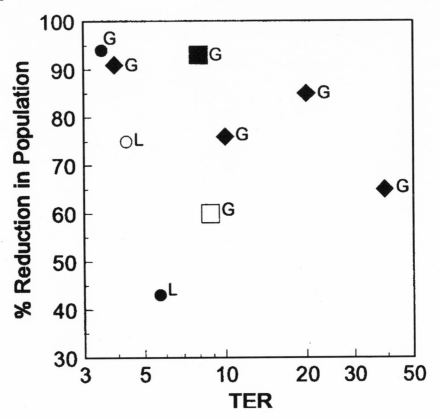

Figure 20-2   Field effect on earthworms after carbofuran application versus toxicity/exposure ratio (TER). Legend: Data are maximum percentage reduction in numbers compared to controls. G = granular formulation, L = liquid formulation. Data from: ● Tomlin and Gore [20], ● Martin [24], ◆ Clements et al., ■ Thompson [23] and ■ Finlayson [26].

be dependent on the prevailing environmental conditions, especially rainfall, and differences in behavior between species.

Ability of populations to recover

The ability of earthworm populations to recover from adverse effects is an important factor in assessing the ecological importance of such effects. Some of the trials reviewed here were continued for long enough to permit an assessment of the population recovery. These are detailed in Figures 20-3 through 20-8. In general, the recovery ability varies between chemicals and is dependent on the persistence of the chemicals in soil. The data in Table 20-4 are from the Pesticide Manual [11].

With low-persistence chemicals carbaryl and thiophanate-methyl, earthworm populations of some species are capable of recovery within 6 months. With the persistent chemicals chlordane, benomyl, and to a lesser extent carbofuran, recovery is possible at some concentrations but is increasingly impeded at higher concentrations. This suggests that, for nonpersistent chemicals (e.g., with a DT50 of < 30 d, or < 60 d as in the EPPO/CoE guideline), a reduction in the population of a particular species > 90% may be considered acceptable, provided no other stresses are placed on the earthworm population during the recovery period.

It should be noted that the trials presented in Figures 20-3 through 20-8, with the exception of Figure 20-3A, were performed on relatively small plots of 4.3 m² to 36 m² with no barriers. It is possible that recovery of earthworm populations may have been influenced by migration of earthworms from surrounding areas. If this were a factor, then the rate of population recovery in large fields may be slower than is suggested by the results presented here.

Derivation of safety factors for acute toxicity exposure ratios

Safety factors may be applied to triggers for further testing (field or semi-field tests) and/or to the final classification of risk. Their purpose is to protect against unacceptable effects in the field by allowing for uncertainties in extrapolation from experimental results. The appropriate magnitude for safety factors therefore depends on what is regarded as an acceptable effect and on the degree of uncertainty in extrapolation.

Nonpersistent pesticides

As some earthworm populations are capable of recovery from impacts of over 90% within 1 y, even large effects might be considered acceptable, provided they were temporary. However, where populations were also subject to other impacts, e.g., from cultivations or other pesticides or repeat applications of the same pesticides, then temporary impacts of this magnitude might not be acceptable.

If short-term effects are of concern, then one might apply the EPPO/CoE guidelines that suggest a reduction of 50% or more in the total population should be considered unacceptable. This seems a sensible figure, especially when it is remembered that a reduction in total numbers may hide more severe effects on particular species. On this basis, it may

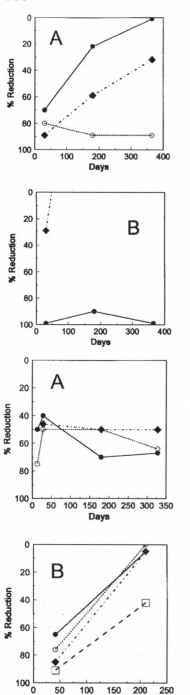

Figure 20-3   Recovery ability of earthworm populations after pesticide applications. Data are percentage reduction in numbers compared to controls. (A) after application of benomyl [19] at:● 2.0 kg(a.i.)/ha, ◆ 5.0 kg(a.i.)/ha and O 10.0 kg(a.i.)/ha. (B) after application of benomyl at 5.0 kg(a.i.)/ha,● *L. terrestris*, ◆ *L. festivus*.

Figure 20-4   Recovery ability of earthworm populations after pesticide applications. Data are percentage reduction in numbers compared to controls. (A) after application of benomyl [18] at:● 0.56 kg(a.i.)/ha, ◆1.12 kg(a.i.)/ha and O 2.24 kg(a.i.)/ha. (B) after application of carbofuran [25] at:● 0.5 kg(a.i.)/ha, ◆1.0 kg(a.i.)/ha, O 2.0 kg(a.i.)/ha and □ 5.0 kg(a.i.)/ha.

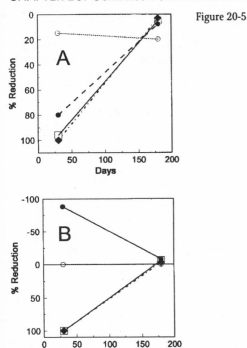

Figure 20-5   Recovery ability of earthworm populations after pesticide applications. Data are percentage reduction in numbers of 2 earthworm species compared to controls after application of carbaryl [22] to 2 different sites, A and B. Results for: *A. longa* at ● 2.5 kg(a.i.)/ha and ◆ 25 kg(a.i.)/ha, *L. terrestris* at ○ 2.5 kg(a.i.)/ha and □ 25 kg(a.i.)/ha.

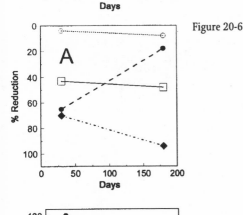

Figure 20-6   Recovery ability of earthworm populations after pesticide applications. Data are percentage reduction in numbers of 2 earthworm species compared to controls after application of chlordane [22] to 2 different sites, A and B. Results for: *A. longa* at ● 5.0 kg(a.i.)/ha and ◆ 10.0 kg(a.i.)/ha, *L. terrestris* at ○ 5.0 kg(a.i.)/ha and □ 10.0 kg(a.i.)/ha.

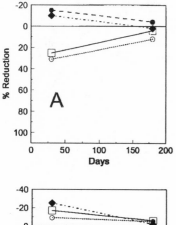

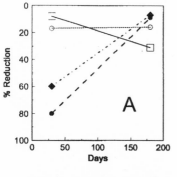

Figure 20-7   Recovery ability of earthworm populations after pesticide applications. Data are percentage reduction in numbers of 2 earthworm species compared to controls after application of pentachlorophenol [22] to 2 different sites, A and B. Results for: *A. longa* at ●12.5 kg(a.i.)/ha and ◆75 kg(a.i.)/ha, *L. terrestris* at○ 12.5 kg(a.i.)/ha and □ 75 kg(a.i.)/ha.

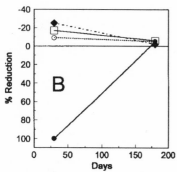

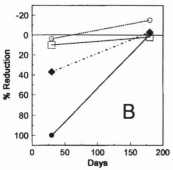

Figure 20-8   Recovery ability of earthworm populations after pesticide applications. Data are percentage reduction in numbers of 2 earthworm species compared to controls after application of thiophanate-methyl [22] to 2 different sites, A and B. Results for: *A. longa* at ● 3.0 kg(a.i.)/ha and ◆6.0 kg(a.i.)/ha, *L. terrestris* at ○ 3.0 kg(a.i.)/ha and □ 6.0 kg(a.i.)/ha.

be seen in Figure 20-1 that with 1 exception, applications with TER > 40 led to acceptable reductions (< 50%). This suggests using a safety factor of 40 to protect against unacceptable effects in the field for nonpersistent chemicals for TERs based on the LC50 and initial PEC.

Persistent pesticides
For persistent chemicals (identified on the basis of DT50 > 60 or DT90 > 100, as in the EPPO/CoE schemes), population reductions persist longer, and therefore the maximum acceptable impact might be lower, perhaps 25%. Reductions of less than 25% may be regarded as being within the limits of normal variation in earthworm populations (Heimbach, personal communication). However, Figure 20-1 shows no obvious relationship with TER for population reductions below 50%, so there is no obvious point at which to set a threshold. Possible solutions include a) obtain field data for more applications with TER > 40 and attempt to identify a further threshold, and b) for persistent pesticides with TER > 40, refine the risk assessment using a comparison between a sublethal test no-observed-effects concentration (NOEC) and the long-term PEC to decide whether a prolonged low-level impact is likely.

Design and interpretation of field and semi-field studies

The data in Figure 20-1 suggest that there is a role for field studies in pesticide risk assessment, as a significant proportion of those with TER in the range 0.1 to 40 caused population reductions of < 50%. However, this review has highlighted the variability in the results of field tests and the degree of influence on this variability of the conditions under which the test is performed. It is desirable to carry out pesticide risk assessment using a realistic worst-case approach, as required by EC Directive 91/414/EEC Annex VI. It thus seems sensible to ensure that, as far as possible, field tests are conducted under realistic worst-case conditions. From this review, a clearer picture has emerged on the desirable design and interpretation of field or semi-field trials, and a number of recommendations can be made to ensure that such trials are as representative as possible of realistic worst-case conditions.

1. Earthworm density should be at least $100/m^2$. [12,13]. The exact variety of species present may be less important than whether they are representative of the ecological categories usually found in agricultural soils. Ideally, the earthworm population of the test site should contain at least 1 and preferably 2 species from each of the categories, aneciques and endoges, identified by Bouché [6,14,15]. The primary objective of field studies should be to ensure that each ecological category is protected; thus the ecological functions performed by each category are protected.

2. The size of the plots needs to be big enough to ensure that significant immigration does not occur, unless the plots are physically separated by a barrier. Since very large plots are inadvisable because of excessive spatial variability, it would seem sensible to ensure that plots are separated by a barrier.

3. Ideally, the soil present should contain the least amount of OM capable of sustaining a reasonable earthworm population. In practice, an OM content of 4% is suitable. The soil OM content should be reported along with the pH and any other relevant information.

4. The amount of crop cover on the test site should be the minimum expected in commercial use: often this may be bare soil.

5. The weather conditions prevalent during the trial are extremely important. The chemical should be applied during a period when earthworm activity is at its maximum level, taking into account the recommended use of the pesticide. This means applying the pesticide in the spring or the autumn or as close to spring or autumn as is consistent with recommended use. However, this does not guarantee that earthworm activity will be at its maximum. A cold spring or a dry autumn could mean that activity is not at a maximum when the chemical is applied, and the conditions are therefore far from a realistic worst case. Some thought should also be given to the effect of rainfall. Mobile chemicals may be leached into the soil under conditions of heavy rainfall, thus changing the exposure pattern to earthworms markedly. Also, some species of earthworm are attracted to the surface following rainfall. This should be regarded as a possible realistic worst case, and it may be necessary to water the site soon after application. However, if it is unclear whether rainfall could reduce exposure, for example due to rapid leaching, it may be desirable to include 2 treatments, 1 with watering and 1 without. Information on temperature and weather should be recorded so that an assessment can be made of whether conditions were close to worst case.

6. The duration of the study and the frequency of sampling depend on the properties of the chemical under consideration. In order to assess acute toxic effects, the first samples should be taken about 1 month after application. The evidence of this review is that, if the chemical is of low persistence, no further effects will occur and earthworm numbers will recover within about a year from any effects that have occurred. This suggests that for nonpersistent chemicals, a 1-month duration is sufficient in appropriate conditions. However, the information on recovery is based on only a few observations, and it is uncertain whether this always happens. A sensible compromise for nonpersistent chemicals would seem to be as follows: if no effects are observed after 1 month, then the trial can be stopped. If effects do occur, then another sample would be taken after 3 to 4 months. If recovery is occurring as expected, the trial can be stopped. If the expected recovery is not observed, then the experiment would be extended to 1 y and a final sample taken at the end of this period. For persistent chemicals, trials should last at least 1 y. Samples should be taken after 1 month, 4 to 6 months, 1 y, and at the end if the trial is longer than 1 y.

7. An assessment of what is an acceptable effect is only possible in the context of other adverse effects on populations. Effects on total numbers can be mislead-

ing, and all earthworms sampled need to be identified according to species. This will then allow an assessment of effects on the 2 agronomically significant categories, aneciques and endoges. Reductions of around 75% in the numbers of either category might be regarded as acceptable provided that no other activities will cause adverse effects within the recovery period, i.e., 1 y. If there is any possibility of other effects, then a 50% reduction might be considered the maximum acceptable. The threshold of acceptability would be further lowered if populations are subjected to other adverse activities.

8. Great care should be taken in interpreting the relevance of the results of field trials. There is little control over environmental factors and the species present. These must be examined before an assessment can be made of whether the trial is reasonably representative of realistic worst-case conditions.

9. This review suggests that there can be significant limitations on the usefulness of field trials in earthworm risk assessment. These principally arise from the lack of control of the earthworm species present and the prevalent environmental conditions. These limitations could be largely overcome by the use of simulated field systems or mesocosms. Such systems would allow greater flexibility than conventional field trials and could allow particular concerns to be investigated in detail.

Acknowledgments - This work was funded by MAFF Pesticides Safety Directorate. We are grateful to Mrs. S. Quinn for help in collation of data, to Dr. H. Kula and Dr C. Kula for useful discussions, to Dr. F. Heimbach for giving us access to his data, and to Dr. H. Thompson for reviewing an earlier draft.

References

1. [OECD] Organization for Economic Development and Cooperation. 1984. Test guideline 207: Earthworm acute toxicity tests. OECD Guidelines for Testing of Chemicals. Paris, France: OECD.

2. EPPO/CoE. 1993. Decision-making scheme for the environmental risk assessment of plant protection products: earthworms. *EPPO Bulletin* 23:131–149.

3. Kula H. 1995. Comparison of laboratory and field testing for the assessment of pesticide side effects on earthworms. *Acta Zool Fennica* 196:338–341.

4. Heimbach F. 1984. Correlations between three methods for determining the toxicity of chemicals to earthworms. *Pestic Sci* 15:605–611.

5. Heimbach F. 1985. Comparison of laboratory methods for the assessment of the hazard of chemicals to earthworms. *Z Pflanzenrankh Pflanzenshutz* 92:186–193.

6. Bouché MB. 1992. Earthworm species and ecotoxicological studies. In: Greig-Smith PW, Becker H, Edwards PJ, Heimbach F, editors. Ecotoxicology of earthworms. Andover UK: Intercept. p 20–35.

7. Heimbach F. 1992. Correlation between data from laboratory and field tests for investigating the toxicity of pesticides to earthworms. *Soil Biol Biochem* 24:1749–1753.

8. van Gestel CAM. 1992. The influence of soil characteristics on the toxicity of chemicals for earthworms: A review. In: Greig-Smith PW, Becker H, Edwards PJ, Heimbach F, editors. Ecotoxicology of earthworms. Andover UK: Intercept. p 44–54.

9. Gerard BM. 1967. Factors affecting earthworms in pastures. *J Anim Ecol* 36:235–252.

10. Ebing W, Pflugmacher J, Haque A. 1985. Survey of the suitability of certain plants and animals for use as index organisms in monitoring soil pollution by organic chemicals [with special reference to earthworms]. *Plant Res Devel* 22:37–84.

11. Tomlin CA. 1994. The pesticide manual. 10th ed. Farnham, Surrey UK: Crop Protection Publications.

12. Edwards CA. 1992. Testing the effects of chemicals on earthworms: the advantages and limitations of field tests. In: Greig-Smith PW, Becker H, Edwards PJ, Heimbach F, editors. Ecotoxicology of earthworms. Andover UK: Intercept. p 75–84.

13. Lofs A. 1992. Measuring effects of pesticides on earthworms in the field: effect criteria and endpoints. In: Greig-Smith PW, Becker H, Edwards PJ, Heimbach F, editors. Ecotoxicology of earthworms. Andover UK: Intercept. p 85–89.

14. Bouché MB. 1972. Lombriciens de France. Ecologie et systematique. Paris, France: Institut National de la Reserche Agronomique.

15. Bouché MB. 1977. Strategies lombriciennes. *Ecol Bull* 25:122–132.

16. Haque A, Ebing W. 1983. Toxicity determination of pesticides to earthworms in the soil substrate. *J Plant Dis Prot* 90:395–408.

17. van Gestel CAM, Dirven-van Breemen EM, Baerelman R, Janssen JAM, Postuma R, Van Vliet PJM. 1992. Comparison of sublethal and lethal criteria for nine different chemicals in standardized toxicity tests using the earthworm *Eisenia andrei. Ecotoxicol Environ Saf* 23:206–220.

18. Tomlin AD, Tolman J, Thorn GD. 1981. Suppression of earthworm populations around an airport by soil application of the fungicide, benomyl. *Prot Ecol* 2:319–323.

19. Edwards PJ, Brown SM. 1982. Use of grassland plots to study the effect of pesticides on earthworms. *Pedobiol* 24:145–150.

20. Tomlin AD, Gore FL. 1974. Effect of six insecticides and a fungicide on the numbers and biomass of earthworms in pasture. *Bull Environ Contam Toxicol* 12:487–492.

21. Potter DA, Spicer PG, Redmond CT, Powell AJ. 1994. Toxicity of pesticides to earthworms in Kentucky bluegrass turf. *Bull Environ Contam Toxicol* 52:176–181.

22. Goats GC, Edwards CA. 1988. The prediction of field toxicity of chemicals to earthworms by laboratory methods. In: Edwards CA, Neuhauser EF, editors. Earthworms in waste and environmental management. The Hague, Netherlands: SPB Academic. p 283–294.

23. Thompson AR. 1970. Effects of nine insecticides on numbers and biomass of earthworms in pasture. *Bull Environ Contam Toxicol* 5:577–585.

24. Martin NA. 1976. Effect of four insecticides on the pasture ecosystem. *New Zeal J Agric Res* 19:111–115.

25. Clements RO, Bentley BR, Jackson CA. 1986. The impact of granular formulations of phorate, terbufos, carbofuran, carbosulfan and thiofanox on newly sown Italian ryegrass. *Crop Prot* 5:389–394.

26. Finlayson DG, Campbell CJ, Roberts HA. 1975. Herbicides and insecticides: their compatibility and effects on weeds, insects and earthworms in the minicauliflower crop. *Ann Appl Biol* 79:95–108.

27. Barker GM. 1982. Short-term effects of methiocarb formulations on pasture earthworms. *New Zeal J Exp Agric* 10:309–311.

28. Saunders DG, Forgie CD. 1977. Some effects of phorate on earthworm populations. Proceedings, 30th New Zealand Weed and Pest Control Conference; 8–10 Aug 1977; Johnsonville, New Zealand. p 222–226.

29. Stenersen J. 1979. Action of pesticides on earthworms. Part I: The toxicity of cholinesterase-inhibiting insecticides to earthworms as evaluated by laboratory tests. *Pestic Sci* 10:66–74.

30. Kula H. 1994. Species-specific sensitivity differences of earthworms to pesticides. In: Donker MH, Eijsackers H, Heimbach F, editors. Ecotoxicology of soil organisms. Boca Raton FL: Lewis. p 241–250.

31. Edwards PJ, Coulson JM. 1992. Choice of earthworm species for laboratory tests. In: Greig-Smith PW, Becker H, Edwards PJ, Heimbach F, editors. Ecotoxicology of earthworms. Andover, UK: Intercept. p 36–43.

32. Ma W, Bodt J. 1993. Differences in toxicity of the insecticide chlorpyrifos to six species of earthworms (oligochaeta, lumbricidae) in standardized soil tests. *Bull Environ Contam Toxicol* 50:864–870.

33. Neuhauser EF, Loehr RC, Malecki MR, Milligan DL, Durkin PR. 1985. The toxicity of selected organic chemicals to the earthworm *Eisenia fetida. J Environ Qual* 14:281–296.

Chapter 21
Development and evaluation of triggers for earthworm toxicity testing with plant protection products

Ian Barber, J Bembridge, P Dohmen, P Edwards, F Heimbach, R Heusel, K Romijn, H Rufli

In regulatory ecotoxicology, a tiered (or iterative) approach is necessary to determine what amount of toxicity testing of a compound will enable an adequate hazard or risk evaluation to be performed. Such an approach allows resources to be better focused on those compounds of concern; initial short-term tests can be utilized to screen out low-risk compounds, i.e., those with a high margin of safety in terms of the calculated toxicity/exposure ratio (TER). This allows resources to be directed toward compounds with potentially higher risk (those with a low margin of safety in terms of the TER). When higher-tier (longer-term) tests are used, the nature, extent, and degree of risk can be more adequately defined.

For any tiered-testing system, the triggers for further testing must be appropriate, avoiding false-negative assessments and at the same time minimizing the number of false-positive assessments. This chapter considers the current status of earthworm toxicity testing as adopted in European Community (EC) Directive 91/414/EEC [7] and proposes revised triggers for assessing the toxicity (and so aiding risk assessment) of plant protection products to earthworms in Europe.

Background

Earthworms are 1 of the major components of soil biomass in all but the driest and coldest land areas of the world, and in a variety of habitats, they are important in the maintenance of soil structure and fertility through their involvement in the cycling of organic matter that reaches the soil surface [1]. Indeed, some of the early Hg and Cu fungicides caused significant reductions in orchard earthworm populations, and the observed increase in leaf litter on the soil surface clearly illustrated the important role of earthworms in removing surface plant material [2].

In addition to their role in the maintenance of soil structure and nutrient cycling, earthworms are also an important part of the terrestrial food web, and they can constitute a significant component of the diet of many species of birds, small mammals, reptiles, amphibians, fishes, and invertebrates [3,4].

Despite the important role that earthworms perform and their ecological relevance as representative soil macrofauna, only a handful of regulatory authorities have previously required hazard or risk assessment of plant protection products for earthworms [e.g., 5,6]. However, following the recent harmonization of the registration process for plant protection products in Europe, the evaluation of risk to earthworms has become a requirement in the assessment procedure prior to inclusion of any compound in Annex I of EC Directive 91/414/EEC [7].

Guidance for the type of testing required for active substances is given in Annex II and for products in Annex III, and the acceptability of the risk that the active substance or any particular product may pose to earthworm populations following use is given in Annex VI. Within the guidance given in Annex II and Annex III, 3 test designs have been recommended to evaluate the potential risk of substances to earthworms; these are a 14-d acute toxicity test [8,9], a sublethal chronic test [10,11], and the option for a semi-field or field test.

Since the implementation of the Directive in 1995 and its adoption by the EU member states, it has become apparent that interpretation of both the level of testing required and the way in which data are evaluated differ significantly between member states to the extent that the expert judgment of one member state may contradict that of a neighboring state. Therefore, in spring 1996, a working group of representatives from the plant protection industry with experience in earthworm toxicity testing and risk assessment constructed a framework for evaluating data requirements and appropriate trigger endpoints, which may be helpful in the hazard or risk assessment of plant protection products to earthworms.

Framework for toxicity testing

As in other areas of regulatory ecotoxicology, data generation and evaluation should follow a tiered approach so that the need for and type of further testing necessary can be evaluated with the best available information at each decision point in the testing process. Any tiered system should start with base-set data (i.e., core information before any hazard or risk assessment can be made), which is then followed, if necessary, by additional testing, usually under increasingly relevant exposure conditions. Ideally, core data should be rapidly generated and of sufficient quality to be able to identify the low hazard compounds at a very early stage of the testing process. Additional data should be necessary only where the risk is considered to be uncertain and should be designed to provide information of added value to the risk assessment process.

The risk assessment procedure should take into account the intrinsic toxicity of the substance and the exposure experienced by the organism (sometimes referred to as the toxicity/exposure ratio [TER] approach) in order to establish the risk to individual organisms. It is appropriate to consider both the intensity and the duration of exposure of organisms to the substance, since the period during which the substance exceeds sublethal or lethal concentrations will influence the ability of individuals to recover and the immigration of new individuals into the treated area.

In the framework detailed below (and summarized in Table 21-1), at each stage of the assessment procedure both the toxicity and the presence or persistence of the substance in soil are considered together for the purpose of triggering further work. For compounds that are toxic but do not persist in soil, there is sufficient evidence to indicate that such

Table 21-1   Proposed triggers for earthworm toxicity testing

Annex II - Active substances	Annex III - Products
Mortality	
Point 8.4.1	Point 10.6.1.1
A 14-d mortality (LC50) test must be performed where products containing the active substance are applied to soil such that earthworm populations may be exposed.	A 14-d mortality (LC50) test should be performed where the ratio of the 14-d LC50 to the initial soil PEC for the active substances is less than 100 unless the toxicity of the product can be adequately assessed on the basis of studies with the active substances or similar products.
Sublethal	
Point 8.4.2	Point 10.6.1.2
Tests to investigate sublethal effects should be performed where:	Tests to investigate sublethal effects should be performed where:
(i) The ratio of the 14-d LC50 to the maximum initial soil PEC for the active substance is less than 20 and the soil DT50 < 30 d OR (ii) The ratio of the 14-d LC50 to the maximum initial soil PEC is less than 100 and the soil DT50 ≥ 30 d Furthermore, the need for a sublethal test should be considered where there are indications of sublethal effects in the mortality test at soil concentrations significantly (i.e., > 10-fold) lower than the LC50 and within an order of magnitude of the initial PEC.	(i) The ratio of the 14-d LC50 to the maximum initial soil PEC for the active substance is less than 20 and the soil DT50 < 30 d OR (ii) The ratio of the 14-d LC50 to the maximum initial soil PEC is less than 100 and the soil DT50 ≥ 30 d Furthermore, the need for a sublethal test should be considered where there are indications of sublethal effects in the mortality test at soil concentrations significantly (i.e., > 10-fold) lower than the LC50 and within an order of magnitude of the initial PEC unless data generated under point 8.4.1 and 10.6.1.1 indicate that the toxicity of the product can be reliably assessed on the basis of studies conducted with the active substance or similar product.
Semi-field/Field	
	Point 10.6.1.3
Semi-field/field tests with the active substance should not be performed.	Semi-field/field studies should be performed where the ratio of the sublethal NOEC (or equivalent ECx) to the maximum initial PEC for the product is < 2.

compounds will be of negligible long-term risk to earthworm populations; equally, compounds that persist in soil but are not toxic will be of negligible risk to earthworm populations.

While intrinsic toxicity can be determined experimentally at a relatively early stage of compound development, exposure is highly dependent on the use pattern of the product and will vary between crops and countries and over time. Therefore, for the purpose of evaluating the potential risk of a compound to earthworms and the need for further work, an initial worst-case calculation of exposure should be used. This initial estimate should be based on the initial predicted environmental concentration of the compound in soil (PEC$_S$) immediately following application, assuming direct contamination of soil to a depth of 5 cm with a soil density of 1.5 g/cm^3 [12,13]. However, the PEC$_S$ should also take into account whether bare soil or crop is sprayed, and for repeat application products, the PEC$_S$ should be the concentration immediately following the final application. In the case of products applied to foliage, it can be assumed that only ½ of the applied material reaches the soil, and the test concentrations should be adjusted accordingly.

Therefore, the initial assessment of the PEC$_S$ should take into account the maximum application rate, number of applications, time between applications, ground cover, and the rate of degradation of the compound (e.g., half-life [DT50] in soil derived from laboratory studies). Where repeat applications are recommended for a product, a simple first-order degradation model should be used to take into account the rate of degradation of the compound between applications (see Figure 21-1). Figure 21-1 illustrates the influence of repeat applications and compound degradation on the maximum initial soil concentration. For a compound applied repeatedly with a DT50 of 5 d and a 14-d interval between applications, the maximum PEC$_S$ after 5 applications will be only 20% greater than the concentration following only 1 application. Therefore, it would be a gross overestimate to compare the toxicity of the compound to a cumulative PEC$_S$ based on the sum of 5 individual applications. Indeed, the same compound would need to have a relatively slow degradation rate (DT50 > 200 d) before the PEC$_S$ would approach the equivalent of 5 cumulative applications. Occasionally, in the case of very slowly degraded compounds in permanent crops (e.g., orchards), it may be necessary to consider soil accumulation over more than 1 season.

Tier I test: acute toxicity

Tier I of the assessment involves measurement of the intrinsic acute toxicity of the active substance or formulated product to a representative test species. For this, the acute (14-d) toxicity test of the compost worm *Eisenia fetida* is a suitable test [9]. Because the test method using an artificial soil has been an acceptable standardized test for the last 12 y, a large comparative database is available within the agrochemical industry, and it has been utilized previously to assess the success of this test in predicting risk to earthworm populations under field conditions.

(a)

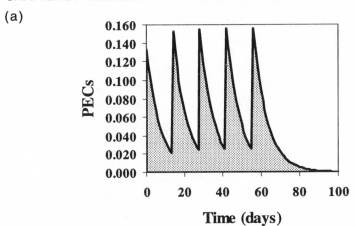

(b)

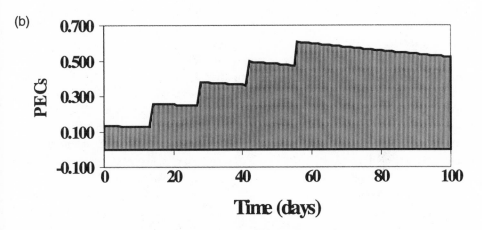

Figure 21-1   Model calculations of PEC$_s$ for repeat application products with different degradation rates (DT50) and timing between applications. The model assumes first-order degradation kinetics. (a) Cumulative PEC$_s$ for a 100 g/ha, 5-application product assuming a DT50 of 5 d and a minimum application interval of 14 d, with no crop interception. The initial PEC$_s$ after 1 application is 0.13 mg/kg soil, while the maximum PEC$_s$ is 0.16 mg/kg soil (i.e., < 20 % greater than the initial concentration following 1 application). (b) Cumulative PEC$_s$ for a 100g/ha 5-application product assuming a DT50 of 200 d and a minimum application interval of 14 d, with no crop interception. The initial PEC$_s$ after 1 application is 0.13 mg/kg soil, while the maximum PEC$_s$ is 0.61 mg/kg soil (i.e., almost equivalent to the sum of 5 applications).

The purpose of the acute toxicity test is to determine the LC50 value (i.e., the concentration that is lethal to 50% of the exposed population) of the test material (active substance or product) to earthworms and, where possible, to determine the highest concentration causing no mortality and the lowest concentration causing 100% mortality. The test should also include observation for effects on gross morphology and behavior of the test organisms. However, the aim of this study should not be to determine a no-observed-effect concentration (NOEC) and lowest-observed-effect concentration (LOEC); such studies in which these parameters are not obtained can still be used for hazard or risk assessment purposes.

For active substances, an acute toxicity test should be performed in which products containing the active substance are applied to soil such that earthworm populations may be exposed. This will apply to most agricultural applications, although there will be some exceptions, e.g., products with limited use for indoor/glasshouse applications, topical plant treatments, etc. (Table 21-1).

Experience with active substances and formulation inerts indicates that the active substance will in most cases be the most significant contributor to the toxicity of any product. Therefore, for products, an acute toxicity test should be necessary only when the active substance may be considered a possible risk to earthworms or when the toxicity of the product cannot be adequately predicted on the basis of data generated for the active substance. When the acute toxicity (i.e., LC50) of the active substance is greater than 1000 mg/kg soil, it is reasonable to assume that any modern plant protection product is unlikely to present a significant risk to earthworms [14]. Furthermore, taking into account the application rate of the substance, if the ratio $LC50/PEC_s$ (i.e., acute toxicity exposure ratio [TER_a]) is greater than 100, the substance is unlikely to present any risk to earthworm populations, and any further testing with the active substance or product should be unnecessary [13]. Therefore, an acute test with the product should be necessary only where the ratio of the 14-d LC50 to the PEC_s for the active substances is less than 100, unless the risk of the product can be assessed on the basis of studies with the active substances or similar products. This may include situations in which the product contains several active substances or contains inerts that may influence the toxicity or availability of the product to earthworms (Table 21-1).

The acute toxicity of a substance to *Eisenia fetida* should always be determined using the artificial soil test according to Organization for Economic Cooperation and Development (OECD) guidelines [9]. However, under certain circumstances, it may be appropriate to modify the test design to evaluate toxicity to other earthworm species or to test using a natural soil (e.g., in cases where degradation of the compound may be significantly affected by soil parameters) before considering whether a Tier II (chronic-reproduction study) is appropriate.

Tier II test: sublethal effects

Longer-term studies (Tiers II and III) should be triggered only for those compounds of moderate and high risk to earthworms, i.e., those compounds that are highly toxic to earthworms although nonpersistent or those that are moderately or highly toxic and persistent.

Annex II of the EU Directive 91/414/EEC states that for active substances "Where on the basis of the proposed manner of use of products containing the active substance or on the basis of its fate and behavior in soil (DT90 > 100 d), continued or repeated exposure of earthworms to the active substance, or to significant quantities of metabolites, degradation, or reaction products can be anticipated, *expert judgement is required to decide whether a sublethal test can be useful.*" This has been interpreted by some authorities to mean that such tests should be performed for all plant protection products that are considered to be persistent (i.e., DT90 in soil > 100 d) or that may be applied on more than 3 occasions per year. However, this does not take into consideration the intrinsic toxicity of the compound and as a consequence has resulted in studies being triggered for compounds that may be considered toxicologically inert. Therefore, longer-term sublethal studies should be triggered only after both toxicity and exposure to the active substance are taken into account, e.g., using the worst-case TER_a.

Where the LC50 of the active substance or product is greater than 1000 mg/kg soil or the TER_a is greater than 100, it is reasonable to assume that compound or product is unlikely to present any significant risk to earthworms [13]. Indeed, data published by Heimbach [15] for a selection of compounds have indicated that where the TER_a was greater than 20, the compound was unlikely to have any significant impact on earthworm populations in the field, irrespective of persistence. Only when the TER_a is less than 8 is a reduction of earthworm field populations likely to occur, although long-term reductions in earthworm populations were observed only where the TER_a was less than 2. However, in addition to the acute TER_a, any sublethal effects observed in the acute toxicity test (e.g., change in weight of surviving worms, gross morphology, or behavior) should also be considered and may be sufficient to trigger longer-term studies. A recent review of the relationship between lethal (LC50) and sublethal (LOEC) effects for a random selection of 44 plant protection products indicated that, for some compounds, sublethal effects may be observed at exposure concentrations several orders of magnitude lower than the LC50 (Table 21-2). The data review indicated that for a small number of insecticides (2 carbamates, 2 organophosphates, and 1 organochlorine), consideration of sublethal effect endpoints in the acute test may be of value for identifying compounds with a very shallow dose-response relationship. Finally, it is important that a distinction is made between nonpersistent (i.e., DT50 < 30 d) and persistent compounds, since the overall assessment

Table 21-2 Ratio of LC50/LOEC for 44 plant protection products

Substance	n =	Ratio LC50:LOEC	
		Minimum	Maximum
Insecticide	19	1.8	216
Fungicide	8	1.3	7.7
Herbicide	15	1.2	4.5
Other	2	1.2	6.9

of acceptability of effect on natural earthworm populations assumes that a reduction in juvenile abundance of less than 20 to 30% after 1 y is acceptable [16].

Taking into account the toxicity of the compound, the initial PEC_s, and the persistence of the compound, tests to investigate sublethal effects should be performed only when the TER_a is less than 20 for compounds that are not persistent. However, because the database published by Heimbach was for a limited number of compounds, an additional margin of safety is proposed for persistent compounds until further data are available. Therefore, for more persistent (e.g., aerobic soil DT50 > 30 d) compounds, a safety margin equivalent to a TER_a of 100 is proposed. Sublethal tests should also be considered where there are indications of sublethal effects at soil concentrations significantly (i.e., greater than 10-fold) lower than the LC50 (i.e., where there is a flat dose-response relationship) and within an order of magnitude of the initial PEC. The same triggers should also apply for products unless mortality data generated for the active substance or product indicate that the risk can be reliably assessed without further study (Table 21-1).

At present, there is limited experience with testing the sublethal effects of plant protection products on *Eisenia fetida*. However, there is no doubt that sublethal tests can be of value for evaluating the long-term effects of compounds and can provide an intermediate step before resources are committed to field testing. Therefore, it is recommended that for the moment a sublethal test be carried out in accordance with Biologische Bundesanstalt Für Land-Und Forstwirtschaft (BBA) Guideline VI, 2-2 [11], using formulated test material applied to the soil surface. However, for repeat application products, the exposure concentrations used in a study should be selected based on the maximum initial PEC_s and not simply as multiples of the single application rate, for the reasons previously outlined (and illustrated in Figure 21-1). Furthermore, the value of the test should be reviewed when further data are available to establish whether effects observed in the laboratory can be demonstrated to be predictive of the risk of compounds to earthworm populations in the field ([17]; Heimbach, Chapter 19).

Tier III: semi-field/field effects

There is little information available on the relationship between the sublethal NOEC in the laboratory sublethal test and effects on field populations. However, data for a limited number of compounds indicate that the sublethal test is approximately 5 × more sensitive than field tests (Heimbach, Chapter 19), such that where the ratio of the sublethal $NOEC/PEC_s$ (TER_{lt}) is greater than 0.2, products are unlikely to have any significant impact on earthworm populations in the field. Therefore, by incorporating an additional safety margin of 10, semi-field or field studies should be necessary only for those products in which the TER_{lt} is less than 2.

Thus, where the TER_{lt} indicates that the active substance or a particular product may pose an unacceptable risk to earthworm populations, semi-field or field studies should be performed to evaluate effects on earthworm populations and communities under natural environmental conditions (Table 21-1).

However, since there are no internationally accepted guidelines available for semi-field or field testing, it is recommended that such tests be carried out in accordance with the principles outlined in BBA Guideline VI, 2-3.

Summary

This chapter presents a framework that incorporates the concept of a tiered-testing strategy using toxicity and exposure (through the use of TERs) to determine the need for further testing in order to be able to perform a risk assessment. The framework is very similar to that previously proposed by the European Plant Protection Organization (EPPO) [13], with some slight modifications to the triggers based upon more recent data comparing the relationship between laboratory acute and sublethal data, laboratory acute and field data, and laboratory sublethal and field data. Using this framework, resources can be focused on those compounds of concern, while undue risk of false-negative assessments can be avoided.

References

1. Edwards CA, Lofty JR. 1977. Biology of earthworms. London UK: Chapman & Hall.

2. Hassall KA. 1992. Biochemistry and use of pesticides. London UK: Macmillan.

3. Cooke AS, Greig-Smith PW, Jones SA. 1992. Consequences for vertebrate wildlife of toxic residues in earthworm prey. In: Greig-Smith PW, Becker H, Edwards PJ, Heimbach F, editors. Ecotoxicology of earthworms. Hants UK: Intercept. p 139–155.

4. Lee KE. 1985. Earthworms: their ecology and relationship with soils and land use. London UK: Academic.

5. MAFF. 1986. Data requirements for approval under the Control of Pesticides Regulations 1986. London UK: Her Majesties Stationary Office.

6. Bode E, Brasse D, Kokta C. 1988. Auswerkungen von Pflanzenschutzmittel auf Nutzorganismen und Bodenfauna: Uberlegungen im Zulassungsverfahren. Gesunde Planzen, 40, p 239–244.

7. European Community. 1991. Council Directive 91/414/EEC concerning the placing of plant protection products on the market. Official Journal of the European Communities No L 230, 19, European Commission, p 8.

8. [EEC] European Economic Community. 1985. Directive 79/831 Annex V, Part C: Methods for the determination of ecotoxicity – level 1, Earthworms—artificial soil test. Commission of the European Communities, European Commission, DG XI/128/82.

9. [OECD] Organization for Economic Cooperation and Development. 1984. Organization for Economic Cooperation and Development guidelines for testing of chemicals: earthworm acute toxicity test. OECDGuideline No. 207. Paris, France: OECD.

10. Van Gestel CAM, Van Dis WA, Van Breemen EM, Sparenburg PM. 1989. Development of a standardized reproduction toxicity test with the earthworm species *Eisenia fetida andrei* using copper, pentachlorophenol, and 2,4-dichloroaniline. *Ecotox Environ Safety* 18:305–312.

11. Kula C. 1994. Auswirkungen von Pflanzenschutzmitteln auf die Reproduktion und das Wachstum von Eisenia fetida/Eisenia andrei - Biologische Bundesanstalt, Richtlinien fur die Prufung von Pflanzenschutzmitteln Tiel VI, 2-2.

12. Beyer WN. 1992. Relating results from earthworm toxicity tests to agricultural soils. In: Greig-Smith PW, Becker H, Edwards PJ, Heimbach F, editors. Ecotoxicology of earthworms. HantsUK: Intercept. p 109–115.

13. [EPPO] European Plant Protection Organization. 1993. Decision making scheme for the environmental risk assessment of plant protection products. Chapter 8, Earthworms. *OEPP/EPPO Bulletin* 23:131–149.

14. Kokta C. 1992. Measuring effects of chemicals in the laboratory: effect criteria and endpoints. In: Greig-Smith PW, Becker H, Edwards PJ, Heimbach F, editors. Ecotoxicology of earthworms. Hants UK: Intercept. p 55–62.

15. Heimbach F. 1992. Effects of pesticides on earthworm populations: comparison of results from laboratory and field tests. In: Greig-Smith PW, Becker H, Edwards PJ, Heimbach F, editors. Ecotoxicology of earthworms. Hants UK: Intercept. p 100–106.

16. Lofs A. 1992. Measuring effects of pesticides on earthworms in the field: effect criteria and endpoints. In: Greig-Smith PW, Becker H, Edwards PJ, Heimbach F, editors. Ecotoxicology of earthworms. Hants UK: Intercept. p 85–89.

17. Reinecke AJ. 1992. A review of ecotoxicological test methods using earthworms. In: Greig-Smith PW, Becker H, Edwards PJ, Heimbach F, editors. Ecotoxicology of earthworms. Hants UK: Intercept. p 7–19.

Modeling earthworm populations: an approach to field ecotoxicology

Chapter 22

Simulation of pesticide effects on populations of 3 species of earthworms

Jørgen Aagaard Axelsen and Martin Holmstrup

The validity of recommendation #19 from the First International Workshop on Earthworm Ecotoxicology, "To allow consistent interpretation until the significance of population level effects is better understood, values of 30% to 50% reduction in population density, and failure to recover to untreated control levels within 12 months can be used as criteria for defining a serious effect in field tests," [1] was investigated by aid of a simulation model.

This model, a stage-structured, energetically driven, metabolic-pool population-dynamic model, is driven by soil temperature and population food demand. The model simulates the population dynamics of 3 species of earthworms, *Lumbricus terrestris* (Ude), *Aporrectodea longa* (L.), and *A. rosea* (Sav.), which were assumed to compete for the same food source. The population growth is dependent on the acquired food, and part of the population is simulated to die if the acquired food is not sufficient to satisfy maintenance requirements. Values for most parameters were found in the literature, but a few were estimated by fitting the model to data from a field survey.

The model was used to simulate a population reduction in all 3 species of 25%, 50%, and 75% in spring and to see how this simulation influenced the population size the following autumn. Soil temperatures (5-cm depth) from a meteorological station near the field site were used as input in the simulations.

The effect of a spring density reduction on a species' autumn density was generally dependent on the relative amount of available food in the field, but the response differed between life stages. When the earthworm population was food limited before the spring reduction, it was able to recover from the reduction and even, in some cases, to overcompensate before the autumn timeframe. The reason for the recovery and overcompensation was the relief from competition. The same simulations did not show recovery for adults and cocoons. When the endpoint "total biomass of a species" was evaluated, an almost complete recovery was found for all 3 species after reductions of 25 and 50%.

The important role of earthworms in the functioning of soil ecosystems is well established. Therefore laboratory protocols for assessing toxicity (acute and chronic) of chemicals to earthworms have been developed [2,3] and are widely used in the ecotoxicological screening of new chemicals. These standardized tests are carried out with the species *Eisenia fetida* in a standardized soil at constant temperature and humidity. Extrapolation of the results from these tests to a field situation, where earthworms live under fluctuating temperature and humidity regimes, may therefore not always give reliable estimates of effects in the field. The available test methods may give an indication of the inherent toxicity of substances, but the ecological consequences of the mortality inflicted by a given toxicant are not clearly identified by such test results. In the recommendations from the First International Workshop on Earthworm Ecotoxicology [1] it is suggested that, until the significance of population-level effects is better understood, populations reduction of 30 to 50% and failure to recover within 12 months should be defined as serious effects of a toxicant [1]. The recommendations from the workshop also encourage development of models that can be used to define the scope of effects that can be tolerated by earthworm populations [1].

Models that can calculate population effects of toxicants and pesticides have been published by Klok and de Roos [4] and Baveco and de Roos [5]. These models adopt a mechanistic approach based on energetics and size-dependent growth, including the effect of toxicants on growth and reproduction rates. The models are useful to calculate effects on populations and the recovery time of the population of 1 species at constant temperature, i.e., the temperature where the laboratory input parameters have been measured. The simulation results are not directly transferable to field conditions where temperatures fluctuate and available food resources influence the development of a population.

In this chapter, a model is presented that is capable of estimating both the long-term effect of a reduction in population size at a given time of year and the recovery time for populations of 3 earthworm species under natural climatic conditions. In the present study, the model was calibrated with data from a survey of the earthworm population in a Danish grass field. The model also takes the possibility of intra- and interspecific competition for food into consideration, which makes it possible to run scenarios with various levels of available food resources. To as large an extent as possible, the model is founded on biological input data from the literature, but a few parameters were estimated during the process of fitting simulation output to field data.

Material and methods

The simulation program was constructed to simulate the temperature-dependent population development from spring to autumn of the 3 species *Aporrectodea longa* (Ude), *Lumbricus terrestris* (L.), and *A. rosea* (Sav.) in a Danish grass field (Figure 22-1). In the model, it was assumed that the 3 species competed for the same food source and that the consumed food was replaced daily. This assumption is based on the fact that they all ingest large amounts of soil and must feed on organic matter (OM) therein. Both *A. longa* and *L. terrestris* are known to feed on surface litter, but in the present field study, surface litter was scarce because the field was used for lea.

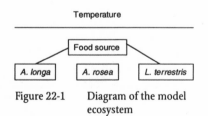

Figure 22-1   Diagram of the model ecosystem

Modeling approach

The model is a 2-trophic, metabolic-pool population-dynamic model [6] that simulates the available food source and the populations of the species based on information on the average animal. The model is stage structured and tracks both numbers and dry weight of cocoons, juveniles, subadults, and adults by means of a series of time-invariant distributed delays [8]. When distributed delay is used in this model, each life stage is split into a number of substages (k), and the dynamics of the system are described by

$$\frac{dQ_1(t)}{dt} = x(t) - r_1(t) - \mu_1 Q_1$$

$$\frac{dQ_2(t)}{dt} = r_1(t) - r_2(t) - \mu_2 Q_2$$

$$\frac{dQ_{k-1}(t)}{dt} = r_{k-1}(t) - y(t) - \mu_k Q_k$$

Equation 22-1,

where Q_i is the quantity (mass or number) in substage i = 1,2,3 ... k, t is the time, $x(t)$ is the input to the life stage, $y(t)$ is the output from the life stage, and r_i is the flow from substage i. Attrition (growth and mortality) is performed within the delays [9] by adding the term $\mu_i Q_i$. The flow through the system is described by

$$r_i(t) = \frac{k}{m} Q_i(t)$$

Equation 22-2,

where m is the average duration of the life stage. The delay process adds variation to the transition time $f(t)$, which is described by the Erlang density function

$$f(t) = \frac{(k/m)^k t^{k-1} \exp(-kt/m)}{(k-1)!}$$

Equation 22-3.

In order to incorporate the effect of temperature on developmental time, the time scale used in the model was a physiological one. Time was transformed into a physiological time scale by aid of hourly temperature measurements, and all rates in the model are related to this time scale. The physiological time scale used in this model was suggested by Taylor [10] and converts temperature into developmental units (percent development) by a Gauss distribution

$$R(T) = R_m \exp\left[-\frac{1}{2}\left(\frac{T_m - T}{T_\sigma}\right)^2\right]$$

Equation 22-4,

where $R(T)$ is the temperature-dependent developmental rate, R_m is the maximum developmental rate, T_m is the optimal temperature, T is the temperature, and T_σ is the variance of the distribution. This physiological time model, as opposed to the most commonly used degree-day model, takes into account that the worms also develop at the low temperatures prevailing in early spring and late fall. The parameters of Equation 22-4 for cocoon development of all 3 species were found by fitting the Gaussian curve to previously published data [11,12]. The fitting process was carried out by aid of graphical illustrations of calculations in a spreadsheet, and the fit to data was evaluated by eye. The T_m and T_σ values from cocoons were assumed also to apply for juveniles, subadults, and adults. This assumption is supported by the results of Tsukamoto and Watanabe [13], who investigated the developmental rate of *Eisenia fetida*. A plot of their measurements of daily developmental rates of cocoons and juveniles (based on the time it took to reach a bodyweight of 100 mg) were found to be almost parallel. The R_m values for juveniles, subadults, and adults of the *A. longa*, *L. terrestris*, and *A. rosea* were found by inserting developmental time measurements from 1 temperature in Equation 22-3 and solving the equation with respect to R_m. The reproduction and mortality rates were related to the physiological time scale, i.e., expressed in units of cocoons and mortality per percent development.

The driving force of the model is the temperature dependent food demand (D) necessary for respiration (D_{resp}), excretion (ß), growth (D_G), reproduction (D_R), and reserves (D_S):

$$D = D_{resp} + D_G + D_R + D_S + \beta \qquad \text{Equation 22-5.}$$

The calculations of the demands for growth and reproduction are based on rate data from investigations where the animals presumably had optimal conditions. The rates may therefore be regarded as maximum rates. The calculations of the demand for growth were based on the growth rate per physiological time unit (b), which was found by solving Equation 22-6 with respect to b.

$$W_e = W_i (1 + b)^{100} \qquad \text{Equation 22-6,}$$

where W_e is the weight by the end of the stage, W_i is the initial weight of the stage, and b is the growth rate. The value 100 is the number of developmental time units (100% development) required to complete the stage. The growth rate for adults was reduced as they approached the maximum weight of the species:

$$b_{adult} = b \left(1 - \frac{\overline{W}}{W_{max}} \right) \qquad \text{Equation 22-7,}$$

where b_{adult} is the growth rate of the adult stage, W is the simulated average dry weight of the stage, and W_{max} is the maximum dry weight of the species.

The excretion rates per unit dry weight were assumed to be equal for the 3 species. The excretion rates were given a random value because the food value of the unknown food source is unknown, i.e., the amounts of food given to the 3 earthworm populations in the model can be regarded as net amounts. The calculation of the respiration demand was based on the temperature-dependent respiration rate ($z(t)$) (h^{-1}), which was calculated according to

$$\text{Equation 22-8,}$$

$$z(t) = z_0 2^{0.1\bar{t}}$$

where z_o is the basic respiration rate and was estimated from the data of Phillipson and Bolton [14] on *A. rosea*. These authors did not find any difference in the respiration of *A. rosea*, *A. caliginosa*, *L. rubellus*, and *A. chlorotica*. Therefore, we assume that the value for *A. rosea* is also representative for *A. longa* and *L. terrestris*.

The demand for reproduction was based on the reproductive output (in numbers) per physiological time unit multiplied by the average weight of a cocoon, and the demand for reserves was calculated by

$$D_s = S_{max} - S$$

Equation 22-9,

where S_{max} is the maximum amount of stored mass, which for all species was assumed to be 10% of the body weight, and S is the actual reserve.

The acquisition of food is made demand dependent and is described by the Gutierrez-Baumgärtner functional response model [7]:

$$M_{(x,y)}* = D_{(x,y)}\left[1-\exp\left(\frac{-s_{(x,y)}qM}{D_{(x,y)}}\right)\right] \qquad 0 \le q \le 1$$

Equation 22-10,

where $M_{(x,y)}*$ is the amount of food acquired by species x stage y, $D_{(x,y)}$ is the demand of species x stage y, $s_{(x,y)}$ is the constant search rate of species x stage y, M is the total amount of food in the system, and q is a factor that takes care of not placing more food at the earthworms' disposal than is available in the system. The value of q was calculated by

$$q = \frac{\min(M,D_{(x,y)})}{\sum_{x=1}^{N}\sum_{y=2}^{K}\min(M,D_{(x,y)})}$$

Equation 22-11,

where N is the number of species, K is the number of life stages, and min(a,b) is a function returning the least of a and b. The summation does not include y = 1 because the cocoon stage does not require food.

The allocation of resources for the different physiological needs follows this priority: excretion, respiration, growth, reproduction, and reserves. This means, in case of shortage of food, the reproduction ceases before the growth, and in case of severe shortage of food, there may not be resources for the respiration. In the model, the latter causes a fraction (*d*) of the population to die, which is described by

$$d = 1 - \left(\frac{M^*}{D}\right)$$

Equation 22-12.

Simulations

The population effect of an acutely toxic pesticide was simulated by reducing the density of juveniles, subadults, and adults found in the field on 6 May by 25, 50, or 75% and then let the model simulate the population development until 24 November. These scenarios were run at different food abundances. The food abundances were expressed in percentage of the optimal food abundance, which is the level at which the growth and reproduction of all 3 species are not clearly food limited in the period from May to November. In the simulations, cocoon mortality as a result of pesticide was set to 0 because cocoons were assumed not to take up the pesticide. In order to concentrate on the effect of the population reduction in spring, sublethal effects were ignored.

Survey of field populations

The field survey was performed on a second-year grass field on a sandy loam soil belonging to The Danish Institute of Agricultural Sciences, Research Centre Foulum, Jutland. The soil consisted of 33.1% coarse sand (> 200 m), 33.3% fine sand (63 to 200 m), 8.5% coarse silt (20 to 63 m), 12.2% silt (2 to 20 m), 8.5% clay (< 2 m), 4.4% humus, and it had a pH-H_2O of 7.3. The earthworm population was monitored in four 15×15 m^2 plots randomly distributed in an area of approximately 2 ha. At 7 sampling occasions from 6 May to 5 November 1996, the densities of cocoons and worms were estimated. At the first 3 sampling occasions, two $25 \times 25 \times 25$ cm^3 soil samples from each plot were hand sorted by washing and sieving the soil through a series of 4 box sieves with decreasing mesh size (10, 4, 2, and 1 mm). At the remaining 4 samplings, only 1 soil core from each plot was taken. Worms and cocoons were classified according to species and stage (i.e., cocoon, juvenile, subadult, or adult) in order to obtain estimates of density m^{-2}.

Hourly soil temperature measurements in 5-cm depth under grass cover and daily precipitation was obtained from a meteorological station about 1 km away from the study site. Only temperature data were used directly in the simulation model. Details of the climatic conditions at the study site can be found in Holmstrup et al. (Chapter 27).

Results and discussion

The dominant species in the grass field were *A. longa*, *A. rosea*, and *L. terrestris*. The density of earthworms was about 175 specimens m^{-2} at the time when the survey was initiated and increased to about 220 m^{-2} in the autumn. In the middle of the summer, the adults of all 3 species almost disappeared (Figure 22-2), which at least for *L. terrestris* and *A. longa* must be explained by a migration to deeper soil layers (beneath sampling depth) due to dry conditions [15]. For *A. longa*, this must have been the case because the steep increase in density of adults between late August and late September could not have originated from the very low number of subadults found in the field in summer. Furthermore, very few adult and subadult *L. terrestris* were found throughout the sampling period, which could not have reflected the real population, since there were rather large numbers of both cocoons and juveniles. This is probably due to the behavior of large *L. terrestris* and to the fact that the sampling depth used in the present study was too shallow to capture this species effectively [15].

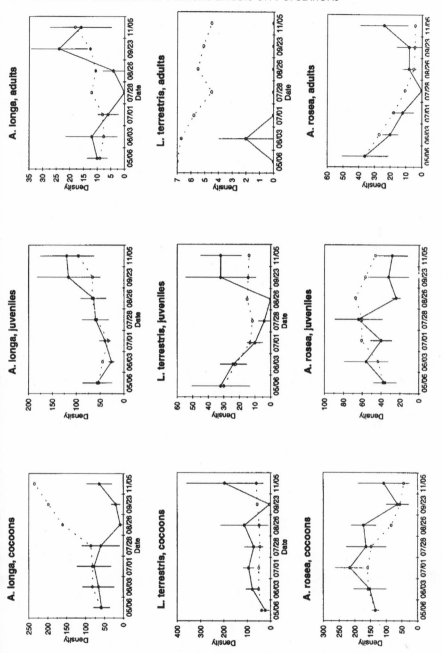

Figure 22-2   Comparison between field data and simulations of densities of cocoons, juveniles, subadults, and adults of the 3 earthworm species *L. terrestris*, *A. longa*, and *A. rosea* in a Danish grass field (May to November 1996). Closed circles are the observed densities (± SEM; N = 4), and open circles are the simulated densities.

There are some strange patterns in the observed field density (Figure 22-2) of *A. longa* cocoons and juveniles. The increase in the density of juveniles between late August and late September could not have occurred with almost no cocoons present in the field in August. One explanation may be that adult *A. longa* had a deeper distribution in summer than in spring and had deposited cocoons in soil layers deeper than the sampling depth. Gerard [15] reports that about 20% of *A. longa* cocoons in pastures are found in 22.5 to 30 cm depth. This makes it difficult to simulate field data. Another possible reason for discrepancies between field data and simulation is that simulation of the seasonal development in available food resources is not attempted. Instead, the same amount is available every day, which is most likely far from the real field situation in which OM from the previous growth season decays through the year, and new food sources (dead roots, fresh OM) are continually produced at seasonally varying rates. However, other unknown circumstances may also contribute to the inconsistencies between simulated and observed data.

In order to estimate unknown parameters (relative food abundance, search rates, and mortality rates), the model was fitted to the field data, ignoring the summer disappearance of *A. longa* adults and the very low abundance of *L. terrestris* adults and subadults. The model fitting was an iterative process, with the parameters of food abundance, search rates and stage mortality rates (given as survival over the average duration of the stages in Table 22-1). The mortality rates describe the parts of the populations that died for unknown reasons. Lack of food could also cause mortality, but a food shortage first affects reproduction; only in cases of severe starvation does it cause mortality. The search rate describes the efficiency of food acquisition. The best obtainable fit to field data (Figure 22-1) was found at a food abundance of 56% of the optimal level, and the survivals during the average time of a stage are presented in Table 22-1. These survival rates are maximum estimates, since the reduced survival from starvation that occurs when the total earthworm density is high is not included in the figures. At the food abundance level 56%, all 3 species are food limited. Other modeled parameters are documented in Table 22-2.

Table 22-1   Estimated (from the model-fitting process) survival (%) over the average duration of the 4 stages of *L. terrestris*, *A. longa*, and *A. rosea*

Species	Estimated survival (%)[1]			
	Cocoons	Juveniles	Subadults	Adults
L. terrestris	37	13	61	61
A. longa	90	78	36	8
A. rosea	90	5	90	2

[1] The survival is the part of the population surviving mortality due to reasons other than starvation and old age (for the adult stage)

The model was used to create scenarios with combinations of different population reductions and food abundance (Figure 22-3). The effect of reducing the population in spring is clearly dependent on both the life stage and the amount of food available to the earthworms. For all 3 species, the density of adults and cocoons in late November is reduced as a consequence of a population reduction in spring. For the juveniles, the impact of a

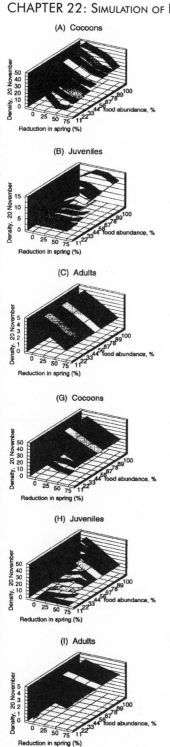

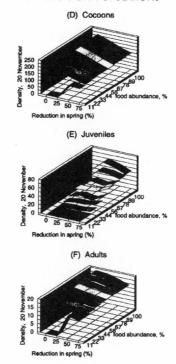

Figure 22-3   Simulated effect in late autumn of reducing the earthworm population by 25, 50, or 75% in spring under different food abundances. 100% food abundance is the level at which earthworm population development is not clearly food limited

Table 22-2   Input parameters used in the simulation model and the sources from which they have been taken or estimated

Input parameter	Model results and sources[1] by earthworm species		
	L. terrestris	*A. longa*	*A. rosea*
R_m (d[-1])			
Cocoons	1.45% [17]	2.3% [12]	3.0% [11]
Juveniles	1.2% [18]	2.4% [18]	2.1% [18]
Subadults	6.0% [EG][2]	7.5% [EG]	9.0% [EG]
Adults	0.66% [19]	0.75% [19]	1.1% [EG]
T_m	20.5 °C [18]	25.0 °C [12]	24.0 °C [11]
T_σ	8.0 [17]	9.5 [12]	8.0 [11]
Growth rate (mg/mg)	0.0461 [18,20]	0.00365 [20]	0.038 [18]
Weight of cocoon (mg)	10.0 [21]	14.0 [21]	2.7 [21]
Reproduction rate (mg/mg/DU)[3]	0.011 [20]	0.015 [20]	0.006 [18,20]
Search rate			
Juveniles	0.025 [FP][4]	0.035 [FP]	0.005 [FP]
Subadults	0.055 [FP]	0.045 [FP]	0.025 [FP]
Adults	0.085 [FP]	0.065 [FP]	0.045 [FP]
z_0	0.0085 d[-1] [14]	0.0085 d[-1] [14]	0.0085 d[-1] [14]

[1]When 2 authors are mentioned as the source of a parameter, the parameter has been estimated by aid of information from both authors.
[2]EG = Educated guess
[3]DU (developmental unit) = percent of average duration of the stage
[4]FP = fitted parameter

reduction in spring is dependent on both the severity of the reduction and the available amounts of food. All 3 species are able to compensate numerically in the juvenile stage for the spring reduction, and there is even an overcompensation at medium levels of food abundance. At high levels of food abundance, *A. rosea* cannot compensate. The mechanism behind the overcompensation after the spring reduction is explored in Figure 22-4, where the simulated population development of *A. longa* at a medium food abundance of 0 and 75% initial reduction in spring is illustrated. In the simulation with the spring population reduction, the number of adults is clearly lower throughout the summer and begins to increase in September but does not reach the level of the control simulation. Keeping the differences in adult population between the 2 cases in mind, it is somewhat surprising that the simulated density of cocoons is higher in mid summer for the spring-reduction group than for the no-spring reduction group. The reason is 2-fold. First, the reduction that is supposed to be caused by a pesticide does not reduce the initial cocoon density, and second, the adults surviving the reduction have relatively more food at their disposal. In the simulation without reduction in spring, the adults suffer such intense inter- and intraspecific competition that they cannot fulfil their food demand (Figure 22-5), which causes the cocoon production to cease. This is not the case in the simulations that include the density reduction in spring, which clearly reduces the competition. The higher cocoon production in the simulation with the reduction is reflected in the density of juveniles, which increases more in late summer than in the simulation without the spring reduction. The strong fluctuations in satiation of food demand (Figure 22-5) is due to both fluctuations in temperature, which changes demand (but not the amount of food that was simulated as a constant daily amount independent of temperature), and mortality due to severe starvation, which eases the competition for the survivors.

Instead of looking only at the numerical response to a density reduction, it makes sense to look at the total biomass of the species, i.e., the mass of cocoons, juveniles, subadults, and adults (Figure 22-6). In terms of biomass, the simulations show a remarkable capacity to compensate for the reduction in spring for all 3 species. The ability to compensate seems to be lowest for *A. rosea* at a high relative food level, where food is not a limiting factor for the population development. The relative food level does not depend on whether the habitat is a rich or a poor one; rather, it depends on the balance between the amount of food available and the density of worms.

The simulations suggest that a reduction of the population's size by 30 to 50%, previously suggested as acceptable [1], can be recovered within 1 y (at least under Danish temperature conditions) if the biomass is considered, but not if the earthworm numbers at the different stages are considered. Thus, a reduction in spring of up to 50% still has an effect on the stage structure in autumn, but not on the biomass. The biomass in autumn is mostly dependent on the food resources.

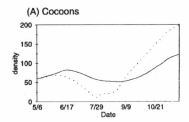

(A) Cocoons

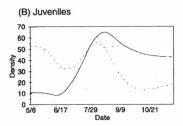

(B) Juveniles

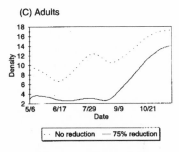

(C) Adults

| · · · No reduction —— 75% reduction |

Figure 22-4　Comparison of the population development from spring to autumn for *A. longa* at medium food abundance (60% of the level where the population development is not clearly food limited) with and without a 75% reduction of the spring population

The model presented here is founded on as much relevant biological information as possible and takes climatic data as the driving input. Therefore, it is to some extent capable of simulating real field data, although information on behavior and physiology in relation to soil humidity is still needed and should be incorporated into the model once these data become available. However, we believe that this disadvantage of the model does not affect the conclusions pointed out here because they are based on comparisons of output from the same model given different input densities of the 3 involved species.

The survival rates estimated by the model-fitting process are rather low, especially for adult *A. longa* and *A. rosea*. This may be due to the difficulty in distinguishing mortality

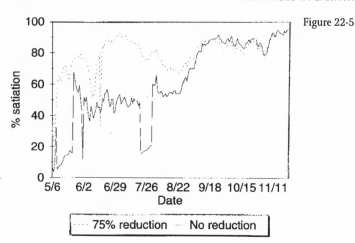

Figure 22-5   Satiation (% of food demand satisfied) of adult *A. longa* over the period from spring to autumn at medium food abundance with and without a 75% reduction of the spring population

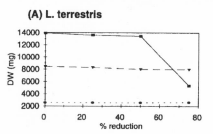

(A) L. terrestris

Figure 22-6   Simulated effect on the total species biomass (dry weight of cocoons, juveniles, subadults, and adults) in late autumn of reducing the earthworm population by 25, 50, or 75% in spring under different food abundances. 100% food abundance is the level at which earthworm population development is not clearly food limited.

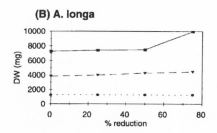

(B) A. longa

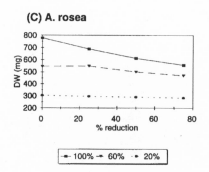

(C) A. rosea

due to unknown reasons (predation, parasitism, etc.) from death due to senescence. Maybe the input values for average duration of the adult stage are too high, thus making it necessary to reduce survival rates in the model to make the simulations fit field data. Furthermore, parts of the mortality of all stages may be caused by starvation.

Acknowledgments - The authors wish to thank Karen Kjær Jacobsen, Gorm Diernisse, John Geert Rytter, Zdenek Gavor, and Lotte Børresen for technical assistance in the earthworm sampling. This project has received financial support from the Danish Pesticide Research Program.

References

1. Greig-Smith PW. 1992. Recommendations of an international workshop on ecotoxicology of earthworms, Sheffield UK (April 1991). In: Greig-Smith PW, Becker H, Edwards PJ, Heimbach F, editors. Ecotoxicology of earthworms. Hants UK: Intercept. p 247–263.

2. [OECD] Organization for Economic Cooperation and Development. 1984. Organization for Economic Cooperation and Development guidelines for testing of chemicals: earthworm acute toxicity test. OECD Guideline No. 207. Paris, France: OECD.

3. Van Gestel CAM, Van Dis WA, Van Breemen EM, Sparenburg PM. 1989. Development of a standardized reproduction toxicity test with the earthworm species *Eisenia fetida andrei* using copper, pentachlorophenol, and 2,4-dichloroaniline. *Ecotoxicol Environ Safety* 18:305–312.

4. Klok C, de Roos AM. 1996. Population level consequences of toxicological influences on individual growth and reproduction in *Lumbricus rubellus* (Lumbricidae, Oligochaeta). *Ecotoxicol Environ Safety* 33:118–127.

5. Baveco JM, de Roos AM. 1996. Assessing the impact of pesticides on lumbricid populations: an individual-based modelling approach. *J Appl Ecol* 33:1451–1468.

6. Gutierrez AP. 1996. Applied population ecology. A supply-demand approach. New York: Wiley. 300 p.

7. Gutierrez AP, Baumgärtner JU, Summers CG. 1984. Multitrophic level models of predator-prey energetics. *Can Entomol* 116:923–963.

8. Manetsch TJ. 1976. Time varying distributed delays and their use in aggregate model of large systems. *IEEE Trans Sys Man Cybern* 6:547–553.

9. Vansickle J. 1977. Attrition in distributed delay models. *IEEE Trans Sys Man Cybern* 7:635–638.

10. Taylor F. 1981. Ecology and evolution of physiological time in insects. *Am Nat* 117:1–23.

11. Holmstrup M, Østergaard IK, Nielsen A, Hansen BT. 1991. The relationship between temperature and cocoon incubation time for some Lumbricid earthworm species. *Pedobiologia* 35:179–184.

12. Holmstrup M, Østergaard IK, Nielsen A, Hansen BT. 1996. Note on the incubation of earthworm cocoons at three constant temperatures. *Pedobiologia* 40:477–478.

13. Tsukamota J, Watanabe H. 1977. Influence of temperature on hatching and growth of *Eisenia fetida* (Oligocheata, Lumbricidae). *Pedobiologia* 17:338–342.

14. Phillipson J, Bolton PJ. 1976. The respiratory metabolism of selected Lumbricidae. *Oecologia (Berl)* 22:135–152.

15. Gerard BM. 1967. Factors affecting earthworms in pastures. *J Animal Ecol* 36:235–252.

16. Lee KE. 1985. Earthworms—their ecology and relationships with soils and land use. Sydney, Australia: Academic. 411 p.

17. Butt KR, Frederickson J, Morris RM. 1992. The intensive production of *Lumbricus terrestris* L. for soil amelioration. *Soil Biol Biochem* 24:1321–1325.

18. Lofs-Holmin A. 1983. Reproduction and growth of common arable land and pasture species of earthworms (Lumbricidae) in laboratory cultures. *Swedish J Agric Res* 13:31–37.

19. Butt KR. 1993. Reproduction and growth of three deep-burrowing earthworms (lumbricidae) in laboratory culture in order to assess production for soil restoration. *Biol Fertil Soils* 16:135–138.

20. Hansen BT, Holmstrup M, Nielsen A, Østergaard IK. 1989. Regnormestudier i Laboratoriet og i felten. Specialerapport. Aarhus, Denmark: Zoologisk Laboratorium, Aarhus Universitet.

21. Christensen O, Mather JG. 1990. Dynamics of lumbricid earthworm cocoons in relation to habitat conditions at three different arable sites. *Pedobiologia* 34:227–238.

Chapter 23

Impact of toxicants on earthworm populations: a modeling approach

Chris Klok and André M. de Roos

Structured population models provide ways to evaluate population level consequences of sublethal toxicant stress. These models formulate predictions about the population, which are firmly based on the individual biology. Species-specific, population-threshold concentrations ($[tox]_{thr}$), above which the population becomes extinct, derived from the structured population models, set the upper boundary for allowable toxicant concentrations. How long and to what depth population density might be reduced if population survival is certain cannot be answered from the perspective of the earthworm species alone. Since earthworm populations in the field are in biotic interaction with their environment, allowable toxicant levels should be based on the risk for the earthworms as well as on the secondary risk for species that are in biotic interaction with the earthworms. Furthermore, the model results indicate that earthworm populations are most sensitive to toxicants that retard maturation. Therefore laboratory tests on the toxicity of chemicals should include chronic tests for pre-adult growth.

Chemical substances are widely distributed in the environment. The treating of agricultural land with pesticides, fungicides, and insecticides and the impact of industrial activities can lead to water, soil, and air pollution. These activities may threaten to pollute soil ecosystems in both rural and natural areas. In these soil ecosystems, earthworms play an important role. They make up a large part of the total biomass, constitute 1 of the central nodes in the food web, and have a large influence on soil processes [1,2]. Earthworms are therefore often used in studies to assess the impact of chemical substances on the soil ecosystem [3–5]. Furthermore, earthworms readily accumulate various chemical substances from the soil [6–8] and make up a substantial part of the diet of over 200 vertebrate species in northwest Europe [9]. Earthworms are hence considered to play a major role in bringing chemical substances from the soil into the above-ground food chain.

Field populations of earthworms live in interaction with their biotic and abiotic environments. These interactions are dynamic in both time and space. Changes in the number of individuals and total biomass of a field population following an application of a toxicant, i.e., a chemical substance in toxic concentrations, therefore cannot be interpreted simply as effects of the substance alone, but the changes will depend on the chemical's properties (degradability, toxicity), on the life history of the earthworm species, and the on interaction of the earthworm population with the environment, both biotic (competition, predation) and abiotic (pH, temperature, etc.). This can make interpretation of the risk that toxicants pose for field populations difficult. Thus, when can we define effects of toxicants on earthworm populations as biologically significant? How much and for how long can the earthworm population be suppressed before its persistence is seriously endangered? Furthermore, considering the role earthworms play in the diet of many vertebrate species, what are the consequences for predator species, given a certain effect of the substance on the earthworm population?

In this chapter, the effects of toxicants on earthworm populations are studied. The chapter focuses on the impact of toxicants on the population while considering the degradability of the substance, the life history of the earthworm species, and the interaction of the earthworm population with other species. The dynamics in time are considered, but the dynamics in space are ignored. Furthermore, only sublethal concentrations of toxicants are reviewed. The following 4 cases are examined:

1) impact of a nondegradable toxicant on the growth potential of a population,
2) impact of a nondegradable toxicant on a population that is in biotic interaction with the environment,
3) impact of a degradable toxicant on a population that is in biotic interaction with the environment, and
4) impact on a species that feeds on earthworm populations which live in soil polluted with a nondegradable toxicant.

To investigate these 4 impacts, "structured-population" or "individual-based" models are used [10,11]. These mathematical models adopt the biological individual as the central unit in the modeling process. This philosophy readily agrees with the fact that toxicants exert their influence on individual organisms. The toxicants are assumed to change life-history parameters, e.g., the rates of individual growth, development, reproduction, and survival. Ultimately these changes in the life history of the individuals induce toxicological consequences at the level of the entire population. In both models and reality, the dynamics of the population are thus firmly based on the life history of the individuals. Detailed descriptions of the models, which deal with the impacts mentioned above, can be found in Baveco and de Roos [12], Klok and de Roos [13], and Klok et al. [14,15]. A broad overview of the results is presented here.

Methods

All models developed to investigate the effects of toxicants on earthworm populations [12–15] use the Dynamic Energy Budget (DEB) model [16] to describe processes at the individual level. The DEB model gives a mechanistic description of how energy available to the individual is allocated to individual growth, development, and reproduction (see Figure 23-1). Assimilated energy is allocated in a fixed proportion K to growth and maintenance and a proportion 1 – K to reproduction. Energy requirements for maintenance always take precedence over growth. Development is assumed to depend on size. The DEB model thus accounts for and integrates the life-history parameters of individual growth, development, and reproduction. At enhanced toxicant levels, individual growth and reproduction can be reduced. The DEB model provides a way to make explicit assumptions about the working mechanisms of the toxicants. For instance, reduction in individual growth can result from reduction in the assimilation of energy or increased maintenance requirements due to detoxification (see Figure 23-1). A decrease in assimilation of energy also directly reduces the reproduction because less energy will be available for reproduction. Both working mechanisms, decrease in assimilation and increased maintenance requirements, reduce reproduction in an indirect manner as well, since smaller individuals will reproduce less.

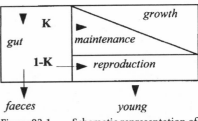

Figure 23-1 Schematic representation of energy channeling in model of individual behavior of *L. rubellus* (after [16]).

The models at the population level use the DEB model for the individual life history as a building block. Depending on the impact examined, the population dynamics are described by discrete-time matrix models [10] or continuous-time structured population models [17]. A fundamental principle of all these structured population models is that the model of the population dynamics is arrived at using bookkeeping-like operations to keep track of all individuals within the population (see Figure 23-2). For the description of the matrix model applied to evaluate the impact of a nondegradable toxicant on the growth potential of a population (impact 1, see introduction), refer to Klok and de Roos [13] and Klok et al. [14]. The description of the differential equations model that is applied to study the 3 other cases (impacts 2, 3, and 4) is given in Baveco and de Roos [12].

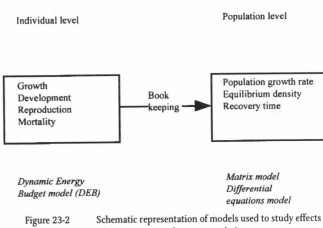

Figure 23-2 Schematic representation of models used to study effects of toxicants on earthworm populations.

Results

Impact of a nondegradable toxicant on the growth potential in *L. rubellus* populations

To investigate effects of nondegradable toxicants on earthworm populations, the impact of Cu on the growth potential of a population is determined. If it is assumed that the population lives in a constant environment (temperature, humidity, food conditions, etc.), is isolated from other earthworm populations, and has no interaction with other species, the effects of a toxicant on the population can be evaluated by the population growth rate. The population growth rate integrates the effect of the toxicant on all life-history processes during the whole life-span of the individuals constituting the popula-tion. In Klok and de Roos [13] and Klok et al. [14], a size-structured matrix model is used to assess the effects of Cu on the population growth rate in *Lumbricus rubellus*. Since de-velopment in *L. rubellus* relies more on size than on age [18–20], a model with distinct size categories was chosen. The model was parameterized with experimental data [8,19] that show reductions in individual growth and reproduction in *L. rubellus* under en-hanced Cu concentrations. In these laboratory experiments [8,19], earthworms were kept at a constant temperature (15 °C) and in abundance of food. The changes in indi-vidual growth apparent in the experimental data were interpreted as resulting from in-creased maintenance requirements for detoxification or decreased assimilation of energy (2 possible working mechanisms of the toxicant in the DEB model) [13,14], and they were captured successfully by parameter changes in the DEB model. Unfortunately, the DEB model cannot accurately capture the large reduction in reproduction observed by Ma [8,19]. These reproduction experiments, however, are not carried out with individu-als that were stressed with Cu throughout their entire lives and might hence not be appli-cable for parameterization of the model. The population growth rates under different Cu concentrations for the modeled population are presented in Figure 23-3.

Figure 23-3 indicates that the popu-lation-level consequences are simi-lar whether the individual earthworm increases its mainte-nance costs or reduces its assimila-tion to cope with enhanced Cu levels. A population exposed to 362 mg Cu kg⁻¹ soil dies out. The extinc-tion results from the severe reduc-tion in individual growth under this high Cu concentration, such that individuals do not reach adult size within their life span [13], and hence they never mature. Reduc-tion in individual growth thus turns out to have a higher impact on the population than does reduc-tion in reproduction. A sensitivity analysis demonstrated furthermore that the population growth rate is most sensitive to changes in growth

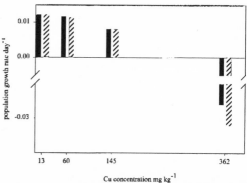

Figure 23-3   Effect of Cu on population growth rate per d in *L. rubellus*, under optimal conditions in loamy, sandy soil, for 3 Cu concentrations and a background concentration of 13 mg kg⁻¹ soil under 2 working mechanisms of the toxicant. Black bar, increased maintenance; stripped bar, decreased assimilation (after [13]).

during the pre-adult stages, while cocoon production is far less important [14]. There-fore, laboratory tests of the toxicity of chemical substances to earthworms should include chronic tests for pre-adult growth. The size-structured model predicts extinction of the population at Cu concentrations around 300 mg Cu kg⁻¹, as its growth rate drops to 0

[14]. This result is in reasonable agreement with empirical field data that show a steep decline in the size of field populations when Cu concentrations rise to 200 mg kg^{-1} [20]. The threshold value of 300 mg Cu kg^{-1} soil is determined for an isolated *L. rubellus* population, which apart from the Cu stress, lives under optimal conditions, i.e., in abundance of food and absence of predation. Under field conditions, *L. rubellus* face a combination of stress factors, and hence population extinction in the field is likely to occur at lower Cu concentrations.

Impact of a nondegradable toxicant on a population that is in biotic interaction with the environment

When the population is considered to be in biotic interaction with its environment and is regulated by this interaction, the equilibrium density is a good estimator for the impact of nondegradable toxicants on the population. In Baveco and de Roos [12], a partial differential equation model was used to assess the effects of an arbitrary nondegradable toxicant on the equilibrium density in *L. rubellus* and *L. terrestris* populations. In this model, earthworm populations are considered to be regulated by predation. Development in *Lumbricidae* relies more on size than on age [18–20]. Therefore the partial differential equation model follows the size distribution of the population in time. This size distribution can change due to changes in individual growth, reproduction (both toxicant induced), and mortality (natural or predator induced) of the individuals. Although food regulation seems to be more important than regulation by predation [21], food regulation is ignored because quantitative data on the relation between food and individual growth, development, and reproduction are virtually absent in the literature. The modeled populations are assumed to be kept in equilibrium at a density of 100 individuals m^{-2} by predation under nonstressed conditions [12], which is low for *L. rubellus* [22–24] and rather high for *L. terrestris* [21].

Toxicants can, for example, 1) decrease individual assimilation, 2) increase metabolic demands, and 3) increase the energetic costs of cocoon production. These physiological effects will reduce individual growth and reproduction and hence have an impact on the population. Figure 23-4 depicts the effect of relative change in these physiological parameters, caused by increase in toxicant concentration, on the equilibrium density of a modeled *L. terrestris* population. The model results (Figure 23-4) show evident extinction (population density = 0) of the *L. terrestris* population at toxicant levels that either increase the maintenance costs by more than 20% or decrease the assimilation of energy by a similar amount. If the toxicant increases the costs for cocoon production, the population can cope with much higher toxicant levels. Both increase in maintenance costs and decrease in the assimilation of energy lead to a decline in individual growth, and thus retard the maturation of the individual. For *L. rubellus*, similar results were obtained (not shown). These results are consistent with those in Klok and de Roos [13] and Klok et al. [14]. Table 23-1 indicates that *L. terrestris* is more sensitive than *L. rubellus* to toxicants that reduce individual growth and thus retard the maturation. While *L. rubellus* populations can cope with toxicants up to levels at which maintenance or assimilation is changed by 30%, *L. terrestris* dies out if this level is above 20%. This difference in sensitiv-

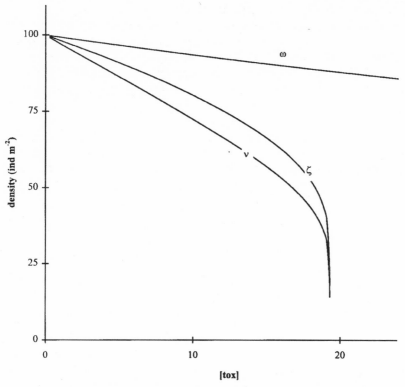

Figure 23-4   Effect of relative increase in an arbitrary toxicant concentration [tox] on equilibrium
               density of a *L. terrestris* population. With a background toxicant concentration ([tox] =
               0), the population is held at an equilibrium density of 100 individuals m^{-2}. υ, toxicant
               increases the reproduction costs; ζ, toxicant increases the maintenance costs; ω,
               toxicant decreases the assimilation of energy (after [12]).

ity results from a difference in maturation time of the species [12]. The maturation time
is about 190 d for *L. terrestris* and 125 d for *L. rubellus* under optimal conditions. *L
terrestris* and *L. rubellus* have comparable population growth rates: 0.011 and 0.014 d^{-1}
(Table 23-2), respectively. Under the assumption that the populations are predation controlled, it necessarily follows that this slightly positive population growth rate is balanced
by an identical death rate due to predation. Both species hence suffer similar predation
rates. The expected lifespan of the species is the reciprocal of this predation rate, which,
therefore, is also comparable for both species. With comparable expected lifespans, *L.
rubellus*, with its shorter maturation time, can cope with higher concentrations of toxicant that retard maturation than can *L. terrestris*.

Impact of a degradable toxicant on a population that is in biotic interaction with the environment

Recovery time is a useful indicator for the impact of a degradable toxicant on earthworm populations that are in biotic interaction with their environment. The effects of degradable toxicants that increase the maintenance costs on the dynamics of *L. terrestris* populations are investigated in Baveco and de Roos [12]. As in the case of nondegradable toxicants (impact 2), the population is assumed to be held at an equilibrium density of 100 individuals m^{-2} by predation under nonstressed conditions. The model results of a single toxicant application on the recovery of a *L. terrestris* population are depicted in Figure 23-5. Initially the toxicant increases the maintenance costs of the individuals by 45%, far above the threshold of 20% at which extinction should occur if this toxicant level were chronic (see Table 23-1). After application of the toxicant, all reproducing individuals

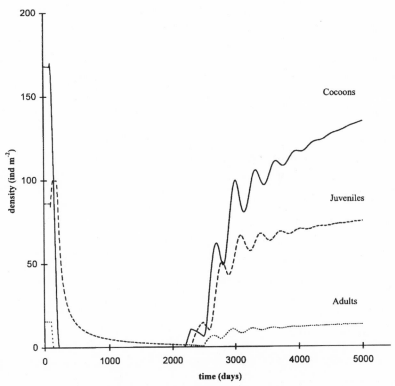

| Figure 23-5 | Recovery of a *L. terrestris* population after application of a sublethal degradable arbitrary toxicant. At time t = 0, the population is held in equilibrium by density-dependent predation at 100 individuals and 170 cocoons m^{-2}. The toxicant is applied at time t = 100 d. The exponential decaying toxicant initially increases the maintenance costs of the individuals by 45% and decays with a half-life of 1680 d. Solid line, cocoons; dashed line, juveniles; and dotted line, adults (after [12]). |

disappear from the population. The increased maintenance costs lead to weight loss of the individuals, which causes adults to shift to the juvenile stages when they shrink to a size smaller than the maturation size. Adults thus disappear from the population. Without adults, cocoons are no longer produced and disappear from the population with a time delay of the cocoon development time. The juveniles suffer mortality resulting from aging and predation and show numerical reduction in time. The toxicant decays exponentially in the subsequent period. When the toxicant has decayed to a level below $[tox]_{thr}$ (Table 23-1), juveniles show positive growth, mature, and start reproducing, which leads to recovery of the population. The time it takes the *L. terrestris* population to recover from the point where the toxicant concentration is lower than $[tox]_{thr}$ to a density of 100 individuals m^{-2} will depend on the population growth rate of the species (see Table 23-2). As can be inferred from Figure 23-5, the recovery time of a population after an application of a sublethal toxicant equals the time it takes the toxicant to degrade from the initial concentration to the species-specific threshold level, where the population can attain positive growth ($[tox]_{thr}$ in Table 23-1), added to the time it takes the population to recover from this point to its former population density. Species sensitivity for the impact of a degradable sublethal toxicant will therefore depend on the species-specific $[tox]_{thr}$, which is mainly determined by the maturation time of the species and the population growth rate. Thus *L. terrestris* with a maturation time of 190 d and a population growth rate of 0.011 d^{-1} under optimal conditions (Table 23-2) is more sensitive than *L. rubellus* with a maturation time of 125 d and a comparable population growth rate of 0.014 d^{-1} (Table 23-2).

Table 23-1   Toxic threshold levels above which population extinction is inevitable. The threshold levels ($[tox]_{thr}$) are given as the relative change in maintenance, assimilation, or reproduction. (after [12]).

Earthworm species	Toxic threshold levels		
	Maintenance increase (%)	Assimilation decrease (%)	Reproduction decrease (%)
L. terrestris	20	20	>100
L. rubellus	30	30	>100

Population extinction obviously takes place if the degradation time (time it takes the toxicant to degrade from the initial concentration to the species-specific $[tox]_{thr}$ (Table 23-1) is longer than the lifespan of the individuals [12]. If the degradation time of the toxicant to the species-specific $[tox]_{thr}$ is shorter than the lifespan of the individuals, and the individuals mature, the high reproductive capacity of earthworms ensures population recovery. Table 23-2 indicates that the minimal recovery time (time it takes the population to recover if it is living under optimal conditions) of *L. terrestris*

Table 23-2   Recovery time under optimal conditions after a drop in density from 100 to 2 individuals m^{-2} (after [12]).

Earthworm species	Population growth rate	Recovery time	
		50% recovery	90% recovery
L. terrestris	0.011 day^{-1}	300 days	350 days
L. rubellus	0.014 day^{-1}	230 days	270 days

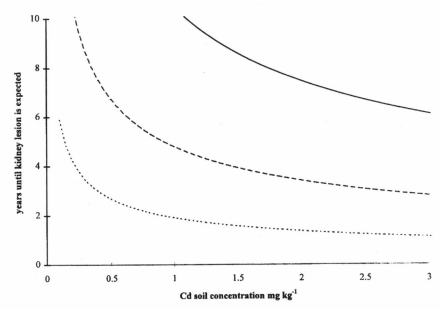

Figure 23-6   Time in y until reaching the threshold level for kidney lesion (200 mg g.⁻¹ dry weight kidney) in badgers in dependence of the Cd concentration and pH of the soil, with a diet composition of 30% earthworms. Solid line, pH 7.2; dashed line, pH 5.5; and dotted line, pH 3.5. (after [15]).

and *L. rubellus* populations, following an initial drop in density of 98% is almost a year. Under field conditions, population recovery is expected to take longer.

Impact on a species that feeds on earthworm populations living in soil contaminated with non-degradable pollutant

Earthworms play a major role in the diet of many organisms and have a high capacity to accumulate toxicants from the soil. Therefore, next to the risk that sublethal toxicant concentrations pose for earthworm populations, the risk posed for their predators also should be considered. Non- and slowly degradable toxicants might constitute a high risk for predator species. This risk may be caused either by decrease in the food availability or by secondary poisoning. Reduction in food availability is expected if the toxicant reduces the individual growth and reproduction of earthworms. On the other hand, if the individual growth and reproduction of earthworms are not influenced by the toxicant but accumulation of the toxicant by earthworms takes place, the predator species might suffer from secondary poisoning.

Cu and Cd are common pollutants in Dutch soil ecosystems. The current Cu and Cd concentrations in some areas in the Netherlands are likely to induce decreased biomass production by earthworms and secondary poisoning of badgers. Cu is expected to reduce the population density and biomass of earthworms, which might lead to food shortage for

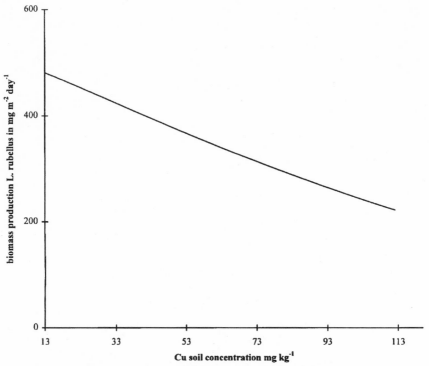

Figure 23-7   Reduction in modeled biomass production of *L. rubellus* as a function of Cu
                soil concentrations (after [15]).

badgers, a protected species in the Netherlands. A variant of the model formulation employed to investigate the impact of a nondegradable chemical substance (impact 2) [12] was used to assess the effect of Cu on the biomass production of earthworms [15]. Figure 23-6 shows the expected reduction in biomass production of the modeled earthworm population that is available for consumption by badgers at different Cu soil concentrations. As indicated in Figure 23-6, even at rather low Cu soil concentrations, the decrease in available earthworm biomass is substantial. Badgers might respond to this decrease in their food levels by foraging over larger areas. In the Netherlands, with its dense infrastructure and already high badger mortality resulting from vehicular traffic (20%) [25], this is likely to make them more vulnerable to traffic mortality.

Cd, on the other hand, decreases earthworm population growth at Cd levels that are very uncommon in Dutch soils, and thus poses little risk for earthworm populations [15]. Earthworms accumulate Cd from the soil and are able to concentrate Cd from the soil by 20-fold [20]. With a diet of 30% earthworms, badgers consume large amounts of Cd. Only a small fraction of the consumed Cd (0.5%) is assumed to be absorbed [15] and deposited mainly in the kidneys. Moreover, excretion of Cd is assumed to be negligible [15]. Kidney lesion in vertebrate species is expected if the Cd concentration in the kidneys is above 200 mg g.$^{-1}$ dry weight [5]. The expected time until badgers suffer from kidney lesion at different Cd and pH values of the soil is given by Figure 23-7. As indicated by Figure 23-7, badgers might reach kidney lesion within their maximum life span, which is about 10 to 15 y [26,27], although only a few badgers reach ages above 6 y in the field. Furthermore, Figure 23-7 shows that the age at which kidney lesion is expected increases with pH. At low pH values of the soil, accumulation of Cd by earthworms is high [20]. Thus Cd consumption by badgers is high and the age at which kidney lesion is expected is low. These results [15] are in agreement with empirical data on Cd concentrations in kidneys of perished badgers [28]. Cd levels exceeded 200 mg g.$^{-1}$ dry weight in the kidneys of badgers aged 4 to 6 y living in polluted areas [28].

Discussion

This chapter discusses the effects of sublethal toxicant concentrations on earthworm populations, given the degradability of the toxicant, the life history of the earthworm species, and the interaction of the earthworm species with a predator species. The results obtained from Baveco and de Roos [12], Klok and de Roos [13], and Klok et al. [14,15] indicate that growth during the pre-adult stages, which determines maturation, is the most

Table 23-3   Review of results: modeled effects of toxicants on earthworm populations

Impact considered	Interaction with other species	Population estimator	Critical life-history process	Population extinction certain if toxicant concentration
Nondegradable toxicant on population growth potential	No	Population growth rate	Maturation	> $[tox]_{thr}$
Nondegradable toxicant on population	Yes	Equilibrium density	Maturation	> $[tox]_{thr}$
Degradable toxicant on population	Yes	Recovery time	Maturation	> $[tox]_{thr}$ during maximal life span of individuals of species

$[tox]_{thr}$ = species-specific threshold concentration above which individuals do not reach adult size

sensitive life-history process for earthworm populations (see Table 23-3). A small decrease in individual growth, leading to retarded maturation, has a far higher impact on the population performance does than an equivalent reduction in reproduction. Population extinction of an isolated population is certain to occur if the reduction in individual growth is so severe that the individuals in the population do not reach their adult size within their life spans [13]. Earthworm populations regulated by predation become extinct if the number of juveniles produced per individual earthworm during its whole life is less than 1 (i.e., individuals do not replace themselves). For a nondegradable toxicant, this extinction condition is met if the toxicant concentration is above the species-specific, population-threshold concentration $[tox]_{thr}$ (Table 23-1). In the case of a degradable toxicant, population extinction is expected if the toxicant concentration is above $[tox]_{thr}$ (Table 23-1) during the whole life of the individual.

Differences in species sensitivity for toxicants (Table 23-1) that retard maturation mainly relate to differences in maturation time. *L. terrestris* that reaches the adult stage under optimal conditions in about 190 d is more sensitive than *L. rubellus*, which matures in about 125 d [12].

The species-specific, population-threshold levels [tox]$_{thr}$ are derived from earthworm populations that are stressed by toxicants and predation; all other conditions are optimal and do not fluctuate. Food is not restricted, soil conditions are assumed to be equal to those in the laboratory experiments used to parameterize the models, and no other toxicants nor other stress factors play a role. Under natural conditions, both food and soil conditions might not be optimal and do fluctuate. This implies, obviously, that the [tox]$_{thr}$ in Table 23-1 are overestimates of the toxic load that field populations can actually sustain.

The species-specific, population-threshold levels [tox]$_{thr}$ (Table 23-1) fix the extreme upper boundary of allowable toxicant concentrations. At toxicant concentrations below these threshold levels, earthworm populations can sustain if all other conditions are optimal. Obviously one should not aim at population suppression to this extreme upper boundary. Even if earthworm population survival is secured, the question is when to define the effects of toxicants on earthworm populations as biologically significant. In other words, how long and to what depth earthworm population density may be reduced by toxicants is difficult to answer from the perspective of the earthworm population alone. Safety factors added to the species-specific [tox]$_{thr}$ possibly provide a way to arrive at allowable toxicant concentrations. If, however, small suppressions in earthworm population density might have large impacts on the species that feed on them [15], allowable toxicant concentrations based on safety factors added to the species-specific threshold levels might not lead to adequate protection of the soil ecosystem. The results of the impact of Cu on the badger show that, even though Cu has a small effect on earthworm populations, the risk to the predator may be substantial. Prevalent Cu soil levels are expected to reduce earthworm biomass by 10 to 20%, which is negligible from an earthworm's point of view. For badgers, these Cu levels might lead to an increase in traffic casualties, as they have to compensate for the drop in food availability [15]. Low Cd concentrations also pose a negligible risk to earthworm populations but do lead to a high risk of secondary poisoning.

The question of when to label the effects of toxicants on earthworm populations as biologically significant if population survival of earthworms is secured should not be answered from the perspective of the earthworm populations alone. Since earthworm populations in the field are in biotic interaction with their environment, allowable toxicant levels should be based on the risk for the earthworms as well as the secondary risk for the species that are in biotic interaction with the earthworms. This would lead to ecologically more relevant and allowable toxicant concentrations than would adding safety factors to species-specific [tox]$_{thr}$.

The models developed to investigate the effects of sublethal toxicant concentrations on earthworm populations [12–15] are based on experimental observations of survival, reproduction, and growth over the entire life cycle of individual organisms. Laboratory bioassays that relate the toxicant level to these life-history parameters are essential to arrive at species-specific, population-threshold levels $[tox]_{thr}$. Since earthworm populations seem very sensitive for toxicants that retard maturation, reduction in pre-adult growth under toxicant treatment should be analyzed. Unfortunately only a few experimental studies are directed toward the effects of toxicants on growth of individuals during the pre-adult development stages, e.g., the effects of Cu and Cd on *L. rubellus* [8,19] and the effects of Cu and Zn on a tropical species *Eudrilus eugeniae* were analyzed [29]. Furthermore, complete ecological information on the life cycle of the tested species is often lacking.

The structured-population approach [12–14] makes it possible to formulate predictions about the population that are firmly based on the individual biology. In the structured-population models, the explicit working mechanism of toxicants can be represented. This makes it possible to compare the population-level effects of toxicants with different working mechanisms. Furthermore, it opens a way to evaluate the population-level consequences of toxicant combinations. One of the major bottlenecks in the use of these models is the amount and nature of experimental data on which they are based. Moreover, the scarcity of field studies applicable to validate the models makes it difficult to test them. This restricts the application of the models [12–14] to specific earthworm species and toxicants.

References

1. Edwards CA, Lofty JR. 1977. Biology of earthworms. New York: Wiley.
2. Lawton JH. 1994. What do species do in ecosystems? *Oikos* 71:367–374.
3. Czarnwska K, Jopiewicz K. 1978. Heavy metals in earthworms as a index of soil contamination. *Polish J Soil Sci* 11:57–62.
4. Carter A, Kenney EA, Guthrie TF. 1980. Earthworms as biological monitors of changes in heavy metal levels in an agricultural soil in British Columbia. In: Proceedings of the 7th International Conference of Soil Biology; July 1979; Syracuse NY. Washington DC: Office of Pesticide and Toxic Substances. EPA/560/13-80-038. p 344–357.
5. Ma WC. 1987. Heavy metal accumulation in the mole *Talpa europea* and earthworms as an indicator of metal bioavailability in terrestrial environments. *Bull Environ Contam Toxicol* 39:933–938.
6. Beyer WN, Cromartie EJ. 1987. A survey of Pb, Cu, Zn, Cr, As, and Se in earthworms and soil from diverse sites. *Environ Monit Assess* 8:27–36.
7. Hunter BA, Johnson MS, Thompson DJ. 1987. Ecotoxicology of copper and cadmium in a contaminated grassland ecosystem. II. Invertebrates. *J App Ecol* 24:587–599.
8. Ma WC. 1982. The influence of soil properties and worm-related factors on the concentration of heavy metals in earthworms. *Pedobiol* 24:109–119.

9. Granval Ph, Aliaga R. 1988. Analyse critique des connaissances sur le prédateurs de lombriciens. *Gibier Faune Sauvage* 5:71–94.

10. Caswell H. 1989. Matrix population models: construction, analysis, and interpretation. Sunderland MA: Sinauer Associates.

11. DeAngelis DL, Gross JG. 1992. Individual-based models and approaches in ecology. population, communities and ecosystems. New York: Chapman and Hall.

12. Baveco JM, de Roos AM. 1996. Assessing the impact of pesticides on Lumbricid populations: an individual-based modelling approach. *J App Ecol* 33:1451–1468.

13. Klok C, de Roos AM. 1996. Population level consequences of toxicological influences on individual growth and reproduction in *Lumbricus rubellus* (Lumbricidae, Oligochaeta). *Ecotox Environ Saf* 33:118–127.

14. Klok C, de Roos AM, Marinissen JCY, Baveco HM, Ma WC. 1997. Assessing the impact of abiotic environmental stress on population growth in *Lumbricus rubellus* (Lumbricidae, Oligochaeta). *Soil Biol Biochem* 29:287–293.

15. Klok C, de Roos AM, Broekhuizen S, Van Apeldoorn RC. 1998. Effecten van zware metalen op de Das. Interactie tussen versnippering en vergiftiging. *Landschap* 15:77–86.

16. Kooijman SALM, Metz JAJ. 1984. On the dynamics of chemically stressed populations: The deduction of population consequences from effects on individuals. *Ecotox Environ Saf* 8:254–274.

17. De Roos AM. 1997. Physiologically structured population models. In: Tuljapurkar S, Caswell H, editors. Structured-population models in marine, terrestrial, and freshwater systems. New York: Chapman and Hall. p 117–204.

18. Cluzeau D, Fayolle L. 1989. Croissance et fécondité comparée de *Dendrobaena rubida tenuis* (Eisen, 1874), *Eisenia andrei* (Bouché, 1972) et *Lumbricus rubellus* (Hoffmeister, 1843) (Oligochaeta, Lumbricidae) en élevage contrôlé. *Rev Ecol Biol Sol* 26:111–121.

19. Ma WC. 1984. Sub-lethal toxic effects of copper on growth, reproduction and litter breakdown activity in the earthworm *Lumbricus rubellus*, with observations on the influence of temperature and soil pH. *Environ Pollut Ser A* 33:53–56.

20. Ma WC. 1988. Toxicity of copper to lumbricid earthworms in sandy agricultural soils amended with Cu-enriched organic waste materials. *Ecol Bull* 39:53–56.

21. Daniel O. 1992. Population dynamics of *Lumbricus terrestris* L. (Oligochaeta: Lumbricidae) in a meadow. *Soil Biol Biochem* 24:1425–1431.

22. Van Rhee JA, Nathans S. 1973. Ecological aspects of earthworm populations in relation to weather conditions. *Rev Ecol Biol Sol* 10:523–533.

23. Bengtson SA, Nilsson A, Nordström S, Rundgren S. 1976. Effect of bird predation on lumbricid populations. *Oikos* 27:9–12.

24. Martin NA. 1978. Earthworms in New Zealand agriculture. Proceedings of the 31st N.Z. weed and pest control conference. Wellington: New Zealand Weed and Pest Control Society. p 176–180.

25. Broekhuizen S, Derckx H. 1996. Durchlässe für Dachse und ihre Effektivität. *Z Jagdwiss* 42:134–14227.

26. Graf M, Wandeler AI. 1982. Altesbestimmung bei Dachsen (*Meles meles* L.) *Revue Suisse Zool* 89:1017–1023.

27. Cheeseman CL, Wilesmith JJ, Ryan J, Mallison PJ. 1988. Badger population dynamics in a high density area. In: Harris S, editor. Mammal population studies. *Symposia of the Zoological Society London* 58:279–294.

28. Ma W, Broekhuizen S. 1989. Belasting van Dassen *Meles meles* met zware metalen: invloed van de verontreinigde maasuiterwaarden? *Lutra* 32:139–151.

29. Reinecke AJ, Reinecke SA, Lambrechts H. 1992. Uptake and toxicity of copper and zinc for the African earthworm, *Eudrilus eugeniae* (Oligochaeta). *Biol Fertil Soils* 42:27–31.

SECTION 6
Progress toward understanding field earthworm ecotoxicology

Chapter 24

Principles for the design of flexible earthworm field-toxicity experiments and interpretation of results

Clive A. Edwards

Although there are well-designed earthworm laboratory toxicity tests, they often need confirmation by field experiments that can provide the most realistic, overall assessments of the acute and chronic toxicity of chemicals to earthworms. However, since earthworm populations are exposed to many site-, region- and climate-dependent variables, it is very important to have flexible but well-defined guidelines for the design of field toxicity experiments. These should minimize variability and provide results that can be interpreted readily in terms of the toxicity, indirect effects and potential long-term hazards of chemicals to earthworm populations. Such field experiments can also provide data on the bioaccumulation of chemicals into earthworm tissues under natural conditions and on the resultant uptake into terrestrial food chains.

Flexible but well-designed field experiments should specify acceptable site variables in terms of soil type, species of earthworms, plant cover, and climate. The experiments need well-defined treatment variables, e.g., standardized size of plots, defined distance between treated plots, potential use of barriers between plots, and timing of treatment applications. It is important to define precisely the formulation and methods of chemical application, e.g., type of chemical used, liquid or solid formulations, and surface application or soil incorporation, and to use a standardized range of dosages. The intermediate and endpoint assessments that should be made of the effects of the chemicals include measurements of numbers of cocoons produced, hatchability of cocoons, numbers and growth stages of earthworms, calculations of time to reach the reproductive phase, and counts of surface casts.

Using such well-defined field experiments, which should be necessary only when potential harm is indicated by 1 or more tiers of laboratory screening tests, the interpretation of results in terms of the overall toxicity of chemicals to earthworms and their effects on species diversity, population growth or decrease, and earthworm behavior can be facilitated greatly. Criteria are developed and defined in this chapter that will enable chemicals to be classified in hazard categories for risk assessment based on their effects on earthworms in the field. These criteria are based on indicative laboratory toxicity data, combined with the effects in the field on earthworm populations, reproduction, and growth. A key issue is that in order for a chemical to be rated not hazardous to earthworms, field populations should be able to recover completely within 12 months of exposure.

General review

Earthworms play a major role in promoting soil fertility, through their activities in the degradation of organic matter (OM) and release of nutrients, and they are also very important in turning over soil, burying OM, providing aeration and percolation channels, and improving overall soil moisture-holding capacity through their influence on aggregate formation [1,2]. They are probably the most important soil-inhabiting invertebrates involved in soil formation and in maintaining soil fertility [3].

For these reasons, and also because earthworms make a major contribution to the total invertebrate biomass in many temperate and tropical soils, they have been selected as key indicator organisms for various ecotoxicological testing programs [4] such as those of the European Union (EU) and the Organization for Economic Cooperation and Development (OECD). Earthworms were the subject of the First International Workshop on Earthworm Ecotoxicology held in Sheffield, England in 1991 [5], and this workshop made very important recommendations on the urgent need for additional laboratory and field protocols for testing the effects of chemicals on earthworms.

There have been many reviews and more than 350 papers on the effects of chemicals on earthworms [2,4,6–11], but the design of most experiments has been very variable and extremely difficult to interpret [4,10,12].

In the earlier reports in the literature, there was a strong orientation toward field experimentation, but in recent years, the emphasis has moved much more toward laboratory toxicity testing. The standardized, artificial-soil laboratory test developed and formalized by Edwards [13,14] for the EU [15] was adopted with minor changes by the OECD in their guideline [16]. This test has been developed further by the German Biologische Bundesandstalt für Land und Fortwirtschaft (BBA) as the Earthworm Reproduction Test [17].

A substantial body of literature now reports the use of the earthworm laboratory artificial soil test to assess the acute effects of chemicals on earthworms [13–16], and this has produced an extensive data bank of laboratory earthworm toxicity data that can be used to predict the effects of most major environmental pollutants on earthworms [18]. Although the species used is *Eisenia fetida*, an epigeic species, good correlations have been established between the toxicity of chemicals to this species and other species [19]. There is an urgent need for a standardized and flexible field-toxicity test protocol to confirm possible serious effects that may be indicated in the laboratory tests of chemicals on earthworms, as was pointed out at the First International Workshop on Earthworm Ecotoxicology [5]. The aim of this chapter is to review the variables involved in field earthworm toxicity testing and to make recommendations about the design and interpretation of such a field test. The recommendations will build upon those made by Edwards [12], the suggestions made in the 1991 workshop [5], and the protocol designed by the BBA [17], as well as discussions in a range of papers published between 1992 and 1997 and in the present volume.

Advantages and limitations of field tests for assessing the toxicity of chemicals to earthworms

A considerable number of benefits are associated with using field experiments to assess the toxicity of chemicals to earthworms, but there are also some serious limitations. The benefits and limitations are tabulated here with a brief discussion of each of the issues raised.

Advantages of field tests

Field tests assessing the effects of chemicals on earthworms involve

- Realistic routes of exposure to chemicals: a flexible field experiment should be designed so that earthworms are exposed to toxic chemicals as they would be in the real world, under natural climatic and environmental conditions. In field experiments, the bioaccumulation of chemicals into earthworm tissues occurs under much more realistic environmental conditions and modes of exposure than in laboratory tests.

- Populations of earthworms rather than single species: field tests also take into account the interactions between earthworm species and expose all stages of the earthworm life cycle to the chemical, as well as allowing for effects on natural mortality and predation. Sampling field experiments sequentially can assess the length of time that earthworms are affected by a chemical, whether the chemical is transient or persistent in soil, and can document the rate of population recovery. These are key criteria in assessing potential hazards to earthworms.

- Determination of indirect effects of chemicals: if a chemical influences the parasites, pathogens, or predators of earthworms, the resultant indirect effects on earthworm populations will be demonstrated in field experiments. If a pesticide influences weed cover or OM availability resulting from dead weeds, which in turn influence earthworm populations, such effects may also be revealed by field experiments.

- Exposure to variable climatic factors that can have effects on the toxicity of chemicals to earthworms: laboratory tests cannot simulate these climatic factors or take them into account. Moreover, earthworms that are under climatic or environmental stress may be much more susceptible to some chemicals in the field than they are under optimal conditions in the laboratory.

Limitations of field experiments

There are a number of serious limitations to using field experiments as a screening system for assessing the effects of chemicals on earthworms, including the following.

- Field tests of toxicity to earthworms are expensive. They are much too costly to be used in a routine screening program, but they have considerable use and application when triggered by tiered laboratory screening tests to confirm adverse effects and provide validation studies under more natural conditions.

- Field tests are limited to 1 or 2 soil types. Soil type can have very considerable effects on the toxicity of different chemicals to earthworms that live in the soil. Some chemicals become reversibly, or irreversibly, bound to particles of clay and OM, which can directly influence both the adsorption and bioavailability of chemicals to earthworms. Moreover, chemicals break down at different rates in different soil types, which may contain very different kinds and populations of microorganisms that contribute to chemical degradation.

· Differential plant cover influences the toxicity of chemicals to earthworms. For instance, bare soil accelerates the loss of chemicals through volatilization at a much greater rate than that from soil with a dense plant cover, even when the chemicals applied are relatively nonvolatile. In grassland, the dense mat of OM at the soil surface may adsorb and inactivate chemicals, and thereby decrease the exposure of earthworms to the chemicals. For valid field tests, both sorts of plant cover should be used unless the chemical is likely to reach either bare or covered soil, in which case either bare soil or an appropriate plant cover should be used.

· Variability in field conditions, in terms of soil characteristics, plant cover, wet or dry areas, and drainage, often necessitate a very considerable degree of plot replication to provide statistically appreciable data. The use of large plots, which may be needed to minimize earthworm migration between treated and untreated plots, may increase this variability because they involve large experimental areas.

· Variability in climatic conditions can make it almost impossible to compare toxicity data on effects of chemicals on earthworms between different seasons or regions. Chemicals behave, degrade, and persist very differently under wet or dry, hot or cold conditions, particularly in terms of bioavailability and consequent effects on earthworms.

Design of flexible field tests

Clearly, the high cost and the other limitations of field experiments place considerable restraints on the widespread use of field tests as screening tools in the early stages of assessing the environmental effects of chemicals on earthworms. However, there is an urgent need for a well-designed and standardized but flexible field test, which could be used as a decisive tool to validate and confirm any potentially serious impact of chemicals indicated through tiered laboratory screening tests.

This chapter discusses the various site and treatment variables involved in the design of earthworm toxicity field tests and how they may be standardized as far as possible, as well as which of them should be relatively flexible in order to account for the various routes of exposure of earthworms to chemicals. Some of these variables are relatively easy to standardize, whereas others demand a considerable degree of flexibility. Kula [20] made recommendations on the design of field experiments, and there is a German-recommended protocol that could provide a partial basis for the development of an internationally accepted field-testing protocol; it is referred to here wherever appropriate [17].

Site variables for field tests

There are a number of site-variable requisites that must be considered in the design of standardized field tests. These variables include the following prerequisites:

1) An uncontaminated site is essential to avoid any preconditioning of the earthworm populations, which could affect their susceptibility to particular chemicals or predispose soil microbial populations to be able to degrade certain chemicals. There should be no history of chemical contamination of the site for at least 5 y.

2) The site should have a representative earthworm population in terms of both species diversity and numbers. Preferably the site should have at least 6 representative species of earthworms, and populations should be at least 100 earthworms per m². If a smaller population than this is required, larger plots and more unit samples per treatment may be used. The earthworm species present represents the relevant ecological group, its vertical distribution, and its form of activity as far as possible. There are 3 such major ecological groupings of earthworm species, based on their feeding and burrowing habits [2]:

· Surface soil and litter species—Epigeic species. These species live in or near the surface plant litter; they typically are small and are adapted to the very variable environmental conditions close to the soil surface; *Lumbricus rubellus* is a typical epigeic species.

· Upper soil species—Endogeic species. Some species move and live in the upper soil strata and feed primarily on soil and associated OM (geophages). They do not have permanent burrows and their temporary channels become filled with cast material as they move through the soil, progressively passing it through their intestines. *Aporrectodea caliginosa* is a typical endogeic species.

· Deep-burrowing species—Anecic species. These earthworms, which are typified by the night crawler, *L. terrestris*, inhabit more or less permanent burrow systems that may extend several meters deep into the soil profile. They feed mainly on surface litter, which some species pull down into their burrows. They may leave plugs, OM, or cast and soil and mineral particles at the mouth of their burrows.

Species from these 3 ecological groupings are exposed to chemicals in quite different ways, with the epigeic and anecic deep-burrowing species being much more susceptible to surface chemical residues, whereas the endogeic species are exposed much more to chemicals that become mixed into soil. Hence, it is important that the species that need to be represented at a test site should typically be related to the location or placing of the residues of the test chemical. The BBA protocol [17] recommends that the species *Lumbricus terrestris* and *Aporrectodea caliginosa* should be present, but this may be too restrictive. It is probably preferable to propose that populations of both epigeic and endogeic species should be present in the selected field site.

3) A representative soil type or soil types is important because some chemicals may be much more toxic in sandy soils, where chemicals are not adsorbed, than in loams or heavy clay loams containing considerable amounts of OM, where adsorption of chemicals and a decrease in their bioavailability is much greater.

Moreover, sandier soils do not usually favor the buildup of the large earthworm populations that are needed for experimental sites. It is important that there should be a single soil type that is as homogeneous as possible across the experimental site. Van Gestel [21] emphasized that soil OM content, clay content, and pH influenced the toxicity of chemicals to earthworms quite strongly. The BBA has recommended 6 categories of soils for such experiments, which are based on a range of mineralogical compositions [22]. Other groups have suggested 3 soil types, e.g., a sandy soil, sandy loam, and a loamy sand. Using recommended soil types may have considerable potential to decrease site-to-site variability, and wherever possible, they should be part of any standardized field experiment design.

Treatment variables for field tests

The management of the site, layout of plots, amounts and method of application of the chemical, and timing of experiments should be standardized as far as possible. Such treatment variables are discussed below.

The plant cover present before and after the application of the chemical treatment is very important. If the chemical to be tested could be expected to reach the ground as fallout or in precipitation, it would be appropriate to apply the test chemical as a surface application to a grassland, woodland, or arable site that contains a suitable and representative earthworm population to simulate natural conditions. However, if a chemical that is to be tested would normally be applied to agricultural land, it preferably should be applied to bare soil and left as a surface residue or cultivated into the soil to avoid binding and adsorption of the pesticide on the surface thatch that is commonly found in grassland. The BBA protocol recommends doing field experiments on long-term grassland [17], but strong reservations are expressed about this as a general procedure because of potential adsorption and inactivation of the chemical by the surface thatch. After the chemical has been applied to the soil, the plots can then be seeded to new grass or other appropriate crops. Row crops should be avoided wherever possible, since they complicate the assessment of earthworm populations, which may become aggregated in the rhizosphere. The testing of herbicides in the field is particularly difficult because the indirect effects of the chemical on earthworms, through the changed weed populations and resulting OM from dead weeds, may be greater than the direct effects of the chemicals. Therefore, in order to assess toxicity, plots must be kept fallow manually [23]. When an arable site with an adequate earthworm population is not available, an alternative is to cultivate a grassland site and prepare a seedbed before applying the test chemical and then sowing the plots with a suitable plant cover.

The size of the plot is very important because plots need to be as small as possible to minimize soil variability across the experimental site. However, plots must be large enough to ensure that they contain a representative earthworm population and that repeated destructive population sampling, of up to 1 m² per sampling date, does not have any significant influence on the overall earthworm population in the plots. Such decreases in populations over a long period have been recorded in some field experiments. To avoid such decreases, the absolutely minimum size of plot that could be used would be 5 m², but plots 10 m² were recommended by the previous workshop [5] and the BBA [17] as a compromise between soil variability, minimizing earthworm migrations between plots, provisioning for adequate destructive sampling, and allowing appropriate replications. Such plot sizes have produced consistent results in a broad range of published studies.

Barriers or spaces between plots offer alternative ways to prevent or minimize earthworm migration between plots. It has been generally accepted by many earthworm scientists that horizontal migration by earthworms over the soil surface is very limited and slow in terms of population movement and distribution [2]. However, Stockdill [24] reported that earthworms could spread from an inoculation point at a rate of about 10 m per y. More recently, it has been suggested that surface migration over appreciable distances can occur, particularly in response to application of toxic chemicals such as benomyl, which may have a repellent effect (See Mather and Christensen, Chapter 25). To minimize the movement of earthworms between plots, it is necessary either to have wide paths of the same dimension as the plots between plots or to have physical barriers, buried into the ground, between plots. A machine for setting up such barriers rapidly and with minimal labor was used by Edwards et al. [25] and has been recommended for use in field environmental experiments such as earthworm field-toxicity testing [26].

The degree of replication of treatments needed is correlated with the variability in soil type, earthworm populations, and plant cover at the experimental site. Experience has shown that a minimum of 4 replicates are needed for most sites, and for more variable sites, 6 or even 8 replicates may be required. The BBA protocol [17] recommends 4 replicates per treatment, but the greater the number of replicates, the greater the precision of the experiment in terms of assessing chemical toxicity to earthworms.

The timing of the experiment is very important. In temperate countries, earthworms are usually most active in spring and autumn. It is recommended that, whenever possible, earthworm field toxicity experiments should begin in spring and should be continued for a minimum of 6 months to include the autumn months for chemicals that persist less than 1 season; they would need to be longer (up to 2 y or more) for testing the effects of more persistent chemicals. This recommendation could be modified if the plots were regularly irrigated to minimize seasonal vertical migrations of earthworms.

The nature of chemical to be tested should be defined clearly in terms of structure and mixture of isomers. If the chemical is a pesticide, its common name should be known. Its physicochemical properties such as solubility, water/lipid partition coefficient, and volatility, which can be used in prediction of toxicity, should be described.

The chemical formulation can have considerable effects on the toxicity of a particular chemical to earthworms and should be clearly specified.

The method of application of the chemical should relate closely to the likely location and occurrence of the chemical in the environment, e.g., if earthworm exposure is to some form of surface-pollutant fallout, this can be simulated by spraying the soil surface with the chemical (either bare soil or with plant cover as appropriate). If the exposure is in the form of a pesticide or other chemical that becomes cultivated into the surface layers of soil, this should be simulated as closely as possible.

The dose of chemical should be calculated as the highest rate likely to occur under natural exposure conditions and should always be compared with untreated controls. Preferably, this should be based on a calculation of the initial predicted environmental concentration (PEC), based on the assumption of a uniform distribution of the applied chemical dosage into the top 5 cm of soil and an average soil density of 1.5 kg/L. For a typical pesticide application of 2 kg/ha active ingredient this would be the equivalent of 2.33 mg/kg dry weight soil (see Barber et al., Chapter 21). This calculation must also involve factors associated with the chemical and soil characteristics; and the dose, method, and number of applications; amount of ground cover; and rate of degradation of the chemical. An additional exposure dose could be $4 \times$ the PEC, as was recommended by Greig-Smith [27]. It is important that the risk assessment must account for both the intrinsic toxicity of the chemical from laboratory tests and the period of exposure to the chemical, expressed as a calculated earthworm toxicity-exposure ratio (TER) expressed as LD50/PEC. Heimbach [28] suggested that when the TER is greater than 20, a chemical is unlikely to have a significant impact on field populations, and long-term reductions in earthworm populations have been recorded only when the TER was less than 2.

It is recommended that a standard chemical be included in the trial. This was suggested by Greig-Smith [27] and is also recommended in the BBA protocol [17]. Such a treatment can act as a reference toxicity point, which enables much more valid comparisons between laboratory and field tests to be made, as well as facilitating comparisons between field experiments with different site and treatment variables. Benomyl was used as a standard during the development of the artificial soil test, and many workers have used this fungicide, or alternatively the insecticide carbaryl, as a standard in earthworm toxicity testing; these chemicals would be appropriate standards.

Selections of suitable endpoints and assessment criteria

A number of appropriate endpoints or other assessment criteria can be used in earthworm field-toxicity tests, including the following:

1) The effect of a chemical on overall earthworm populations in terms of both numbers and biomass is one of the most important criteria to be used in assessing the field toxicity of a chemical. Where possible, earthworms should be identified to species levels.

2) A suitable method of assessing earthworm populations must be used. Ideally, this should include all stages of the earthworm population (cocoons, immatures, resting and active adults). The best method is by hand sorting soil samples from a quadrat (0.25 to 1.0 m²) to a depth of 30 to 60 cm, preferably in

the strata of 10 cm in the field or laboratory. However, this is extremely laborious and time consuming, and for this reason, it is seldom feasible.

An alternative method of assessing the active phases of earthworm populations in the field is by using quadrats (0.25 m^2) and pouring dilute formalin (50 cc of 40% formaldehyde in 9 L of water) on to the soil surface gradually until it is saturated. Earthworms will come to the surface within 15 to 20 minutes, from where they can be picked up with forceps and stored in 5% formalin until they can be identified and counted [29]. Earthworms stored in formalin can be dried on paper tissues and weighed for biomass measurements, since their dry weight is very similar to their live weight. This method of bringing earthworms to the soil surface works relatively quickly but has the drawback that it works best when earthworms are at peak seasonal activity and is most effective for those earthworms with permanent burrows, e.g., *L. terrestris* and *Allobophora longa*.

The method recommended for maximum efficiency is to hand sort earthworms in the top 5 to 10 cm of soil in the field quadrats on polythene sheets and then pour dilute formalin on to the surface of the quadrat. This method recovers both those earthworm species that spend most of their lives close to the soil surface (epigeic and endogeic species) as well as those that penetrate as much as 2 m deep into the soil through burrows (anecic species) and that would not be recovered by only hand sorting the surface soil layer.

It has been suggested that an electrical method of sampling by passing a current from a 220 V generator through the soil is a suitable sampling method [2,6,30]. However, this method does not sample a sharply delimited area of soil, and it is influenced greatly by soil pH and moisture content, so it is not to be generally recommended, even on a comparative basis.

3) A method of sampling numbers of cocoons is necessary. To do this, take soil cores from the plot, dry the soil, hand sort on plastic sheets, and count full and empty cocoons under a magnifying lens. Holmstrup et al. (Chapter 27) described an alternative method of sampling cocoons by washing and sieving.

4) An assessment of bioaccumulation of chemicals into earthworm tissues is an essential endpoint for a field toxicity test. Such chemicals can be taken up into the earthworm tissues, in solution through the cuticle, as a function of lipid/water partition for lipophilic chemicals, or through contaminated food or soil as it passes through the earthworm's guts. The types of chemicals that tend to accumulate in earthworms' tissues are mainly heavy metals and certain pesticides. Representative samples of earthworms must be taken and a standard weight must be analyzed for chemical residues, using the appropriate methodology for the particular pollutant. Care should be taken to keep the earthworms on moist filter paper for 24 h to allow them to void their gut contents prior to analyses.

5) The timing of earthworm sampling in field experiments should be related to the persistence of the test chemical in soil, as far as it is known. For chemicals that persist less than 6 months in soil, monthly sampling is adequate. For chemicals

that persist longer, it is recommended that samples should be taken 1, 2, 4, 12, and 24 months after treatment, with sampling done every 6 months subsequently for very persistent chemicals. The recommendations in the BBA protocol [17] were 1, 4 to 6, and 12 months after application.

6) The number of earthworm population samples that need to be taken per replicate plot can differ with populations of earthworms, variations in soil type, and size of plot. However, it is recommended that a minimum of 4 quadrats per plot be sampled on each sampling date if a statistical change in population on the order of 25% is to be identified. With small earthworm populations, more samples may be needed.

7) Earthworm community analysis could be 1 additional endpoint particularly applicable to field experiments. It has been suggested that redundancy analysis that has been used to assess the effects of perturbations, e.g., chemicals on complex soil communities, could be a useful tool for examining earthworm field-toxicity data (see Bembridge et al., Chapter 26).

Potential of semi-field and mesocosm experiments

Considerable attention has been focused on the use of semi-field or microcosm experiments to assess the environmental effects of chemicals. Large intact and relatively deep soil cores, taken from a relatively small area of field site, can be much less variable than whole plots in a large-scale experiment, and such mesocosms can still maintain most of the functions that occur in full-size field plots [31–33].

Mesocosms can be taken from the field, either before or after treatment with chemicals, and studied under standardized conditions of temperature and moisture, or they can be replaced into the field where they are exposed to normally fluctuating climatic conditions. A number of smaller mesocosm studies reported in the literature also have included assessments of the effects of chemicals on earthworms [33,34], and one such test [35] was specifically aimed at ecotoxicological studies on earthworms. Another mesocosm system was described at the Second International Symposium on Earthworm Ecotoxicology by Olesen and Weeks (Chapter 30). Mesocosm and other semi-field tests also offer the opportunity to manipulate populations of earthworms in field populations, e.g., by adding key species that were absent from the test site to the mesocosm before returning it to the field. It is outside the scope of this chapter to define a mesocosm methodology, but it clearly holds considerable potential as an alternative to field experiments.

Need for earthworm field-ecotoxicity experiment and interpretation of results

At the First International Workshop on Earthworm Ecotoxicology, it was suggested that successive tiers of testing, each invoked only when potential hazards may be indicated, should be used to clarify progressively whether a chemical is substantially harmful to earthworms [27]. A hierarchical system would allow an assessor to classify effects from high risk to no risk, and only if the initial screening indicated a potential for substantial effects on earthworm populations, would it be necessary to do higher-tier field studies.

Among the recommendations made were that the Artificial Soil Test [13–16] using *Eisenia fetida* should be used for preliminary screening, with acute toxicity determined as mortality as the main endpoint expressed as an median lethal concentration (LC50) and no-observed-effect concentration (NOEC). It was concluded that this could be followed by 1 (or more) higher-tier laboratory tests, which could be more flexible than the artificial soil test, in order to tailor tests to particular routes of exposure. These artificial soil tests could include a reproductive test or even a small mesocosm test [33]. Such tests could lead to yet a higher-tier-level field test if potentially serious hazards were indicated (see Heimbach, this volume). It was recommended that, although a flexible protocol for such a field test was needed, more standardization than that used in past field-ecotoxicity experiments was considered very desirable.

In the interpretation of results, it was pointed out that care should be taken to distinguish between the statistical significance of changes and their biological significance, which is often hard to identify. In particular, it was suggested that the results could be considered in terms of the scale of population decrease in response to a chemical and the time taken for a population to recover. Population reductions of 30 to 50% below control levels and the rate at which numbers or biomass recover were suggested as simple threshold criteria. Heimbach [28] suggested a classification based on the criterion that a reduction in an earthworm population of 75% was heavy, 40 to 75% medium, and 25 to 40% slight or weak. Such mortalities could be used as indicators for defining effects on earthworms and as risk assessment thresholds in field tests. However, in a review of laboratory and field studies on effects of pesticides on earthworm populations, Jones' and Hart's (Chapter 20) results suggested that for nonpersistent chemicals, recovery of earthworm populations was possible within a year after reductions up to 90%. Similarly, Lofs [36] reported that after cultivation of a lucerne lea, 70% of the earthworm population and 97% of its biomass were destroyed; however, 12 months later, the population had completely recovered, and it was assumed that this was through population multiplication rather than immigration.

Thus, defining population decreases of 30 to 50% in a single season as possibly serious effects on earthworm communities may be based on an underestimate of the potential of earthworm populations to decrease and still recover quite rapidly. It is not at all unusual for seasonal changes to decrease earthworm populations to as little as 10%, with full recovery achieved by the following season [2]. There is also good evidence that populations can double in 12 months in response to suitable food or environmental factors [37]. Based on these data, it seems that setting a critical level of population decrease as an index of potential hazard due to chemicals may be difficult or even impossible. A much more important criterion for a serious earthworm population hazard due to a chemical is whether any significant decrease in population persists for longer than 12 months. If the chemical has no significant residual effect after 12 months, the chemical should not be classified as potentially hazardous to earthworms.

References

1. Edwards CA, Bohlen PJ, Linden D, Subler S. 1995. Earthworms in agroecosystems. In: Hendrix P, editor. Earthworm ecology and biogeography in North America. Boca Raton FL: Lewis. p 185–213.

2. Edwards CA, Bohlen PJ. 1996. Biology and ecology of earthworms. 3rd ed. New York: Chapman and Hall. 426 p.

3. Hendrix PF. 1995. Earthworm ecology and biogeography in North America. Boca Raton FL: Lewis. 244 p.

4. Edwards CA, Bohlen P. 1992. The assessment of the effects of toxic chemicals upon earthworms. In: Ware G, editor. *Rev Env Contam Toxicol* 125:23–99.

5. Greig-Smith PW, Becker H, Edwards PJ, Heimbach F. 1992. Ecotoxicology of earthworms. Filey, Yorkshire UK: Intercept. 269 p.

6. Satchell JE. 1955. The effects of BHC, DDT and parathion on the soil fauna. *Soils Fert* 18:279–285.

7. Davey SP. 1963. Effect of chemicals on earthworms: a review of the literature. *USDI Spec Scien Rep Wildlife* No. 74, Washington DC. Patuxent MD: U.S. Fish and Wildlife Service. p 20.

8. Edwards CA, Thompson AR. 1973. Pesticides and the soil fauna. *Residue Rev* 45(1):1–73.

9. Thompson AR, Edwards CA. 1974. Effects of pesticides on non-target invertebrates in freshwater and soil. In: Pesticides in soil and water. Soil Sci. Soc. Amer. Spec. Publ. 8, 1974. p 341.

10. Lofs-Holmin A, Boström U. 1988. The use of earthworms and other soil animals in pesticide testing. In: Edwards CA, Neuhauser EF, editors. Earthworms in waste and environmental management. The Hague, Netherlands: SPB Academic. p 303–313.

11. Edwards CA, Neuhauser EF. 1988. Earthworms in waste and environmental management. The Hague, Netherlands: SPB Academic. 391 p.

12. Edwards CA. 1992. Testing the effects of chemicals on earthworms: the advantages and limitations of field tests. In: Greig-Smith PW, Becker H, Edwards PJ, Heimbach F, editors. Ecotoxicology of earthworms. Filey, Yorkshire UK: Intercept. p 75–84.

13. Edwards CA. 1983. Development of a standardized laboratory method for assessing the toxicity of chemical substances to earthworms. Environment and quality of life commission of European communities. Brussels, Belgium: Commission of the European Communities. EUR 8714 EN.

14. Edwards CA. 1984. Report of the second stage in development of a standardized laboratory method for assessing the toxicity of chemical substances to earthworms. Environment and quality of life commission of European communities. p 99; 8714 EN. Brussels, Belgium: Commission of the European Communities. EUR 9360 EN. p 141.

15. [EEC] European Economic Community. 1985. Directive 79/831 Annex V, Part C. Methods for the determination of ecotoxicity- Level 1, earthworms. Artificial soil test. Brussels, Belgium: Commission of the European Communities. DG XI/128/82.

16. [OECD] Organization for Economic Cooperation and Development. 1984. Organization for Economic Cooperation and Development guidelines for testing of chemicals: earthworm acute toxicity test. OECD Guideline No. 207. Paris, France: OECD.

17. [BBA] Biologische Bundesandstalt für Land und Fortwirtschaft. 1994. Richtlinienvorschlag für die Prüfung von Pflanzenshuctz-mitteln Auswirkungen von Pflanzenshutzmitteln auf die Reproduktion und Wachstrum von *Eisenia fetida/Eisenia andrei*. Biol. Bundesanstalt Land. und Först. No VI, 2-2.

18. Beyer WN. 1992. Relating results from earthworm toxicity tests to agricultural soil. In: Greig-Smith PW, Becker H, Edwards PJ, Heimbach F, editors. Ecotoxicology of earthworms. Filey, Yorkshire UK: Intercept. p 109–115.

19. Heimbach F. 1985. A comparison of laboratory methods using *Eisenia foetida* and *Lumbricus terrestris* for assessment of the hazard of chemicals to earthworms. *Zeit Pflanzenkrank Pflanzenschutz* 92(2):186–193.

20. Kula H. 1992. Measuring effects of pesticides on earthworms in the field: test design and sampling methods. In: Greig-Smith PW, Becker H, Edwards PJ, Heimbach F, editors. Ecotoxicology of earthworms. Filey, Yorkshire UK: Intercept. p 90–99.

21. van Gestel CAM. 1992. The influence of soil characteristics on the toxicity of chemicals for earthworms: a review. In: Greig-Smith PW, Becker H, Edwards PJ, Heimbach F, editors. Ecotoxicology of earthworms. Filey, Yorkshire, UK: Intercept. p 44–54.

22. Kuhnt G, Muntau H. 1992. EURO-soils, identification, collection treatments, characterization. Brussels, Belgium: Joint Research Centre, Commission of the European Communities. ISPRA 1-84.

23. Edwards CA, Stafford CJ. 1978. Interactions between herbicides and the soil fauna. *Ann Appl Biol* 89(2)91:125–146.

24. Stockdill SMJ. 1982. Effect of introduced earthworms on the productivity of New Zealand pastures. *Pedobiol* 24:29–35.

25. Edwards CA, Sunderland KD, George KS. 1979. Studies on polyphagous predators of cereal aphids. *Appl Ecol* 16(3):811–823.

26. Edwards CA. 1979. Tests to assess the effects of pesticides on beneficial soil organisms. Proc Umweltbundsamt Sem on Ecol Tests Relevant to Envir Chem: Evaluation and Research Needs. December 1977. Berlin, Germany: Umweltbundesamt. p 240–253.

27. Greig-Smith PW. 1992. Risk assessment approaches in the UK for the side-effects of pesticides on earthworms. In: Greig-Smith PW, Becker H, Edwards PJ, Heimbach F, editors. Ecotoxicology of earthworms. Filey, Yorkshire UK: Intercept. p 159–168.

28. Heimbach F. 1992. Effects of pesticides on earthworm populations: comparison of results from laboratory and field tests. In: Greig-Smith PW, Becker H, Edwards PJ, Heimbach F, editors Ecotoxicology of earthworms. Filey, Yorkshire UK: Intercept. p 100–108.

29. Edwards CA. 1991. Methods for assessing populations of soil-inhabiting invertebrates. *Agric Eco Envir* 34:145–176.

30. Theilmann U. 1986. Elektrischer Regenwurmstrant mit der Oktett-Methods. *Pedobiol* 29:295–302.

31. Van Voris P, Tolle DA, Arthur MF, Chesson J. 1985. Terrestrial microcosms: applications, validation and cost-benefit analysis. In: Cairns J, editor. Multispecies toxicity testing. New York: Pergamon. p 117–142.

32. Kuehle JC. 1988. Radiological studies on earthworms and their value for ecotoxicological risk assessment. In: Edwards CA, Neuhauser EF, editors. Earthworms in waste and environmental management. The Hague, Netherlands: SPB Academic. p 377–388.

33. Edwards CA, Knacker T, Pokarzhevskii A, Subler S, Parmelee R. 1997. The use of soil microcosms in assessing the effects of pesticides on soil ecosystems. In: Environmental behavior of crop protection chemicals. Proc. Int. Symp. Use of Nuclear and Related Techniques for Studying Environmental Behavior of Crop Protection Chemicals. Vienna, Austria: International Atomic Energy Agency. SM 343/3. p 435–451.

34. Bogomolov DM, Chen S-K, Parmelee RW, Subler S, Edwards CA. 1996. An ecosystem approach to soil toxicity testing: a study of copper contamination in laboratory microcosms. *Appl Soil Ecol* 4:95–105.

35. Svendsen C, Weeks JM. 1997. A simple low-cost field mesocosm for ecotoxicological studies on earthworms. *Comp Biochem Physiol* 117(1):31-40.

36. Lofs A. 1992. Measuring effects of pesticides on earthworms in the field: effect criteria and endpoints. In: Greig-Smith PW, Becker H, Edwards PJ, Heimbach F, editors. Ecotoxicology of earthworms. Filey, Yorkshire UK: Intercept. p 85–89.

37. Edwards CA, Lofty JR. 1982. Nitrogenous fertilizers and earthworm populations in agricultural soils. *Soil Biol Biochem* 14:515–521.

Chapter 25

Earthworm surface migration in the field: influence of pesticides using benomyl as test chemical

Janice G. Mather and Ole M. Christensen

Migration by soil organisms in the field in relation to pesticides has generally been disregarded, even though this behavior influences population density, biomass, and species diversity. Earthworm movement over the soil surface was investigated using trap units in grassland. Three concentrations of benomyl were considered: 0.5, 1.0, and 2.0 kg a.i. ha^{-1}. Traps were tended daily for 35 d after spraying. Captured earthworms were counted, identified, and weighed, resulting in 4258 individuals and 7 lumbricid species. Significant increases in surface migration were evident by day 2 for 1.0 and 2.0 kg a.i. ha^{-1} benomyl and by day 4 for 0.5 kg a.i. ha^{-1} benomyl, indicating that migration is a comparatively quick and sensitive parameter for measuring chemical side effects in the field. By the end of the study, 0.5 kg a.i. ha^{-1} benomyl gave a clear 2.5-fold increase in surfacing, 1.0 kg a.i. ha^{-1} benomyl gave a 1.4-fold increase, whereas for 2.0 kg a.i. ha^{-1} benomyl, a 1.3-fold decrease occurred; the latter 2 trends probably were confounded by mortality and downward migration. Numerical changes were apparent at the species level, with significant increases for *Lumbricus terrestris* at all benomyl concentrations. Benomyl caused changes in species composition, which was most evident for the highest concentration that significantly altered the proportions of the 2 anecic species, increasing *L. terrestris* and decreasing *Aporrectodea longa*. Weight-class frequency distribution for the dominant species *A. longa* indicated that all benomyl concentrations induced a higher proportion of smaller-sized individuals to surface migrate. Results are discussed in relation to both ecological consequences for habitats and conventional ecotoxicological testing.

The migratory behavior of soil animals as a consequence of pesticide application in the field has so far tended to be overlooked in ecotoxicological studies; these organisms generally are considered relatively sedentary. More recently, however, it has been proposed that migration is an ideal behavioral parameter for assessing side effects of chemicals because the measured response is relatively sensitive and immediate and provides an early warning of environmental hazard [1–3].

Migration as a study parameter has high ecological relevance because the movement of individuals influences population dynamics, which in turn affect species diversity, age structure, and density and biomass. Also, although a chemical-induced change in migration is a sublethal effect, if all individuals in a given area migrate and recolonization is inhibited, then the ecological consequence for the habitat is equivalent to 100% mortality [compare 2,3].

Furthermore, migratory behavior—both horizontally over the soil surface and vertically within the soil profile—can influence the degree of exposure to chemicals, unfavorable climatic conditions, and predation. Thus, a pesticide assessed as nonlethal to nontarget organisms in the laboratory (where the test animal is confined in a container, thereby restricting movement) may prove to have detrimental or lethal consequences under field conditions. A potential environmental hazard remains hidden, therefore, if chemical screening is based exclusively on laboratory tests.

The aim of the present investigation is to assess the influence of the fungicide benomyl on the surface migratory behavior of lumbricid earthworms in the field.

Materials and methods

Study site
The investigation was conducted in a 2-y clover–grass field at Askhøj, Foulum Research Centre, Denmark, during spring 1996. Within this field, the study site covered an area of 3.3 ha (130 m × 250 m). The site has a sandy-loam soil, the soil texture being 67.5% sand, 20.5% silt, 7.4% clay, and 4.6% humus. Soil pH (H_2O is 7.3. Records of climatic conditions (precipitation and soil temperature) during the investigation were obtained from the nearby climatic station at Foulum (data from a similar grass field approximately 1 km away).

Soil sampling of earthworms (on 23 April) just prior to the investigation period gave mean population estimates of 122 (± 3.8 SE; n = 4) individuals m^{-2}. Based on number, species composition was 40% *Aporrectodea rosea* (Savigny), 30% *Aporrectodea longa* (Ude), and 30% *Lumbricus terrestris* (Linnaeus). Life-stage composition was 79% juveniles and 21% mature individuals (adults plus sub-adults).

Trap units
Earthworm surfacing and movement over the soil surface were recorded using specially designed trap units that allow assessment of migration degree and direction (Figure 25-1). A trap unit consists of 2 troughs separated by an H-shaped barrier; this barrier ensures that earthworms enter each trough from only 1 main direction. The apparatus is constructed of 2-mm thick, gray PVC.

A steel frame driven into the ground produced a clearly defined outline for digging the hole into which the trap unit was to be fitted. The trap unit was then carefully positioned within the ground so that the top of each trough was flush with the soil sur-

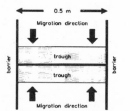
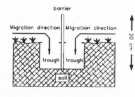

Figure 25-1   Trap unit used for recording earthworm surface migration. Plan and cross section of a trap unit within the soil profile are portrayed.

face, with no gaps between the trough edges and grass sward. The H-shaped barrier extended approximately 12 cm above and 8 cm below ground. A small amount of water (depth approximately 1.5 cm) poured into each trough stopped earthworms from escaping and helped to keep them alive (preventing desiccation and rinsing chemical-exposed individuals) for subsequent examination. A wooden stick acted as a ladder, allowing other accidentally caught organisms (e.g., insects, small mammals) to escape from the troughs.

Experimental design

The field was divided into 3 blocks, each comprising 3 experimental plots (Figure 25-2). Each experimental plot (30 m × 15 m) consisted of a sprayed (treatment) and a nonsprayed (control) subplot (each 15 m × 15 m), the treatment areas having their own abutting control areas in order to minimize any effect of possible variation in earthworm spatial distribution pattern (density or species) across the study site. Sprayed and nonsprayed areas alternated between N- and S-ends of the experimental plots to control for possible directional preferences due to site characteristics and other orientational cues. Within each experimental plot, trap units were placed in 3 rows (4 trap units per row at 1-m intervals) positioned centrally within the treated and control areas, as well as along the mid borderline between the 2 (Figure 25-2). There were thus 12 troughs per subplot.

The carbamate fungicide benomyl was chosen as a test chemical because it is known to be toxic to earthworms in the field [4–8]. Based on this literature, 3 benomyl concentrations aimed at less radical treatment were considered. The 3 blocks allowed 3 replicates per chemical concentration and their respective controls. Treatment type was allocated randomly within each block.

Spraying took place on 1 May 1996 (day 0); the fungicide was applied by a tractor-mounted boom spray unit (low-drift nozzles) at 203 kPa and a tractor speed of 3.8 km h^{-1}. The 3 benomyl treatments were 0.5, 1.0, and 2.0 kg a.i. ha^{-1} (+ 400 L

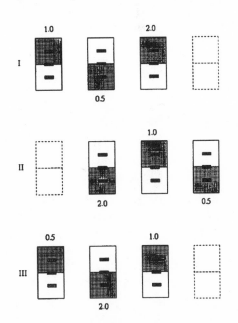

Figure 25-2   Experimental design. Present study involved 3 experimental blocks (I-III), each comprising 3 experimental plots (rectangles), which were divided into sprayed (hatched) and nonsprayed subplots. Each horizontal bar represents a row of 4 trap units, with 1-m intervals between trap units. Numbers denote benomyl concentration in kg a.i. ha^{-1}. Plots outlined by dashed lines were part of another investigation.

H_2O ha⁻¹). Weather conditions (wind-still and dry) were ideal for spraying according to the experimental design, producing a well-defined borderline between treated and control subplots. During spraying, the trap units were temporarily covered by plastic sheeting to prevent chemical contamination. In order to ensure that the fungicide reached the soil substrate relatively quickly, the entire area was artificially irrigated (40 mm precipitation) over 2 and 3 May (days 1 and 2 after spraying).

The 108 trap units (216 troughs) were examined daily (09.00 to 12.00 h) from the morning prior to spraying (day 0) to 35 d after spraying. Earthworms collected from each trough were subsequently counted, identified according to species and life stage, and weighed (individual fresh weight, after storage in glass jars for 3 d at 10 °C in order to empty the gut). Throughout the study, care was taken to hold disturbance of the plots to a minimum; in particular, it was ensured that the ground in front of the troughs remained untrampled.

Results

In all, 4258 earthworms representing 7 species were collected from the trap units over the investigation period, all specimens being alive and active. The species found were *Aporrectodea longa* (Ude), *Lumbricus terrestris* (Linnaeus), *Aporrectodea rosea* (Savigny), *Aporrectodea caliginosa* (Savigny), *Allolobophora chlorotica* (Savigny), *Octolasion cyaneum* (Savigny) and *Lumbricus festivus* (Savigny).

During the investigation period, daily mean soil temperature at 10 cm depth ranged from 7.5 °C to 13.7 °C and at 30 cm depth from 7.1 °C to 12.2 °C, and daily natural precipitation ranged from 0 mm to 12.5 mm (Figure 25-3).

The daily mean numbers and cumulative daily means of earthworms caught per subplot for the 3 different treatments and their respective controls over the entire investigation period are shown in Figures 25-4 through 25-6. Peak captures coincide with periods of rainfall, this being evident for both treated and control areas. The influence of precipitation on surface movement was particularly clear over days 17 to 20 when rainfall occurred over 4 consecutive days, amounting to 31.5 mm. Irrigation over the first 2 d after spraying appeared to have little effect on surfacing.

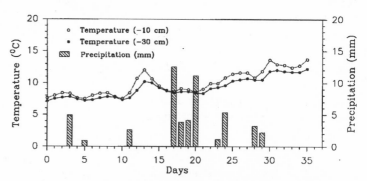

Figure 25-3   Climatic data covering the study period. Daily mean soil temperature at 10-cm and 30-cm depth, and rainfall are given for 1 d prior to spraying (day 0) to 35 d after spraying.

For 0.5 kg a.i. ha⁻¹ benomyl treatment (Figure 25-4), captures tended to be greater in sprayed areas compared with their respective controls over the entire study. The difference between sprayed and nonsprayed areas is particularly evident from the cumulative daily means; each period of rainfall tended to increase the distance between the 2 curves such that a 2.5-fold difference existed by the end of the investigation.

A similar pattern was shown for treatment with 1.0 kg a.i. ha⁻¹ benomyl (Figure 25-5), though the higher level of captures in sprayed areas compared with controls was most evident over the period prior to day 17. Again, the difference between sprayed and nonsprayed areas is clearly seen from the cumulative daily means. The 2 curves tended to diverge until day 16, converge over the next few rainy days, and then parallel each other until the end of the study; the final difference was 1.4-fold.

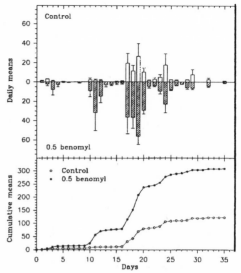

Figure 25-4 Number of earthworms captured per subplot over the study period for areas sprayed with 0.5 kg a.i. ha⁻¹ benomyl and respective control areas. Upper graph: daily means plus SE (n = 3). Lower graph: cumulative means.

A somewhat different pattern was evident for 2.0 kg a.i. ha⁻¹ benomyl treatment (Figure 25-6). Captures were again greater in sprayed areas compared with controls until day 16, but then the 2 curves for the cumulative daily means converge and cross such that from day 19 onward, control captures exceed those from treated areas, thus ending with a 1.3-fold difference between the 2.

A closer examination of the cumulative daily means over the first week at the trough level (n = 36 per treatment/control; Figure 25-7) reveals that the difference in captures between control and treated areas was significant by day 4 after spraying for 0.5 kg a.i. ha⁻¹ benomyl (Mann-Whitney U test P < 0.02), and already by day 2 for 1.0 and 2.0 kg a.i. ha⁻¹ benomyl (Mann-Whitney U test P < 0.002 and P < 0.001, respectively). Although capture levels for the first week's cumulative daily means appear to increase from lower to higher pesticide concentration, relative to the respective controls there is a 2.7-fold difference for 0.5 kg a.i. ha⁻¹ benomyl, a 13.6-fold difference for 1 kg a.i. ha⁻¹ benomyl, and an 8.3-fold difference for 2 kg a.i. ha⁻¹ benomyl.

With regard to species composition based on number (Figure 25-8), in control areas the dominant species was *A. longa*, which ranged from 68.5 to 85.4%, followed by *A. rosea* ranging from 9.2 to 15.5%, then *L. terrestris* ranging from 3.5 to 15.5%, the remaining 4 species combined contributed only 0.2 to 0.9%. The domination of *A. longa* on control

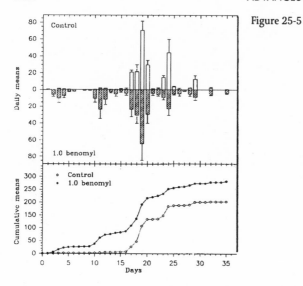

Figure 25-5   Number of earthworms captured per subplot over the study period for areas sprayed with 1.0 kg a.i. ha⁻¹ benomyl and respective control areas. Upper graph: daily means plus SE (n = 3). Lower graph: cumulative means.

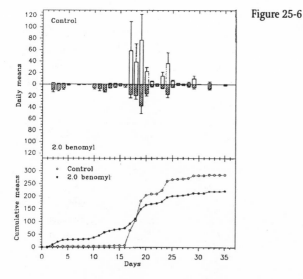

Figure 25-6   Number of earthworms captured per subplot over the study period for areas sprayed with 2.0 kg a.i. ha⁻¹ benomyl and respective control areas. Upper graph: daily means plus SE (n = 3). Lower graph: cumulative means.

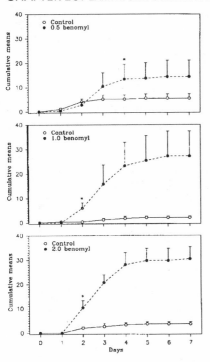

Figure 25-7   Cumulative mean number of earthworms captured per subplot over the first week for areas sprayed with different concentrations of benomyl and respective control areas. Upper graph: 0.5 kg a.i. ha⁻¹ benomyl. Middle graph: 1.0 kg a.i. ha⁻¹ benomyl. Lower graph: 2.0 kg a.i. ha⁻¹ benomyl. SE given (n = 36). (*) denotes the first occurrence of the significant difference between treatment and respective control.

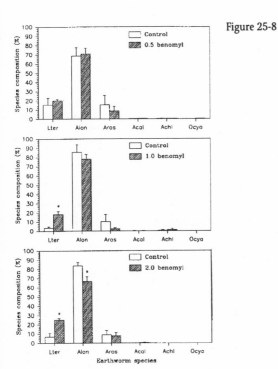

Figure 25-8   Species composition (%) based on number of earthworms captured per subplot for areas sprayed with different concentrations of benomyl and respective control areas. Upper graph: 0.5 kg a.i. ha⁻¹ benomyl. Middle graph: 1.0 kg a.i. ha⁻¹ benomyl. Lower graph: 2.0 kg a.i. ha⁻¹ benomyl. SE given (n = 3). (*) denotes a significant difference between treatment and respective control. Species as follows: Lter = *Lumbricus terrestris*; Alon = *Aporrectodea longa*; Aros = *Aporrectodea rosea*; Acal = *Aporrectodea caliginosa*; Achl = *Allolobophora chlorotica*; Ocya = *Octolasion cyaneum* (data for *Lumbricus festivus* are too few for presentation).

samples was even more evident based on biomass; the range was 83.2 to 95.0%. No significant changes in species composition were evident as a result of spraying with 0.5 kg a.i. ha^{-1} benomyl compared with the respective controls. However, spraying with 1.0 kg a.i. ha^{-1} benomyl resulted in a significant increase in the proportion of *L. terrestris* (18.2%

Table 25-1 Number of earthworms captured per subplot for the 3 dominant species over the study period (35 d) for areas sprayed with different concentrations of benomyl and respective control areas. Mean ± SE (n = 3).

Species	Control	0.5 kg a.i. ha^{-1} benomyl
A. longa	88 ± 52	210 ± 50
L. terrestris	10.3 ± 3.2	61.7 ± 17.3
A. rosea	23.3 ± 12.8	33.7 ± 21.8
Species	Control	1.0 kg a.i. ha^{-1} benomyl
A. longa	206.5 ± 51.9	245.7 ± 49.5
L. terrestris	7.7 ± 2.6	55.8 ± 13.2
A. rosea	17.5 ± 10.8	5.5 ± 3.3
Species	Control	2.0 kg a.i. ha^{-1} benomyl
A. longa	248 ± 151	150.3 ± 39.1
L. terrestris	8.7 ± 2.0	54.0 ± 6.7
A. rosea	27.7 ± 13.4	15.7 ± 5.6

treated versus 3.5% control; Mann-Whitney U test, $P < 0.05$). Also, spraying with 2.0 kg a.i. ha^{-1} resulted in a significant increase in the proportion of *L. terrestris* (25.0% treated versus 6.9% control; Mann-Whitney U test, $P < 0.05$) and a concomitant significant decrease in the proportion of *A. longa* (66.7% treated versus 83.9% control; Mann-Whitney U test, $P < 0.05$). Those significant differences that were based on number were likewise found when based on biomass.

Analysis of mean numbers of trapped earthworms per subplot for the above 3 species (Table 25-1) indicates trends for enhanced surface migration by all species as a result of spraying with 0.5 kg a.i. ha^{-1} benomyl, by *A. longa* and *L. terrestris* for 1.0 kg a.i. ha^{-1} benomyl, and by just *L. terrestris* for 2.0 kg a.i. ha^{-1} benomyl. Only increases for *L. terrestris* reached significance (Mann-Whitney U test, $P < 0.05$).

The results of life-stage composition based on number for all species combined (Figure 25-9) show that in control areas juveniles dominated the samples with 70.2 to 73.0%, whereas subadults comprised only 1.8 to 4.3% and adults 25.1 to 26.7%. Based on biomass, juveniles ranged from 54.1 to 60.1%, subadults from 2.6 to 5.8%, and adults from 37.3 to 42.4%. No significant changes in life-stage composition (based on number or biomass) were evident as a result of spraying compared with respective controls.

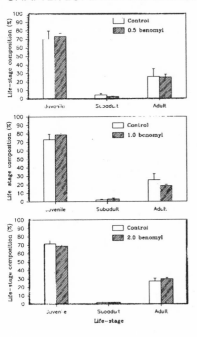

Figure 25-9   Life-stage composition (%) based on number of earthworms captured per subplot for areas sprayed with different concentrations of benomyl and respective control areas. Upper graph: 0.5 kg a.i. ha^{-1} benomyl. Middle graph: 1.0 kg a.i. ha^{-1} benomyl. Lower graph: 2.0 kg a.i. ha^{-1} benomyl. SE given (n = 3).

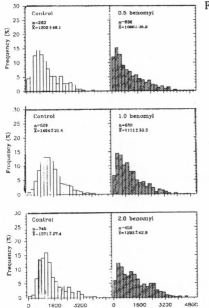

Figure 25-10   Weight-class frequency distribution (%) for *A. longa* for areas sprayed with different concentrations of benomyl and respective control areas. Upper graph: 0.5 kg a.i. ha^{-1} benomyl. Middle graph: 1.0 kg a.i. ha^{-1} benomyl. Lower graph: 2.0 kg a.i. ha^{-1} benomyl. Data are analyzed according to 200 mg freshweight categories.

A weight-class frequency distribution is presented for the dominant species, *A. longa* (Figure 25-10). In control areas the mean individual weight ranged from 1,202 mg final weight to 1,464 mg final weight, whereas in treated areas the mean individual weight ranged from 1086 mg final weight to 1292 mg final weight. For all 3 concentrations of benomyl, mean body weight of captured earthworms was significantly lower in sprayed areas compared with respective controls (Mann-Whitney U test, P < 0.001), i.e., there was a higher incidence of smaller-sized juveniles in benomyl-treated areas.

Discussion

Surface migration by lumbricid earthworms appears to be a commonplace, natural phenomenon involving many species regardless of ecological category [1,9]. During the evening and night, when climatic conditions are favorable (cool and damp), lumbricids often emerge from their burrow systems and move over the surface of the ground, as evidenced by the numerous surface trails left behind in patches of mud and silt sediment [1].

The vertical extension of their soil habitat, which includes the soil and air boundary layer, gives earthworms access to a zone where relatively nonrestricted horizontal movement is possible, so enhancing the efficiency of resource-seeking behavior. The distance traveled during these crepuscular/nocturnal surface forays is often considerable; measurement of trail length depicting a night's activity indicates that larger species such as *L. terrestris* and *A. longa* may crawl over 19 m [1] or over 23 m (unpublished data, Mather and Christensen). Even so, earthworm migration in the field has been considered seldom in ecotoxicological studies, as reflected by the use of small experimental plots (sometimes as small as 3 m^2) with no physical barriers [4,8,10].

The capture of numerous earthworms (1,826 individuals over 35 d) comprising several species in control areas in the present investigation again infers that surface migration is a normal activity. This is further supported by pitfall trapping in a pasture in France where 1,104 earthworms and 8 lumbricid species were found when sampled twicemonthly over a 13-month period [11] and in a pasture in New Zealand where adult *A. caliginosa* were noted to move over the surface during wet weather [12]. Also, direct observations of surface migrating earthworms in Danish grassland have indicated a total of 10 lumbricid species, both litter and soil dwellers [9].

Under conditions of environmental stress (natural or anthropogenic), earthworms can attempt to avoid hostile conditions either by migrating from the habitat, i.e., performing horizontal surface migration, or by remaining within the habitat but moving to a different zone within the soil profile, i.e., performing downward vertical migration. Sudden exposure to noxious substances or interaction with predators seems to induce surfacing as an avoidance reaction to such stimuli; this innate behavior allows a rapid removal from harmful situations [cf. 2,3,9].

Although certain chemicals are known to expel earthworms from the soil, and indeed have been used to estimate earthworm activity and population density (e.g., solutions of mercuric chloride [13], potassium permanganate [14], mowrah meal [15], formalin [16], chloroacetophenone, [17] and mustard [18]), a possible expellant effect of chemicals used in agricultural practice appears not to have been considered until recently [2,3]. The present investigation clearly demonstrates that the carbamate fungicide benomyl likewise induces earthworms to emerge from the soil and move over the surface of the ground. Similar results may be predicted for other pesticides.

In both sprayed and control areas, earthworm surface migration and rainfall tend to be synchronized; bouts of precipitation (particularly prolonged) enhance the level of migration. Rainfall is known to increase the surface activity of lumbricids [1,9,12,19–23]. The low (or lack of) response by control earthworms to both irrigation and rainfall during the first 2 weeks of the present investigation is likely to relate to an extended period of dry weather prior to commencement of the study (totals of merely 5.3 mm during April and 0.7 mm during March). Dry conditions lead to downward vertical migration and inactivity of lumbricids in pasture [24]. In addition, the soil surface environment was probably unfavorable during the first fortnight due to stronger winds (up to 10 m s^{-1}), lower temperatures, and shorter grass height (about 6 cm).

Despite these apparently suboptimal conditions prevailing at the soil surface over the initial period of the study, spraying with benomyl induced surfacing (also during daylight and even sunshine), thereby causing earthworms to become exposed to environmental conditions they would otherwise avoid by remaining within the soil body. The elevated level of surface migration relative to normal behavior became significant already by day 2 for the higher benomyl concentrations (1 and 2 kg a.i. ha^{-1}) and by day 4 for the lowest benomyl concentration tested (0.5 kg a.i. ha^{-1}). This is a comparatively quick sublethal ecotoxicological response for a field-based study, compared with assessments using other parameters affecting population density and biomass, e.g., growth and reproduction.

The different patterns evident for the cumulative mean curves (treatment versus control) relative to benomyl concentration (i.e., "diverging" for 0.5 kg a.i. ha^{-1}, "converging" for 1.0 kg a.i. ha^{-1}, "crossing" for 2.0 kg a.i. ha^{-1}) are likely to be a function of egress from the population due to both active surface migration (non-captures plus experimental removal of captured individuals) and mortality, as well as suppressed surface activity (e.g., downward vertical migration).

Mortality can be expected to be greater in areas with highest benomyl concentrations, not only because of lethal effects of the chemical per se but also because of the expellant effect leading to an increased risk of a) chemical exposure on the soil surface, b) exposure to deleterious environmental conditions such as sun and wind leading to dehydration, and c) predation by both nocturnal and diurnal predators. Direct observations of areas treated with 2.0 kg a.i. ha^{-1} benomyl revealed many dead or dying earthworms on the soil surface, as well as an absence of surface- casting activity. Reductions in lumbricid earthworm populations and surface casting have been reported for orchards (with grass cover) sprayed on 13 occasions with 0.28 kg a.i. ha^{-1} benomyl over a 4-month period (= 3.64 kg a.i. ha^{-1}), the effect being seen not only for density and biomass of all species combined but also for each individual species [5].

A greater mortality leaves fewer individuals available for trap capture. This is the most likely explanation for the crossing pattern whereby control captures exceed treatment captures in areas with highest benomyl concentration, though some suppression of surface activity due to downward migration is also plausible. Decreases in overall captures at highest benomyl concentration appear to be brought about by reductions in the anecic species *A. longa* and the endogeic species *A. rosea*. In contrast, surfacing by the anecic *L. terrestris* is enhanced even at this highest concentration.

At lowest benomyl concentration, on the other hand, where mortality is lower and the chemical has a considerable stimulating effect on surface migration, a diverging pattern results. All 3 main species appear to be influenced regardless of ecological category and associated burrow systems. Enhanced migration probably reflects the lowering of habitat suitability due to benomyl treatment. For example, a reduction in food quality is inferred by reports of a significant decrease in feeding by captive *L. terrestris* when leaf material had spray deposits of benomyl at 0.87 mg cm^{-2}, and there was cessation of feeding when deposits were 1.75 mg cm^{-2} [25]. In addition, benomyl treatment may reduce the amount of soil fungal mycelia, which is an important food source. Thus, migratory behavior in chemical-treated areas is probably aimed at locating a more suitable habitat.

For the intermediate benomyl concentration, a moderate mortality will result in the converging pattern observed for surface migration in sprayed areas versus controls. Here, enhanced surfacing due to treatment still appears for *L. terrestris* and *A. longa*, whereas trap captures for *A. rosea* tend to be reduced; this latter trend probably reflects both mortality and downward migration.

Regarding species composition, according to trap samples in control areas, species dominance ranked *A. longa* > *A. rosea* > *L. terrestris*, the remaining 4 species having only minor contribution. This rank order differs from that found for soil sampling, namely *A. rosea* > *A. longa* = *L. terrestris*, suggesting species-specific differences in behavior. Higher concentrations of benomyl significantly altered the proportions of the species *L. terrestris* and *A. longa*; the chemical induced an increase and a decrease, respectively.

The finding that spraying with benomyl did not induce any significant changes in life-stage composition relative to the controls is perhaps surprising because chemical exposure is likely to be greater for juveniles, owing to their habit of dwelling in the upper soil profile. However, life-stage-specific effects of benomyl may be masked by species differences; a consideration of life-stage for separate species (in preparation) will elucidate this.

The weight-class frequency distribution for the dominant species *A. longa* indicates that benomyl generally induces a higher proportion of smaller-sized individuals to surface migrate. This may arise from a greater chemical exposure of these smaller individuals not only because of their position within the top soil but also because of their higher surface area to volume body ratio.

Chemically induced changes in the migratory behavior of lumbricids in the field, as demonstrated in the present study, can lead to alterations in population density or biomass and species diversity, so affecting the structure and function of earthworm communities. Egress (migration and/or mortality) due to chemical treatment will, in general, have most relevance for low-density populations and small-area habitats, and repeated annual treatment will probably inhibit recolonization through immigration. Reduction of the deep-burrowing anecic species will lower the engineering capacity of the earthworm population, thus affecting soil structure.

Surface migration by free-living earthworms appears to be a sensitive behavioral parameter for assessing side effects of chemicals on field populations, and it is a parameter with a high degree of ecological relevance. The present investigation not only reveals elevated levels of earthworm surfacing due to benomyl treatment but also highlights species- and body-size-specific differences. Further evaluation of the migratory response is necessary in order to elucidate the proportion of the population affected, as well as the distance traveled and the pattern of movement (i.e., whether direct or circuitous). Studies examining these aspects are under way.

Acknowledgment - This investigation is part of a 3-year study aimed at examining the influence of pesticides on the migration and reproduction of earthworm populations and at developing a model to predict side effects. The study is supported by a grant from the Danish Ministry of the Environment. We wish to express thanks to the following people: Erik Pram-Nielsen, Steffen Christensen, Mads Bisgård, and Arne Grud at Foulum Research Centre for the use of Askhøj as an experimental site and invaluable practical assistance with spraying and irrigation; Birgit Sørensen at Foulum Research Centre for providing climatic data; and Martin Holmstrup at the National Environmental Research Institute in Silkeborg for soil sampling data on earthworms.

References

1. Mather JG, Christensen O. 1988. Surface movements of earthworms in agricultural land. *Pedobiologia* 32:399–405.

2. Mather JG, Christensen OM. 1994. Earthworms as bio-indicators of side-effects of fungicides in arable land: A pilot study of surface migration in the field and a laboratory study of mortality, growth and reproduction. *Ministry of Environment, Danish Environmental Protection Agency* 7:1–135.

3. Christensen O, Mather JG. 1994. Earthworms as ecotoxicological test organisms. *Ministry of Environment, Danish Environmental Protection Agency* 5:1–99.

4. Tomlin AD, Gore FL. 1974. Effects of six insecticides and a fungicide on the numbers and biomass of earthworms in pasture. *Bull Envir Contam Toxicol* 12:487–493.

5. Stringer A, Lyons CH. 1974. The effect of benomyl and thiophanate-methyl on earthworm populations in apple orchards. *Pestic Sci* 5:189–196.

6. Tomlin AD, Tolman JH, Thorn GD. 1981. Suppression of earthworm (*Lumbricus terrestris*) populations around an airport by soil application of the fungicide benomyl. *Prot Ecol* 2:319–323.

7. Lofs-Holmin A. 1981. Influence in field experiments of benomyl and carbendazim on earthworms (Lumbricidae) in relation to soil texture. *Swedish J Agric Res* 11:141–147.
8. Edwards PJ, Brown SM. 1982. Use of grassland plots to study the effect of pesticides on earthworms. *Pedobiologia* 24:145–150.
9. Mather JG, Christensen O. 1992. Surface migration of earthworms in grassland. *Pedobiologia* 35:51–57.

Chapter 26

Variation in earthworm populations and methods for assessing responses to perturbations

John Bembridge, T J Kedwards, P J Edwards

Earthworm populations are influenced by a number of biotic and abiotic factors that may affect the population size, biomass, and species composition. Large fluctuations may be observed over time, particularly between seasons and years and in response to specific perturbations. Data from long-term field experiments for periods of up to 10 y are presented to illustrate these points. In an agricultural environment, further factors, e.g., crop, cultivation, and chemical application, may have additional effects. The impact of these factors should be assessed in comparison to natural variation.

As a result of the first International Workshop on Earthworm Ecotoxicology, a 30 to 50% reduction in population density was suggested, and failure to recover to untreated control levels within 12 months could be used as a criterion for defining effects. However, a further understanding of the significant effects on populations was required. This chapter explores the biological significance to earthworm populations of natural and imposed perturbations and suggests some possible methods of identifying significant effects.

To study the effect of an agricultural chemical in a field experiment, it is important to have a base line against which to compare the treated population; this normally takes the form of an untreated control. A method to assess the significance of any differences between treated and untreated plots must be identified. Measurements such as the total number of earthworms, ratio of adults to immatures, total biomass, and individual species comparisons may be useful. Recovery to the same status as an untreated population over a set period of time has been identified as a useful indicator. However, it is often difficult to gain an overall view of the effects of an agricultural chemical on an earthworm population.

Redundancy analysis has been used in other areas of ecotoxicology to assess the effects of perturbations on complex communities and may be a useful statistical tool for examining earthworm data. This multivariate technique allows an overall assessment of the earthworm population as a community by identifying any changes in size and structure and allowing comparison to untreated populations. Examples showing the use of this technique to identify recovery times from field study data are presented and compared to other methods.

Earthworms play a key role in maintaining the fertility of the soil through drainage, aeration, and the incorporation of organic matter (OM) [1,2], and they have therefore been adopted as important indicator species used to assess the safety of agricultural chemicals to soil organisms. A tiered approach to testing has been adopted, starting with relatively simple studies in the laboratory and culminating in studies carried out in the field that aim to mirror the conditions under which the chemical will normally be applied. Such field studies are assumed to be the most realistic indicators of any potential risk to the soil macro-fauna.

Field studies are designed to identify any acute or chronic effects on the overall earthworm community, and this usually is carried out by assessing changes in the population size or biomass in addition to the age and species structure. However, to accurately assess the ecological impact of a perturbation on an earthworm community, it is essential to understand the natural fluctuation of populations and their response to abiotic and biotic factors.

Earthworm populations are dynamic, with constant changes in their size, biomass, and relative species makeup due to seasonal variations, direct perturbations, or a combination of interacting factors. This leads to problems in measuring changes in communities and relating them to particular factors. Without an understanding of how earthworm communities vary in the absence of imposed perturbations, it is impossible to accurately interpret the importance of the effects of agricultural practices on such communities.

In this chapter, the major biotic and abiotic factors that affect earthworm populations are reviewed. Data from a long-term field trial are presented to assess the relative importance of these factors, and a new statistical approach is suggested to allow an ecological interpretation of the data.

Variations in earthworm communities

The size and structure of earthworm communities are known to be affected by a number of factors: moisture, temperature, food availability, pH, soil structure, predation, parasitism, and inter- and intra-specific interactions. These can result in large changes in communities or may act as limiting factors in the growth or biodiversity of the community. The management of agricultural land may also have an influence through manipulation of 1 or more of these factors or by direct toxicity.

The main factors that are believed to influence earthworm populations in the field are moisture and temperature, and these normally interrelated factors result in the seasonal fluctuations seen in the field [3–6]. Lack of soil moisture affects populations by causing increased mortality, reducing cocoon production [3], and changing behavior and activity levels. Deep-burrowing species such as *Lumbricus terrestris* move deeper into the soil, while other species may become quiescent or enter diapause. An added complication is that sampling efficiency is affected by the depth and activity levels of the earthworms, and in periods of drought populations, they may be underestimated [7].

Extremes of temperature can have dramatic effects on the size of populations; periods of cold weather have been shown to reduce populations by 70% and decrease biomass by 80% [8]. The removal of insulating levels of plant cover is quoted by Hopp and Hopkins [9] as the main cause of reduced earthworm populations in arable fields in America. The effects of high temperature are less clear, and although experiments have been conducted in the laboratory [2,10], in the field it is difficult to separate the effects of high temperatures and low moisture contents. There are, however, clear relationships between temperature and the rate of production of cocoons, length of incubation, and development of juveniles [11–13] with earthworms exhibiting an optimum temperature for growth and reproduction.

The supply and distribution of food also has a major influence on earthworm populations. The addition of OM, e.g., as manure, can increase the overall number of earthworms, and this has been exhibited in grassland and arable land [13,14].

Other factors that may affect the seasonal abundance of communities include predation and parasitism. Earthworms are food items for a wide range of birds, mammals, reptiles, and invertebrates [15,16]. The rate of predation is linked to the availability of earthworms and alternative food supplies. However, as with parasitism and disease, they are not considered to be prime factors in the fluctuation of earthworm communities.

Edaphic factors such as pH and soil type do affect earthworm communities, but these are usually limiting factors affecting size and diversity rather than causing direct responses.

The high level of variation in earthworm communities with a number of different species and in a range of soils has been demonstrated by many authors [3–6, 17–24] either within a year or over a longer period. Evans and Guild [3] showed a variation within 12 months of 5 to 300 worms in their samples, while Van Rhee and Nathans [5] reported fluctuations ranging from 150 to 750 worms m^{-2} within the 12 y of their study.

Earthworms in agriculturally managed soil are exposed to further perturbations. These factors can be linked to the main inputs involved with agriculture, which Edwards [25] describes as cultivation, cropping patterns, fertilization, and crop protection. It is clear that 2 of these, cropping patterns and fertilization, relate directly to the availability of food and distribution of OM.

The effects of cultivation on earthworms remain a contentious point with conflicting results in the literature. Early investigators such as Zicsi [26] suggested that the difference in earthworm numbers between arable land and grassland was due to cultivation, particularly plowing. While cultivation can kill earthworms by mechanical damage and increased predation, this is normally only the case where highly disruptive forms of management, such as rotary cultivation are used. In these cases, Lofs has shown that populations may be reduced by as much as 70% [8]. However, work by Edwards and Lofty [27] suggests that the effect of plowing may initially increase the number of earthworms by improving the distribution of OM in the soil, but that repeated cultivation results in declining numbers.

The introduction of direct-drilling or no-till agriculture allowed the comparison of earthworm populations from cultivated and undisturbed areas. Direct drilling is seen to have a highly beneficial effect on earthworm populations with higher numbers and larger biomass. The effects on populations of permanent burrow dwellers like *Lumbricus terrestris* were found [28] to be more pronounced than on nonpermanent burrowers.

A number of agricultural chemicals have been shown to affect earthworm populations either through acute toxicity or by chronic effects on reproduction and development. It is important to distinguish the indirect effects of crop protection chemicals from direct ones when they are tested in the field. For example, an herbicide may kill the surface vegetation, causing an initial increase in OM, which supports a higher earthworm population; however, in subsequent years the amount of new OM introduced into the soil will be less.

Therefore, it is apparent that in arable land earthworms are subjected to a number of pressures that influence the size and structure of the community. These pressures may be due to naturally occurring factors, such as environmental conditions, that are caused by an indirect effect of agricultural management (e.g., cropping patterns and the availability of food) or that are caused by a direct effect such as an agrochemical.

To assess the effects of crop protection products on an earthworm community, it is important to bear in mind the levels of variation that can occur from other factors and hence try to assess the biological and ecological significance of any changes seen. Grieg-Smith [29] stated that a 30 to 50% reduction in population density and failure to recover to untreated control levels should be used as criteria for defining a serious effect in the field. However, he noted a lack of understanding of the tolerance of populations to reduction in size or biomass.

As illustrated, changes in population density of 30 to 50% and considerably higher are common in naturally or agriculturally managed communities. It is the ecological effect on the community that is important, and therefore, the concept of recovery to the status of an untreated community is most useful.

Experimental data

Methods

Long-term studies were carried out to assess the toxicity of agrochemicals to earthworms. The trials were set up using small grassland plots (6m × 6m) as described by Edwards and Brown [30]. Guard rows of 1.5 m were used to minimize the effects of earthworm movement between plots or the effect of treatments used on adjacent plots. Sampling was carried out for a period of 10 y using the formalin expellent method [31] in spring and autumn to coincide with periods of earthworm activity. At each sampling occasion two 60 × 60 cm squares were sampled from each of the 3 replicates for each treatment.

Two untreated control treatments were included in the test design. During this study, the control plots received no chemical applications but were treated in the same way as the treated plots in all other respects.

The fungicide Benlate (containing benomyl), which is known to be toxic to earthworms, was sprayed at rates of 5 or 2 kg a.i. ha^{-1} at approximately yearly intervals for 10 y. Additionally a number of experimental pesticides were included in the trial but the details and results of these applications will not be considered here.

Identification of worms was carried out on live specimens and was to species level following the terminology of the Linnean Society at the time [32]; some of the species groupings used would no longer be considered taxonomically correct [33], but because they were consistent throughout the study, they are sufficient for this purpose. Biomass was recorded in terms of fresh weight.

Results

The total number of earthworms and biomass for the plots are presented in Figures 26-1 and 26-2. The high degree of variation in the earthworm communities in the control plots is illustrated in Tables 26-1 and 26-2, with large fluctuations in numbers and biomass from year to year, season to season, and between consecutive samplings.

The largest decrease in the populations of the control plots was seen in the autumn of 1983, when summer temperatures were very high and rainfall was low at the study site. These conditions may have affected the depth and activity levels of the earthworms at the autumn sampling, leading to an underestimation of the population. Further work will be required to prove a direct correlation of weather conditions to earthworm populations, but it is likely that they were the major factors.

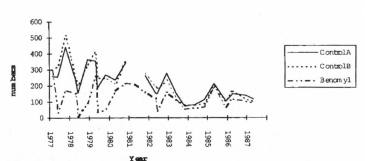

Figure 26-1   Earthworm numbers over a 10-y period for control and benomyl plots

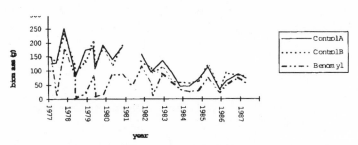

Figure 26-2   Earthworm biomass over a 10-y period for control and benomyl plots

The numbers of earthworms and their biomass in the Benlate-treated plot were clearly correlated to the application of the chemical, with marked declines following each application in the spring and a gradual recovery by the autumn.

The age structure of the community may be a useful indicator of its stability, and this can be examined by calculating a ratio of the number of juveniles to the number of adults. The ratios for the control plots are shown in Table 26-1. These change substantially through-out the study period, with swings from adult- to juvenile-dominated communities.

Table 26-1   Range of earthworm numbers, biomass, age structure, and species ratios in 2 untreated grassland plots over a 10-y period

Endpoint	Plot A	Plot B
Individuals m^{-2}		
Overall range	72–442	80–512
Autumn	72–442	80–512
Spring	82–354	80–410
Biomass (g)		
Overall range	35–253	36–233
Autumn	35–253	36–233
Spring	46–182	57–202
Immature:adult ratio		
Overall range	0.55–5.14	0.57–4.80
Autumn	0.55–4.29	0.57–4.00
Spring	0.84–5.14	0.58–4.80
Non *Lumbricus:Lumbricus* ratio		
Overall range	0.7–7.38	0.75–5.45
Autumn	0.82–7.38	0.75–5.45
Spring	0.70–4.05	1.02–4.72

Table 26-2   Fluctuations in earthworm numbers in 2 untreated grassland plots over a 10-y period

Untreated grasslands	Autumn to autumn		Spring to spring		Consecutive samples	
	Max. increase	Max decrease	Max. increase	Max decrease	Max. increase	Max decrease
Plot A	+52%	−74%	+157%	−45%	+134%	−52%
Plot B	+42%	−65%	+148%	−50%	+122%	−50%

A simple indication of the biodiversity of the community can be gained by calculating the ratio of non-*Lumbricus* species to *Lumbricus* species. The results in Table 26-1 show that the species ratio within the control communities, as with all the other parameters, varies over time.

Analysis

The control plots show a high degree of variation in terms of numbers, biomass, age structure, and species diversity that must be the result of naturally occurring factors. Therefore, earthworm populations in undisturbed land are subject to fluctuations, which can be substantial, in their size, biomass, and age structure. This must be taken into account when analyzing field trial data.

Traditionally, there have been a number of ways in which data produced from field studies have been used to detect effects. The most common methods are to look at numbers or biomass [34], but to gain a fully ecological view as to the significance of any changes, the age structure and species diversity of the population should also be considered.

The large collection of data from a field trial can be analyzed by a number of different methods to allow interpretation of the effects of a compound on the earthworm population. In many cases, it is difficult to interpret these data easily because there may be differences between species or age groups, with some areas appearing to be significantly affected while others are not. It is important to consider the overall effect on the whole community; this is not always possible using traditional univariate statistical methods.

Univariate statistical methods are normally used for the analysis of both laboratory and field data. However, as we recognize that field experiments are highly complex in comparison to laboratory studies, it may be preferable to use different approaches. Univariate methods are well accepted for laboratory studies and allow the easy interpretation of data that are usually based upon just 1 species. In the field, however, there normally are a number of different species and variables. If these are analyzed separately, it may be difficult to weight them and interpret overall effects on a community. Using univariate statistics, it would be necessary to carry out an analysis of variance (ANOVA) on each group for each sampling interval to investigate effects, resulting, in this case, in excess of 600 analyses.

One possible alternative is the use of a multivariate approach to looking at the overall effect of a perturbation on a community. Multivariate techniques allow the use of the whole dataset in 1 powerful statistical analysis to identify a community's status.

A number of ordination techniques have been used to interpret the data. All ordination methods aim to simplify the large, multivariate data matrix derived from such studies into a set of new hypothetical axes that summarize the variation within the dataset. In this way, it is possible to derive a community-level picture of the earthworm population throughout the study period.

The ordination techniques specifically used in this study have been used with considerable success in recent years in aquatic ecotoxicology [35–41]. Redundancy analysis (RDA) is the constrained form of the more well-known principal component analysis (PCA), which summarizes the variation within the collated data in new axes, taking into account the data structure (i.e., which datapoints originate from specific treatments, sampling times, and replicates). The output from such an analysis can be plotted as a biplot on which both species and sampling times are jointly represented on a single graph.

Figure 26-3 provides a graphical interpretation of the response of the treated and untreated communities. The lines start at similar points, representing the similar community structure prior to treatment. Following treatment, the 2 lines diverge rapidly, showing that there has been an effect on the community of the treated plot. It should be noted that the control line, which combines data from both control plots, moves relative to the axes. The movement of the control line represents the response of this community to other factors such as the environment.

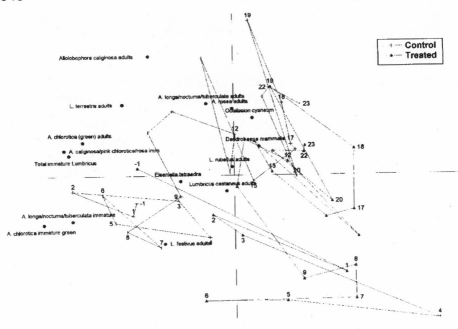

Figure 26-3   Ordination diagram showing response of an earthworm community to repeated
               application of benomyl in comparison to an untreated control

Figure 26-3 also shows that some species were responsible for a larger part of the variation than others. Species groups to the left of the vertical axis can be considered the most
sensitive to treatment effects, with greater distance from the origin representing a higher
level of sensitivity. Species to the right of the axis are considered tolerant to the treatment
effects.

However, the interpretation of such biplots is often complicated by the many sampling
times and treatments to be displayed at once. Another technique called the principal response curve (PRC) [42] can be used to graphically represent the differences between 2
communities at any time. If the community structure of the control plots is set as the
base axis, it allows comparison of the treated plots at any time point by comparing the
proximity of the 2 lines. Further insight into the similarity of the communities can be
gained by carrying out a Monte Carlo permutation test at each sampling occasion.

The PRC (Figure 26-4) allows an easier interpretation of the data, showing that the largest
differences between the communities occurred at the beginning of the study following
the applications of Benlate at 5 kg ai ha^{-1}. There was a gradual recovery of the community
after applications were reduced to 2 kg ai ha^{-1} until the 2 communities became less different and can be considered as similar at 3 sampling points.

Discussion and conclusions

Earthworm populations are dynamic, responding to biotic and abiotic factors with changes in their size, biomass, and diversity. Large fluctuations are possible, particularly in response to environmental conditions and the availability of OM. Agricultural management of the soil exposes earthworms to imposed perturbations such as cultivation and crop protection, which may also result in large changes in the population.

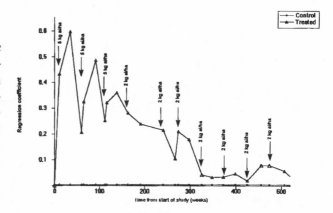

Figure 26-4   Principal response curve demonstrating differences between an earthworm community exposed to repeated applications of benomyl and an untreated control

Any attempt to evaluate the effect of a single perturbation on an earthworm population must relate the importance of the level of effect to the high level of variability in natural populations so that the ecological significance of any such effect can be assessed. This provides challenges to the design and analysis of earthworm field studies.

Recovery to the status of an untreated community within a defined period is the most useful measure of the effect of a perturbation because it takes into account, partly at least, the level of variation in the natural community.

At present, field studies remain the most effective way of gaining realistic data on the response of field populations to imposed perturbations. The complex dataset that is produced from such studies needs to be analyzed in the most appropriate way to allow accurate interpretation. Multivariate statistical methods offer the chance to compare differences at the community level and, therefore, should be considered for future ecological trials.

However, while new statistical methods can be used to improve the design and analysis of field trials, they should be used only as a tool by the ecotoxicologist to allow a greater ecological interpretation of the data produced. Whatever statistical methods are used to analyze field data, an ecological interpretation is still vital to predicting the effects of perturbations.

References

1. Darwin C. 1881. The formation of vegetable mould through the action of worms, with observations on their habits. London UK: Murray.

2. Lee KE. 1985. Earthworms, their ecology and relationships with soils and land use. Sydney, Australia: Academic.

3. Evans AC, Guild WJ McL. 1947. Studies on the relationships between earthworms and soil fertility. I . Biological studies in the field. *Ann Appl Biol* 34:307–330.

4. Gerard BM. 1967. Factors affecting earthworms in pastures. *J Anim Ecol* 36:235–252.

5. Van Rhee JA, Nathans S. 1973. Ecological aspects of earthworm populations in relation to weather conditions. *Rev Ecol Biol Sol* 10:523–533.

6. Ganihar SR. 1996. Earthworm distribution with special reference to physiochemical parameters. *Proc Indian Natl Sc Acad* B62:11–18.

7. Bouché MB, Gardner RH. 1984. Earthworm functions. VIII. Population estimation techniques. *Rev Ecol Biol Sol* 21:37–63.

8. Lofs A. 1992. Measuring effects of pesticides on earthworms in the field: Effect criteria and endpoints. In: Greig-Smith PW, Becker H, Edwards PJ, Heimbach F, editors. Ecotoxicology of earthworms. Andover, Hants UK: Intercept. p 85–89.

9. Hopp H, Hopkins HT. 1946. The effect of cropping systems on the winter populations of earthworms. *J Soil Water Conserv* I:11–13.

10. Reinecke AJ. 1974. The upper lethal temperature of *Eisenia rosea* (Oligochaeta). *Natuurwetenskappe* 62:1–14

11. Butt K. 1991. The effects of temperature on the intensive production of *Lumbricus terrestris* (Oligochaeta: Lumbricidae). *Pedobiologia* 35:257–264.

12. Holmstrup M, Østergaard IK, Nielsen A, Hansen BT. 1991. The relationship between temperature and cocoon incubation time for some lumbricid earthworm species. *Pedobiologia* 35:179–184.

13. Satchell JE. 1955. Some aspects of earthworm ecology. In: Kervan DK Mc E, editor. Soil zoology. London UK: Butterworths. p 180–201.

14. Edwards CA, Lofty JR. 1977. Biology of earthworms. London UK: Chapman and Hall.

15. MacDonald DW. 1983. Predation on earthworms by terrestrial vertebrates. In: Satchell JE, editor. Earthworm ecology. London UK: Chapman and Hall. p 393–414.

16. Cooke AS, Greig-Smith PW, Jones SA. 1992. Consequences for vertebrate wildlife of toxic residues in earthworm prey. In: Greig-Smith PW, Becker H, Edwards PJ, Heimbach F, editors. Ecotoxicology of earthworms. Andover, Hants UK: Intercept. p 139–155.

17. Waters RAS. 1955. Numbers and weights of earthworms under a highly productive pasture. *NZ J Sci Technol* 36:516–525.

18. Marinissen JCY. 1992. Population dynamics of earthworms in a silt loam soil under conventional and "integrated" arable farming during 2 years with different weather patterns. *Soil Biol Biochem* 24:1647–1654.

19. Daniel O. 1992. Population dynamics of *Lumbricus terrestris* L. (Oligochaeta: Lumbricidae) in a meadow. *Soil Biol Biochem* 24:1425–1431.

20. Tomlin AD, McCabe D, Protz R. 1992. Species composition and seasonal variation of earthworms and their effect on soil properties in Southern Ontario, Canada. *Soil Biol Biochem* 24:1451–1457.

21. McCredie TA, Parker CA, Abbott I. 1992 . Population dynamics of the earthworm *Aporrectodea trapezoides* (Annelida: Lumbricidae) in a Western Australian pasture soil. *Biol Fertil Soils* 12:285–289.

22. Baker GH, Barrett VJ, Carter PJ, Williams PML, Buckerfield JC. 1993. Seasonal changes in the abundance of earthworms (Annelida: Lumbricidae and Acanthodrilidae) in soils used for cereal and lucerne production in South Australia. *Aust J Agric Res* 44:1291–1301.

23. Boström U, Lofs A. 1996. Annual population dynamics of earthworms and cocoon production by *Aporrectodea caliginosa* in a meadow fescue ley. *Pedobiologia* 40:32–42.

24. Bennour SA, Nair GA. 1977. Density, biomass and vertical distribution of *Aporrectodea caliginosa* (Savigny 1826) (Oligochaeta, Lumbricidae) in Benghazi, Libya. *Biol Fertil Soil* 24:145–150.

25. Edwards CA. 1989. The importance of integration in sustainable agricultural systems. *Ag Ecosys Environ* 27:25–35.

26. Zicsi A. 1958. Einfluss der Trockenheit und der Bodenbearbeitung auf das Leben der Regenwürmer in Ackerboden. *Acta Agron* 7:67–74.

27. Edwards CA, Lofty JR. 1969. Effects of cultivations on earthworm populations. *Report Rothamsted Experimental Station for 1968*. p 247–248.

28. Edwards CA, Lofty JR. 1982. The effect of direct drilling and minimal cultivation on earthworm populations. *J Appl Ecol* 19:723–734.

29. Greig-Smith PW. 1992. Recommendations of an international workshop on ecotoxicology of earthworms. In: Greig-Smith PW, Becker H, Edwards PJ, Heimbach F, editors. Ecotoxicology of earthworms. Andover, Hants UK: Intercept. p 247–261.

30. Edwards PJ, Brown SM. 1982. Use of grassland plots to study the effect of pesticides on earthworms. *Pedobiologia* 24:145–150.

31. Raw F. 1959. Estimating earthworm populations by using formalin. *Nature* 184:1661–1662.

32. Cernosvitov L, Evans AC. 1947. The Linnean Society of London synopses of the British fauna. *No. 6 - Lumbricidae*. London UK: Linnean Society.

33. Sims RW, Gerard BM. 1985. Earthworms: Keys and notes for the identification and study of the species. London UK: Linnean Society.

34. Lofs-Holmin A, Boström U. 1988. The use of earthworms and other soil animals in pesticide testing. In: Edwards CA, Neuhauser EF, editors. Earthworms in waste and environmental management. The Hague, Netherlands: SPB Academic. p 303–313.

35. Van Breukelen SWF, Brock TCM. 1993. Response of a macro-invertebrate community to insecticide application in replicated freshwater microcosms with emphasis on the use of principal component analysis. *Sci Total Environ* p 1047–1058.

36. Van den Brink PJ, Van Donk E, Gylstra R, Crum SJH, Brock TCM. 1995. Effects of chronic low concentrations of the pesticides chlorpyrifos and atrazine in indoor freshwater microcosms. *Chemosphere* 31:3181–3200.

37. Van den Brink PJ, Van Wijngaarden RPA, Lucassen WGH, Brock TCM, Leeuwangh P. 1996. Effects of the insecticide dursban 4e (active ingredient chlorpyrifos) in outdoor experimental ditches: II. Invertebrate community responses and recovery. *Environ Toxicol Chem* 15:1143–1153.

38. Van Wijngaarden RPA, Van den Brink PJ, Crum SJH, Voshaar JHO, Brock TCM, Leeuwangh P. 1996. Effects of the insecticide dursban 4e (active ingredient chlorpyrifos) in outdoor experimental ditches: I. Comparison of short-term toxicity between the laboratory and the field. *Environ Toxicol Chem* 15:1133–1142.

39. Shaw JL, Manning JP. 1996. Evaluating macroinvertebrate population and community level effects in outdoor microcosms: use of in situ bioassays and multivariate analysis. *Environ Toxicol Chem* 15:608–617.

40. Matthews RA, Matthews GB, Hachmoller B. 1990. Ordination of benthic macroinvertebrates along a longitudinal stream gradient. In: Annual Conference, North American Benthological Society. Blacksburg VA: North American Benthological Society.

41. Warwick RM, Clarke KR. 1994. Changes in marine communities: an approach to statistical analysis and interpretation. Plymouth UK: Natural Environment Research Council. p 1–144.

42. Van den Brink PJ, Ter Braak CJF. 1997. On the analysis of the time-dependent multivariate response of a biological community to stress in experimental ecosystems. *Environ Toxicol Chem*

Chapter 27

Estimation of effects of benomyl application on the reproductive rate of earthworms in a grass field

Martin Holmstrup, Klaus S. Jensen, Erik Kirknel

The aim of this study was to apply a method for estimation of in situ reproduction of earthworms in an attempt to validate laboratory test results of effects of benomyl on reproduction. In a randomized block design with 4 treatment levels of benomyl, samplings for earthworm cocoons and adults were done during the 1996 season. Sampled cocoons were incubated in the laboratory under controlled conditions, and the hatching distribution was recorded. An estimate of the number of cocoons produced per time per adult was then obtained from a calculated age-probability distribution for the incubated cocoons and the corresponding number of adults found in the soil sample. In the field during early summer, adult *Aporrectodea rosea* produced about 1 viable cocoon every 14 d. Benomyl application rates up to 2.0 kg ha^{-1} did not have any adverse effects on reproduction in this species. Adult *Aporrectodea longa* had a much lower reproduction, about 0.4 viable cocoon every 14 d. Reproduction in this species was severely influenced by benomyl application, even in the lowest dose of 0.5 kg ha^{-1}. The difference in sensitivity between the 2 species most likely is due to their different exposures to benomyl. *A. longa* is far more active on the surface than *A. rosea* and therefore may take up more of the pesticide through skin and through ingestion of surface soil and decaying plant material. A comparison of the results from the present investigation with data from laboratory reproduction studies suggests that laboratory test results may predict the effects on reproduction occurring in the field fairly well if the test protocol involves the same species and method of application. However, it is proposed that further parallel laboratory and field experiments be performed to test this hypothesis.

An important aim of earthworm ecotoxicology is the validation of laboratory test results in the field. It is desirable to be able to predict effects of xenobiotics on natural populations under real-world conditions based on laboratory experiments. Well-documented laboratory protocols for earthworm reproduction tests have been developed [1,2], but their validity in the field is not known. For example, it is not known if reproductive rates are affected to the same extent in laboratory earthworms as in free-living earthworms. Even though earthworms readily reproduce in laboratory cultures, this environment is highly unnatural. The spatial and hygro-thermal conditions of free-living earthworms are much different than the laboratory situation. In the field, earthworms may search out hot spots with favorable food and moisture conditions and stay there, and this may cause reproduction to be influenced differently by xenobiotics.

One goal for earthworm ecotoxicology is the protection of earthworm populations and communities. The use of laboratory test methods that focus on a sublethal endpoint to predict how a certain chemical would influence the earthworm population is based on the assumption that this particular parameter (e.g., reproduction) plays a major role in the population dynamics of earthworms. If laboratory reproduction tests should have any relevance to natural populations, at least 2 questions should be addressed: 1) What is the significance of effects on reproduction to earthworm populations? or How severe must effects be before they have any consequences for earthworm populations?, and 2) Is the effect on reproduction observed in the laboratory of the same magnitude under field conditions?

The first question implies that it is necessary to know how significant a given reduction in reproductive rate is to the earthworm population. It could be that a slight decrease in reproduction is compensated for by a less severe intraspecific competition. The best way to investigate the significance of decreased reproduction is probably the use of simulation models, in which the significance of single parameters can be analyzed [3, 4]. The second example is a true field validation of the laboratory reproductive test. To do such a validation, it must be possible to measure the reproductive rate in the field. The present study uses a method proposed by Jensen and Holmstrup [5] to estimate the reproductive rate of earthworms in the field after treatment with a pesticide. The fungicide benomyl was chosen as a stressor because this is one of the best-investigated pesticides with regard to effects on earthworm reproduction. In a randomized block design with 4 treatment levels of benomyl, samplings for cocoons and adults were done in order to estimate the dose–response relationship of effects of benomyl on the reproductive rate. The objective of the study was to compare effects on in situ reproduction rates with effect levels of benomyl found by other authors in the laboratory.

Methods

Experimental site
The field experiment was performed on a second-year grass field on a sandy loam soil belonging to The Danish Institute of Agricultural Sciences, Research Centre Foulum, Jutland. The soil consisted of 33.1% coarse sand (> 200 mm), 33.3% fine sand (63 to 200 mm), 8.5% coarse silt (20 to 63 mm), 12.2% silt (2 to 20 mm), 8.5% clay (< 2 mm), and 4.4% humus and had a pH-H_2O of 7.3. A preliminary survey (washing and sieving of 10 soil samples) showed that the density of earthworms was about 150 specimens m^{-2} at the time when the experiment was initiated. The dominant species were *Aporrectodea longa* (Ude), *A. rosea* (Savigny), and *Lumbricus terrestris* (L.). The experiment was set up as a randomized block design with 4 blocks, each consisting of 4 15 H 15 m^2 plots. Plots were situated 15 m apart from each other, and the 4 blocks were situated 30 m apart from each other. On 1 May, benomyl (Du Pont) was applied by spraying (400 L ha^{-1}) in 4 concentrations, i.e., 0 (control), 0.5, 1.0, and 2.0 kg a.i. ha^{-1}. During the following 2 d, the plots were irrigated with approximately 40 mm precipitation in order to incorporate into the soil as much of the fungicide as possible.

Physical and chemical measurements

Hourly soil temperature measurements at 5-cm depth under grass cover and daily precipitation was obtained from a meteorological station situated about 1 km away from the study site. Soil water potential at the 5-cm depth was measured weekly in the plots by use of Wescor soil psychrometers connected to a portable Wescor HR-33T Dew Point Microvoltmeter (Wescor Inc., Logan UT) operated in the dewpoint mode. In order to document the behavior and fate of benomyl in the soil, soil for chemical analysis was sampled in 2 horizons (i.e., 1.5 to 3.5 and 6.5 to 8.5-cm depth) from plots receiving 0.5 and 2.0 kg ha^{-1}. This was done on 5 occasions, namely at 5, 19, 47, 103, and 175 d after spraying. Benomyl rapidly reacts with water and the principal breakdown product, methyl 2-Benzimidazole carbamic acid (MBC), is the active component of the fungicide [6]. Extraction of MBC from soil samples and high-performance liquid chromatography analysis was done according to Kirkland et al. [7]. The results were corrected for recovery, which was found to be 50% in spiked soil. The quantitative limit of detection with this method was 0.08 mg kg^{-1}. The analysis was replicated twice.

Sampling of earthworms

On 7 sampling occasions after fungicide treatment (5, 33, 61, 89, 117, 145, and 188 d after treatment), the densities of cocoons and worms were estimated. At the first 3 sampling occasions, two $25 \times 25 \times 25$ cm^3 soil samples from each plot were hand sorted by washing and sieving the soil through a series of 4 box sieves with decreasing mesh size (10, 4, 2 and 1 mm). At the remaining 4 samplings, only 1 soil core from each plot was taken. Worms and cocoons were classified according to species and stage (i.e., juvenile, subadult, or adult) in order to obtain estimates of density m^{-2}. The soil cores were kept at 5 °C until hand sorting.

Incubation of cocoons

The cocoons found by washing and sieving of soil cores were incubated on wet filter paper in Petri dishes. Cocoons of each species from each sample were kept separately and incubated at 5 °C until all soil samples from a sampling occasion had been processed. Thereafter, the Petri dishes were placed at 20 °C and the accumulated number of hatchings was recorded 3 times a week until all viable cocoons had hatched.

Estimation of reproduction

An estimate of the number of cocoons produced per time per adult can be obtained from a calculated age-probability distribution for the incubated cocoons and the corresponding number of adults found in the soil sample. In the following section, the method for estimating the number of cocoons produced in a specified time interval from the distribution of observed hatching times is outlined. A more detailed description of the mathematical and numerical aspects is given by Jensen and Holmstrup [5]. By numerical integration of the instantaneous development rate, knowledge about the influence of temperature on the development rate for constant temperatures can be used to obtain the development time of earthworm cocoons incubated under fluctuating temperature conditions. The mathematical description can be expanded by incorporating individual variation in the development rate, expressed as a symmetrical frequency distribution [8]. Doing so, Jensen and Holmstrup [5] demonstrated that the distribution of development times for cocoons incubated under naturally fluctuating temperatures can be reliably predicted for *A. caliginosa*, *Allolobophora chlorotica*, and *Dendrobaena octaedra*. In the same paper, it was shown that the age-probability distribution for a cocoon with known hatching time can be determined by simply reversing the direction of time in the algorithm.

In the present study, constant-temperature data from older studies [9,10] are used to estimate the population mean development rate (r) as a function of temperature (T). Figure 27-1 shows the fitted r-functions (polynomials) for *A. longa* and *A. rosea*. Experimental values were not available for 5 °C. As a qualified guess, $1/r$ equal to 400 d is used, which is the approximate value found for *A. chlorotica* and *D. octaedra* [5].

Since individuals in a population do not develop with identical rates, a symmetrical frequency distribution for the relative development rate (i.e., normalized with respect to the mean rate, $r(T)$, in the population) is specified. As in Sharpe et al. [8], a quadratic distribution is used for reasons of simplicity. The width (SD) of the frequency distribution for relative development rate has not been derived from experimental data for *A. longa* and *A. rosea* but is set to 0.1, as found for *A. caliginosa*, *A. chlorotica*, and *D. octaedra* [5].

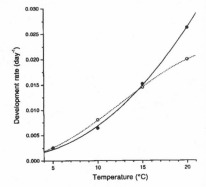

Figure 27-1   Development rate of *A. longa* and *A. rosea* cocoons as a function of temperature. Experimental data were available only for T = 10, 15, and 20 °C [9, 10]. Fitting of functions by least-squares method. Open circles and dashed line, *A. longa*; filled circles and solid line, *A. rosea*

By numerical integration of $r(T(t))$ backward in time and application of the frequency distribution for the relative development rate, a single hatching time observation gives rise to a probability distribution for the production time (Figure 27-2). Here, the temperature time course is an idealization of the experimental conditions (temperature set to 20 °C). The dashed curve in Figure 27-2 shows the raw distribution as calculated by integration. The solid curve shows the corrected distribution that follows from the fact that the cocoon could not have been produced after the soil sample was taken at $t = t_{samp}$. The areas under the 2 curves are identical (= 1).

From the recorded observations of hatchings, a discrete frequency distribution can be made with the same sample period as used in the temperature time series (1 h in the present case).

The integration scheme should then be repeated for each discrete value in the hatching time frequency distribution. After weighing the resulting (corrected) production time distributions according to the hatching time distribution, summation finally yields the overall production time probability distribution. This is illustrated in Figure 27-3 for an experimental sample with 22 observed hatchings. The shaded area under the calculated production time distribution between $t = -14$ d and $t = 0$ d ($= t_{samp}$) is an estimate of the number of cocoons produced in that time interval. The area appears to be 8.6. The upper panel shows the observed frequency distribution of hatching (right side of panel) and how this is transformed into the calculated probability distribution of production time (left side of panel). The shaded area represents the number of cocoons produced during the period $-14 < t < 0$. The lower panel shows the temperature experienced by the cocoons, first in the field ($t < 0$), then incubated in the laboratory, initially at 5 °C, then at 20 °C until hatching.

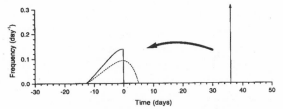

Figure 27-2   Calculated probability distribution for the production time of a single cocoon (A. *rosea*) hatched at t = 36 d. The soil sample is taken at t = 0 d, and the temperature is assumed to be 20 °C during the entire time course. Dashed curve: raw distribution. Solid curve: corrected distribution.

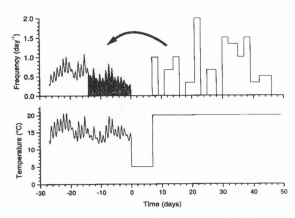

Figure 27-3   Calculated production-time probability distribution for a sample of 22 cocoons (A. *rosea*) taken at t = 0. Also shown are the observed hatching frequency distribution

Results

In this type of investigation, the hygro-thermal conditions of the soil are of particular importance because these data are used directly in the estimation of the reproductive rate. Data for soil temperature, precipitation, and soil water potential (pF) are shown in Figure 27-4. It has been shown that cocoon development may be retarded if pF becomes higher than approximately 3.3 [11,12]. This implies that a correct estimation of cocoon age is only possible for sampling occasions in which the soil moisture in the previous period has been optimal for cocoon development (i.e., pF below 3.3). As a consequence of this, reproductive rates have been estimated only for periods associated with 3 of the 7 sampling occasions, namely at 33, 61, and 188 days, respectively, after benomyl spraying (compare Figure 27-4).

Benomyl residue (MBC) was detected in the upper soil horizon, 1.5 to 3.5 cm, in plots treated with 2.0 kg ha^{-1}, where the concentration of MBC at 5 d after spraying was found to be approximately 0.6 mg kg^{-1} dry soil (Table 27-1). In plots treated with 0.5 kg ha^{-1}, concentrations were almost consistently below the detection limit. In plots treated with 2.0 kg ha^{-1}, there seemed not to be any substantial decrease in the concentration of MBC during the experimental period.

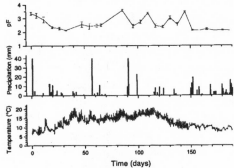

Figure 27-4　　Daily precipitation, soil temperature (5 cm), and soil water potential (5 cm) at the study site. The area was irrigated with approximately 40 mm at 3 occasions, indicated as precipitation

Table 27-1　　Measured concentrations of benomyl residues (methyl 2-benzimidazole carbamic acid) in 2 soil horizons of plots treated with 2.0 kg ha-1 and 0.5 kg ha-1

Time (d after application)	Measured soil concentration (mg kg^{-1} dry soil)			
	2.0 kg ha^{-1}		0.5 kg ha^{-1}	
	1.5 to 3.5 cm	6.5 to 8.5 cm	1.5 to 3.5 cm	6.5 to 8.5 cm
5	0.60	0.08	n.d.	n.d.
19	0.17	n.d.	0.17	n.d.
47	0.73	n.d.	n.d.	n.d.
103	0.33	n.d.	n.d.	n.d.
175	0.51	n.d.	n.d.	n.d.

n.d.: not detectable

Tables 27-2 and 27-3 show the density of adults and cocoons of *A. longa* and *A. rosea* in the experimental plots at the selected sampling occasions. Benomyl in high dose (2.0 kg ha^{-1}) caused a drastic reduction in density of adult *A. longa* (Table 27-2). At 188 d after spraying, still no adults of this species were found in these plots. Lower doses had only transient effects on the density of adult *A. longa*. The density of adult *A. rosea* seemed to be unaffected by benomyl application, even at the highest dose (Table 27-3). The density of *A. longa* cocoons in control plots was about 70 m^{-2} at all 3 sampling occasions (Table 27-2). In benomyl-treated plots, there was a decrease in cocoon density with time, indicating that more cocoons disappeared (by hatching or mortality) than were produced during this period. Cocoons of *A. rosea* were found in somewhat higher densities, ranging from 56 to 288 m^{-2} (Table 27-3).

Table 27-2 Density (mean ± SE m^{-2}) of adults and cocoons of A. *longa* in benomyl-sprayed plots at 3 sampling occasions

Benomyl treatment (kg ha^{-1})	Sampling occasion (d after treatment)	Adult density (mean ± SE m^{-2})	Cocoon density (mean ± SE m^{-2})
0 (control)	33	12.0 ± 9.5	66.0 ± 42.1
	61	6.0 ± 3.8	78.0 ± 45.3
	188	16.0 ± 11.3	64.0 ± 33.9
0.5	33	8.0 ± 4.6	84.0 ± 37.2
	61	6.0 ± 6.0	74.0 ± 55.8
	188	20.0 ± 20.0	32.0 ± 26.9
1.0	33	8.0 ± 4.6	62.0 ± 24.5
	61	2.0 ± 2.0	18.0 ± 13.2
	188	16.0 ± 9.2	28.0 ± 16.5
2.0	33	8.0 ± 5.7	58.0 ± 18.3
	61	0	30.0 ± 15.1
	188	0	16.0 ± 6.5

Table 27-3 Density (mean ± SE m^{-2}) of adults and cocoons of A. *rosea* in benomyl-sprayed plots at 3 sampling occasions

Benomyl treatment (kg ha^{-1})	Sampling occasion (d after treatment)	Adult density (mean ± SE m^{-2})	Cocoon density (mean ± SE m^{-2})
0 (control)	33	20.0 ± 5.2	152 ± 52.5
	61	12.0 ± 6.9	216 ± 46.8
	188	24.0 ± 15.3	108 ± 81.9
0.5	33	32.0 ± 19.9	250 ± 164
	61	18.0 ± 11.5	184 ± 74.1
	188	44.0 ± 20.0	104 ± 36.1
1.0	33	22.0 ± 10.0	156 ± 51.1
	61	6.0 ± 3.8	108 ± 33.1
	188	16.0 ± 6.5	56.0 ± 23.1
2.0	33	42.0 ± 28.4	288 ± 149
	61	26.0 ± 5.0	192 ± 71.2
	188	28.0 ± 23.0	228 ± 129

For each replicate, the probability distribution of production time for the whole cocoon sample was generated (Figure 27-3). When the reproductive rate was calculated, a window of 14 d (corresponding to the shaded area in Figure 27-3) was arbitrarily chosen, assuming that the density of adults during this period was largely the same as the density observed at the sampling time.

The estimated reproductive rates for A. *longa* and A. *rosea* are shown in Figure 27-5. Each value is calculated as the mean of 4 replicates without measures of statistical deviation indicated. This could not be derived because a substantial number of replicates did not contain any adults, and therefore, reproductive rates for those particular replicates could not be calculated. Reproduction in A. *longa* was about 0.4 viable cocoon adult^{-1} in control plots in the period 19 to 33 d after treatment. Reproduction in control plots during the

2 later periods was much lower, approximately 0.1 viable cocoon adult⁻¹. Benomyl treatment caused a drastic reduction in reproductive rate in *A. longa*. Even though adults were present in all treatments 33 d after application, reproduction had ceased in plots treated with 1.0 and 2.0 kg ha⁻¹, and 0.5 kg ha⁻¹ apparently caused a 50% reduction of the reproductive rate (Figure 27-5). Also 61 d after

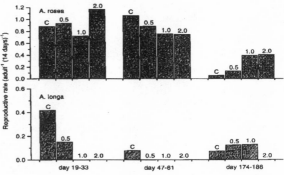

Figure 27-5   Estimated reproductive rate of *A. longa* and *A. rosea* in the field at 3 sampling occasions. Treatments are indicated above the columns.

treatment, an effect of benomyl was seen. After 188 d, the effect had apparently disappeared. The reproductive rate in *A. rosea* was somewhat higher than observed for *A. longa*, about 0.9 viable cocoon adult⁻¹ (14 d)⁻¹ at the 2 first sampling occasions (Figure 27-5). At the last sampling occasion, a much smaller reproduction was observed. In *A. rosea*, there were no observed adverse effects of benomyl on reproduction.

Discussion

The aim of the present study was to assess the effects of benomyl on in situ reproduction of earthworms using a new method developed in a previous study [5]. To our knowledge, specific methods to estimate reproductive rates of earthworms under field conditions have not previously been used, although other authors [13,14] have given rough estimates of cocoon production in another species, *A. caliginosa*, based on the seasonal dynamics of adult earthworms and cocoon densities in the field.

The highest reproductive rate observed in *A. longa* in control plots was about 0.4 viable cocoon adult⁻¹ during 2 weeks with soil temperature ranging between 6 and 15 °C (19 to 33 d after spraying) (Figure 27-4). During the 2 later periods with temperatures ranging between 12 and 20 °C (47 to 61 d after spraying) and in late autumn (174 to 188 days after spraying) when temperature ranged between 7 and 10 °C, the reproduction was only approximately 0.1 viable cocoon adult⁻¹ during 14 d. *A. longa* is known to aestivate during summer, which may explain the low reproduction in controls 47 to 61 d after benomyl treatment. At the sampling in late autumn, the low temperature is probably the reason for a low reproduction in this species. In the laboratory, when cultured at various constant temperatures ranging from 9 to 20 °C and given cow dung as feed, cocoon production in *A. longa* is reported to range from 0.7 to 1.5 cocoon adult⁻¹ (14 d)⁻¹ [15–18]. It is evident that reproduction in this species was much lower in the field than in laboratory cultures. Several factors, e.g., food quality and quantity or soil water status, may be limiting for reproduction in the field. Reproduction in *A. rosea* in control plots was approximately 0.9 viable cocoon adult⁻¹ (14 d)⁻¹ with soil temperature ranging between 6 and

20 °C (late May and June), whereas in autumn (7 to 10°C) it was below 0.1 adult^{-1} (14 d)$^{-1}$, probably due to low temperature. Laboratory studies (similar to those listed for *A. longa*) show that the reproductive rate of *A. rosea* in cultures may range from 0.5 to 1.4 cocoons adult^{-1} (14 d)$^{-1}$ at constant temperatures from 12 to 15 °C [16,17,19,20], results that are quite similar to the estimates for May and June in the present investigation. Phillipson and Bolton [21] used mesocosms buried in the field and estimated the annual cocoon production of *A. rosea* to be about 3 cocoons adult^{-1}. In the light of the present investigation, these estimates seem unrealistically low.

Reproduction in *A. longa* was clearly impaired by benomyl application (even though this could not be statistically tested), whereas reproduction in *A. rosea* seemed to be unaffected (Figure 27-5). This difference between the 2 species probably relates to their differing behavior and spatial distribution in the soil, which causes the 2 species to be differently exposed to benomyl. *A. longa* belongs to the aneciques [22], species that live in more or less permanent burrows in the mineral soil but come to the surface to feed. *A. rosea* belongs to the endogees [22], species that are more confined to the mineral soil horizons. The measured soil concentrations of benomyl residues (Table 27-1) suggest that most of the residues were present in the uppermost soil layers (assuming a homogeneous distribution in the upper 5 cm of soil, 2.0 kg ha^{-1} should be equivalent to about 2.8 mg kg^{-1} dry soil), maybe even on the soil surface, which means that *A. rosea* was probably not exposed to very high concentrations. A parallel investigation of surface activity of earthworms in the same experimental plots as used in this study showed that *A. longa* was much more active on the soil surface than *A. rosea* (see Mather and Christensen, Chapter 25) and therefore would be exposed to much higher concentrations than would be the case for *A. rosea*. In autumn (188 d after spraying), the effects of benomyl application in doses of 0.5 and 1.0 kg ha^{-1} seemed to have disappeared in *A. longa*, but in plots that received 2.0 kg ha^{-1}, adults of this species were still absent. In *A. rosea*, after 188 d there was a tendency to an increased reproduction with increasing application rate. An explanation for this could be that a lowered competition from *A. longa* and *L. terrestris* (data not shown) may have promoted *A. rosea*.

While effects of benomyl on reproduction have not previously been studied in the field, there exist some reports of laboratory studies. Kula [23] studied responses of *Eisenia fetida* and *L. terrestris* to benomyl applied by spraying on the surface of the test soil. It was found that reproduction was not affected by 0.25 kg a.i. ha^{-1}, but an application rate of 1.25 kg ha^{-1} completely arrested reproduction in these species. The behavior of *L. terrestris* and *A. longa* is quite similar (both are aneciques), and these laboratory results may therefore be comparable with the present field study. One should bear in mind, however, that the inherent sensitivity to benomyl may differ between *L. terrestris* and *A. longa*. Other studies [24,25], in which benomyl was mixed into the soil, are difficult to compare with field studies because of the extreme spatial (vertical) variation of pesticide concentrations in field soils. More work involving parallel studies of reproduction in laboratory and field using the same species and method of pesticide application is needed.

The method used here for estimation of reproduction is based on a number of assumptions that need to be fulfilled for estimates to be reliable. First, correct age determination of cocoons is possible only when cocoon development in the field proceeds as observed in empirical studies that form the basis of the model. Moisture must be optimal in the field [5], and this was documented in the present investigation by water potential measurements (Figure 27-4). It should also be documented that development rate of cocoons incubated in soil does not differ from incubation on wet filter paper. No such information exists for the species involved in the present study, but Jensen and Holmstrup [5] reported that development rate of *A. caliginosa* cocoons, tested in 3 different soil types, was apparently not affected by the method of incubation (wet filter paper versus moist soil). The presence of benomyl in the soil must not have any influence on development rate. In order to investigate this, a preliminary experiment in the laboratory with *A. longa* cocoons was done. This study showed that the development rate was unaffected at concentrations up to 1.0 mg benomyl kg⁻¹ dry soil, when the fungicide was mixed homogeneously into the soil (M. Holmstrup, unpublished study). This question needs further investigation, but a comparison of these preliminary results and the benomyl residue concentrations found in the field experiment suggests that the development rate of field-sampled cocoons was probably not affected by benomyl application. Gerard [26] reported that most cocoons of *A. longa* and *A. rosea* in a grass field were found in soil layers deeper than 2.5 cm and that this was most pronounced for *A. longa*. The actual vertical distribution of cocoons therefore further supports the idea that effects of benomyl on development rate are probably of minor importance.

The present study suggests that age determination of field-sampled cocoons is feasible, and earlier work showed that the method is relatively reliable for calculating the production-time probability distribution for samples of cocoons [5]. In the case of *A. longa*, the other component in the estimation of reproductive rate, density of adults, has probably been underestimated. The employed sampling method is not optimal for this species because adult *A. longa* live in more or less vertical, deep-burrow systems and therefore may escape sampling by quick retraction to depths below the sampled soil core. For *A. rosea*, however, the method is probably adequate because of this species' confinement to the upper soil layers. With some modifications of the sampling procedure, including a better estimation of the density of adults by means of deeper and larger soil samples, the method described here may be a valuable tool in the studies of earthworm ecology and ecotoxicology.

Acknowledgments - The authors wish to thank Karen Kjær Jacobsen, Gorm Diernisse, John Geert Rytter, Zdenek Gavor, and Lotte Børresen for technical assistance and endurance in the earthworm sampling. Valuable comments of 2 anonymous reviewers are also greatly acknowledged. This project has received financial support from the Danish Pesticide Research Program.

References

1. [BBA] Biologische Bundesanstalt Für Land-Und Forstwirtschaft. 1997. Auswirkungen von Pflantzenschuzmitteln auf die reproduktion und das wachstum von *Eisenia fetida/Eisenia andrei*. Richtlinien für die prüfung von pflanzenschutzmitteln im zulassungsverfahren, teil VI, 2-2, Jan 1994. Ribbesbüttel, Germany: Aphir-Verlag.

2. Van Gestel CAM, Van Dis WA, Van Breemen EM, Sparenburg PM. 1989. Development of a standardized reproduction toxicity test with the earthworm species *Eisenia fetida andrei* using copper, pentachlorophenol, and 2,4-dichloroaniline. *Ecotoxicol Environ Safety* 18:305–312.

3. Klok C, de Roos AM. 1996. Population level consequences of toxicological influences on individual growth and reproduction in *Lumbricus rubellus* (Lumbricidae, Oligochaeta). *Ecotoxicol Environ Safety* 33:118–127

4. Baveco JM, de Roos AM. 1996. Assessing the impact of pesticides on lumbricid populations: an individual-based modeling approach. *J Appl Ecol* 33:1451–1468

5. Jensen KS, Holmstrup M. 1997. Estimation of earthworm cocoon development time and its use in studies of in situ reproduction rates. *Appl Soil Ecol* 7:73-82

6. Kilgore WW, White ER. 1970. Decomposition of the systemic fungicide 1991 (Benlate). *Bull Environ Contam Toxicol* 5:67–69.

7. Kirkland JJ, Holt RF, Pease HL. 1973. Determination of benomyl residues in soils and plant tissues by high-speed cation exchange liquid chromatography. *J Agr Food Chem* 21:368–371.

8. Sharpe PJH, Curry GL, DeMichele DW, Cole CL. 1977. Distribution model of organism development times. *J Theor Biol* 66:21–38.

9. Holmstrup M, Østergaard IK, Nielsen A, Hansen BT. 1991. The relationship between temperature and cocoon incubation time for some Lumbricid earthworm species. *Pedobiologia* 35:179–184.

10. Holmstrup, M., Østergaard, I. K., Nielsen, A. and Hansen, B. T. 1996. Note on the incubation of earthworm cocoons at three constant temperatures. *Pedobiologia* 40:477–478.

11. Gerard BM. 1960. The biology of certain British earthworms in relation to environmental conditions. [Ph.D. Thesis]. London UK: University of London.

12. Holmstrup M. 1994. Physiology of cold hardiness in cocoons of five earthworm taxa (Lumbricidae: Oligochaeta). *J Comp Physiol B* 164:222–228.

13. Nowak E. 1975. Population density of earthworms and some elements of their production in several grassland environments. *Ekologia Polska* 23:459–491.

14. Bostrom U, Lofs A. 1996. Annual population dynamics of earthworms and cocoon production by *Aporrectodea caliginosa* in a meadow fescue ley. *Pedobiologia* 40:32–42.

15. Butt KR. 1993. Reproduction and growth of three deep-burrowing earthworms (Lumbricidae) in laboratory culture in order to assess production for soil restoration. *Biol Fertil Soils* 16:135–138.

16. Evans AC, Guild WJ McL. 1948. Studies on the relationships between earthworms and soil fertility IV. On the life cycles of some British Lumbricidae. *Annals of Applied Biology* 35:471–484.

17. Hansen BT, Holmstrup M, Nielsen A, Østergaard IK. 1989. Regnormestudier i Laboratoriet og i felten. Specialerapport, Zoologisk Laboratorium, Aarhus Universitet. Aarhus, Denmark: University of Aarhus.

18. Meinhardt U. 1974. Vergleichende Beobachtungen zur Laboratoriumsbiologie einheimischer Regenwurmarten: III. Erfahrungen mit Arten, die sich für die datierte Zucht nicht eigneten. *Zeitschr für Angev Zool* 61:265–299.

19. Graff O. 1953. Die Regenwürmer Deutschlands. *Schrift Forsch Land Braunschweig-Volk* 7:5–79.

20. Lofs-Holmin A. 1983. Reproduction and growth of common arable land and pasture species of earthworms (*Lumbricidae*) in laboratory cultures. *Swedish J Agric Res* 13:31–37.

21. Phillipson J, Bolton PJ. 1977. Growth and cocoon production by *Allolobophora rosea* (Oligochaeta: Lumbricidae). *Pedobiologia* 17:70–82.

22. Bouché MB. 1977. Strategies lombriciennes. In: Lohm U, Persson T, editors. Soil organisms as components of ecosystems. *Ecol Bull (Stockholm)* 25:122–132.

23. Kula H. 1994. Auswirkungen von pflanzenschutzmitteln auf regenwürmer (Oligochaeta: Lumbricidae). Zur problematik der bewertung letaler und subletaler effekte in labor und feldversuchen. [PhD thesis]. Braunschweig, Germany: Naturwissenschaftlichen Facultät der Technischen Universität Carolo-Wilhelmina zu Braunschweig : 151 p.

24. Van Gestel CAM. 1991. Earthworms in ecotoxicology. [PhD thesis]. Utrecht, the Netherlands: University of Utrecht. 197 p.

25. Lofs-Holmin A. 1982. Measuring cocoon production of the earthworm *Allolobophora calignosa* (Sav.) as a method of testing sublethal toxicity of pesticides. *Swedish J Agric Res* 12:117–119.

26. Gerard BM. 1967. Factors affecting earthworms in pastures. *J Animal Ecology* 36:235–252.

Chapter 28

Comparison of acute predicted environmental concentrations of pesticides in soil and earthworms with concentrations determined by residue analysis

Ainsley Jones, Colin McCoy, Andrew D.M. Hart

The Seeking Confirmation about Results at Boxworth (SCARAB) project is a major, long-term investigation of the effects of reduced pesticide use on invertebrates and soil microflora in UK arable cropping systems. As part of the project, soil and earthworms were sampled (from the top 5 cm) between 1 and 5 d after applications of 3 pesticides, chlorpyrifos, propiconazole, and dimethoate, to a variety of crops. Pesticide concentrations were determined by chemical analysis and compared to the initial predicted environmental concentration (PEC) that is used in pesticide risk assessment for earthworms. Maximum concentrations in soil after application to grass were reasonably consistent with initial soil PECs (between 60 and 150%) in 4 out of 5 cases. In one case, the maximum soil concentration was only 27% of the initial soil PEC. Results for pesticides applied to summer cereals were more variable, with maximum concentrations ranging from < 5% to 114% of the initial soil PEC. Only 1 application to cereal in spring was studied, and the maximum soil concentration was 80% of the initial soil PEC. PECs in earthworms were calculated by assuming that all the pesticide in the earthworm is contained in the soil in the gut contents. The results of residue analysis of earthworms suggested that this was not accurate in all cases.

Assessment of the risks of pesticides to earthworms involves comparing the estimated exposure in the field with laboratory toxicity data. Under the European Plant Protection Organization/Council of Europe (EPPO/CoE) guidelines [1], the exposure estimate is calculated as the PEC in the soil. It is assumed that all of the pesticide applied reaches the soil and is evenly distributed in the top 5 cm. If the soil is plant-covered to a large degree (e.g., cereal crops in summer), it is assumed that only 50% of the pesticide reaches the soil.

Earthworms are an important food source for some birds and mammals. Since any contaminated earthworms may be eaten by predators, the EPPO/CoE guidelines allow for an assessment of secondary poisoning risk. The guidelines give no guidance on calculating expected residues in earthworms. In the UK, the Pesticide Safety Directorate calculates earthworm PEC by assuming that all of the pesticide present is in the soil in the earthworm's gut and that this soil is 30% of the weight of the earthworm. Thus earthworm PEC (wet weight) is 30% of soil PEC (dry weight).

The SCARAB project is a major, multi-site study to investigate the effects of reduced pesticide use on invertebrates and soil microflora in UK arable cropping systems. The project was designed to determine whether the pesticide effects found in a previous single-site experiment (The Boxworth Project) [2] could be repeated at other locations on different soil types and with different agricultural rotations. The experiment was conducted on 8 fields (A–H) at 3 locations in England: High Mowthorpe in North Yorkshire, Gleadthorpe in Nottinghamshire, and Drayton in Warwickshire.

As part of the project, samples of soil and earthworms (from the top 5 cm) were taken shortly after applications of a number of pesticides to a variety of crops. The purpose of this investigation was to compare PECs in both soil and earthworms with actual pesticide concentrations, determined by chemical analysis.

Methods

Details of the pesticide applications are given in Table 28-1. Samples of soil and earthworms were taken using a soil corer of 25-mm diameter that was driven into the soil to a distance of 5 cm. Any earthworms present were immediately separated from soil. On each sampling occasion, 10 to 20 samples were taken by this method. The samples were combined and thoroughly mixed, and a single 5-g subsample was taken for chemical analysis. Concentrations of chlorpyrifos and dimethoate were determined by gas chromatography with flame photometric detection after extraction with acetone/methanol (50:50). Propiconazole was determined by gas chromatography with nitrogen-phosphorous detection after extraction with ethyl acetate and cleanup on neutral alumina columns. Limits of detection were 25 to 35 ng/g for chlorpyrifos in soil and earthworms, 25 to 50 ng/g for dimethoate in soil, 10 ng/g for propiconazole in soil, and 35 ng/g for propiconazole in earthworms.

Results

Actual pesticide concentrations in soil are compared to initial soil PECs in Figures 28-1 through 28-3. PECs were calculated (in mg/g) as

$$\text{PEC} = \frac{(\text{Application rate in mg/cm}^2) \times (\text{\% chemical reaching soil})}{500 \times (\text{Bulk density in g/ml})}$$

Application rate in mg/cm^2 = Application rate in kg/ha × 10.

The bulk density of soil was assumed to be 1.4 g/ml.

In some cases, the pesticides were applied to the study areas in summer, when earthworms are likely to retreat deeper into the soil. Consequently, earthworms were not found in the top 5 cm of soil after some applications, and there were fewer earthworm samples than soil samples available for chemical analysis. Actual concentrations in earthworms are compared to PECs in earthworms (calculated on the basis of 30% of soil PEC) in Figure 28-4.

Table 28-1   Pesticide applications for which samples were taken for chemical analysis

App no.	Site	Field	OM (%)	Pesticide	log Kow	Date of application	Crop	Age of crop (d)	Rainfall (mm) on days after application*					
									Day 0	Day 1	Day 2	Day 3	Day 4	Day 5
1	Drayton	A	4.4	Chlorpyrifos	4.70	14-6-94	Grass	62	0	0	0	0		
2	Drayton	A	4.4	Chlorpyrifos	4.70	22-3-95	Grass	339	0	0	0	0.3	3.1	1.8
3	Gleadthorpe	F	1.65	Chlorpyrifos	4.70	19-6-95	Spring wheat	179	0	0	0	0		
4	High Mowthorpe	C	4.4	Chlorpyrifos	4.70	28-6-95	Winter wheat	270	0	0	0			
5	Drayton	B	4.7	Propiconazole	3.72	11-4-94	Grass	397	0.3	2.5	0	0	0	
6	Drayton	B	4.7	Propiconazole	3.72	3-4-95	Grass	755	0	0	0			
7	Drayton	A	4.4	Propiconazole	3.72	3-4-95	Grass	351	0	0	0			
8	Gleadthorpe	G	1.5	Dimethoate	0.70	30-3-95	Spring wheat	103	0.3	0.6	0	0		
9	Gleadthorpe	G	1.5	Dimethoate	0.70	30-6-95	Spring wheat	193	0	0	0			
10	High Mowthorpe	D	5.4	Dimethoate	0.70	17-7-95	Winter wheat	292	0.5	0.2	0	2.8		
11	High Mowthorpe	E	5.4	Dimethoate	0.70	17-7-95	Winter wheat	292	0.5	0.2	0	2.8		

Kow (Octanol/water partition coefficient) values obtained from The Pesticide Manual [3].

* Rainfall data given only for the time period during which samples were taken.

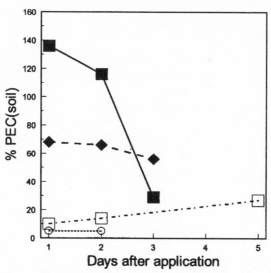

Figure 28-1   Chlorpyrifos soil residues as percentage of PEC. Chlorpyrifos applications: number 1
               to grass at Drayton field A in June 1994 (■), number 2 to grass at Drayton field A in
               March 1995 (□), number 3 to spring wheat at Gleadthorpe field F in June 1995 (◆),
               number 4 to winter wheat at High Mowthorpe field C in June 1995 (○). PECs
               calculated for applications 1 and 2 by assuming 100% of chemical reached soil and for
               applications 3 and 4 by assuming 50% reached soil. No residues were detected after
               application number 4 and the limits of detection are plotted as percentage of PEC.

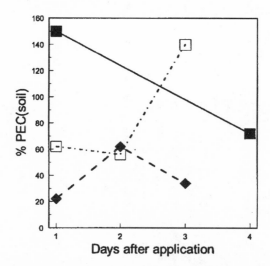

Figure 28-2   Propiconazole soil residues as PEC. Propiconazole applications: number 5 to grass at
               Drayton field B in June 1994 (■), number 6 to grass at Drayton field B in April 1995
               (□), number 7 to grass at Drayton field A in April 1995 (◆). PECs calculated by
               assuming 100% of chemical reached soil.

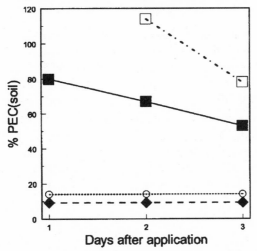

Figure 28-3   Dimethoate soil residues as percentage of PEC. Dimethoate applications: number 8 to spring wheat at Gleadthorpe field G in March 1995 (■), number 9 to spring wheat at Gleadthorpe field G in June 1995 (□), number 10 to winter wheat at High Mowthorpe field D in July 1995 (◆), number 11 to winter wheat at High Mowthorpe field E in June 1995 (O). PECs calculated for applications 1 and 2 by assuming 100% of chemical reached soil and for applications 3 and 4 by assuming 50% reached soil. No residues were detected after the application of chlorpyrifos to winter wheat in June 1995 and the limits of detection are plotted as percentage of PEC. No residues were detected after applications 10 and 11 and the limits of detection are plotted as percentage of PEC.

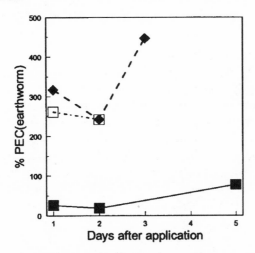

Figure 28-4   Pesticide residues in earthworms as percentage of PEC. Applications: number 2 of chlorpyrifos to grass at Drayton field A in March 1995 (■), number 6 of propiconazole to grass at Drayton field B in April 1995 (□), number 7 of propiconazole to grass at Drayton field A in April 1995 (◆).

Discussion

Considering first those pesticides applied to grass, there is reasonable agreement between predicted soil concentrations and those from residue analysis in 4 cases out of 5, with maximum concentrations between 60 and 150% of the soil PEC. After 1 chlorpyrifos application (#2), however, the initial concentration was less than 10% and the maximum concentration less than 30% of the soil PEC. The reason that such a small proportion of the chemical appeared to reach the soil is unclear. A possible explanation is that most of the chemical was initially deposited on the grass. However, an application of propiconazole to the same field a few days later (#7) resulted in a maximum concentration of 60% of the soil PEC. The reason for the steady rise in the soil concentration over the period studied after application #2 may be that the rainfall on days 3, 4, and 5 caused some wash-off from the grass onto the soil. Grassland plots are normally used in field trials on the effects of pesticides on earthworm populations, and these results suggest that, under some circumstances, exposure of earthworms may be less than predicted.

For cereals in summer, there was less good agreement between predicted and actual soil concentrations, and the maximum determined concentration ranged from < 5% to 114% of soil PEC. Results from applications #10 and #11 suggest that wash-off from summer cereals onto the soil caused by rainfall was not a significant factor, even for a chemical with a relatively low log Kow that is likely to be more susceptible to foliar runoff than the 2 other chemicals studied that have much higher log Kow values. The results suggest that the soil PEC gives a reasonable worst-case estimate of earthworm exposure for cereals in summer. However, in some cases, exposure may be much lower than expected.

One application to cereals in spring was studied, and measured concentrations were in good agreement with the soil PEC, but caution must be exercised in drawing conclusions from only 1 application.

Considering pesticide concentrations in earthworms, it is difficult to draw conclusions from such a small data set. However, there was evidence that the simple assumption that pesticide concentrations in earthworms are around 30% of soil concentrations is not accurate in all cases. The maximum concentration of chlorpyrifos in earthworms was 86% of the maximum soil concentration. Maximum measured earthworm concentrations after the 2 propiconazole applications were 56% and 218% of the maximum soil concentration. It appears that earthworms may be absorbing these chemicals into body tissues, something that is not surprising for chemicals such as these with relatively high log Kow values. Unfortunately, in this study, no data were available for dimethoate with a much lower log Kow, so a comparison between chemicals with very different log Kow values was impossible. This work does, however, indicate that the crude calculation currently used to predict the risk to animals that eat earthworms may not be reliable. Further work is required to provide a more reliable basis for calculating PEC in earthworms.

Acknowledgment - This work was funded by the Pesticides Safety Directorate of the Ministry of Agriculture, Fisheries and Food.

References

1. [EPPO/CoE] European Plant Protection Organization/Council of Europe. 1993. Decision-making scheme for the environmental risk assessment of plant protection products: earthworms. *EPPO Bulletin* 23:131–149.

2. Greig-Smith PW, Frampton GK, Hardy AR. 1992. Pesticides, cereal farming and the environment, the Boxworth project. London UK: Her Majesties Stationary Office.

3. Tomlin CA. 1994. The pesticide manual. 10th ed. Farnham, Surrey UK: Crop Protection Publications.

Chapter 29
Ecological relevance of ecotoxicological risk assessment

Herman A. Verhoef

In the framework for environmental risk assessment, 4 components can be distinguished: hazard identification, exposure and exposure-response assessment, risk characterization, and risk management. Science has the task to identify the resources at risk and to define the desired or acceptable conditions of the resources. Science, however, is not able to define optimum conditions for most ecosystems. It is proposed that the methods to assess risk of chemical contaminants include examination of the biota from the ecosystem for contaminant effects and examination of the characteristics of the ecosystem under study. This can successfully be performed using microcosms for toxicity testing. Ecosystem functions such as mineralization and decomposition can be studied in microcosms in a realistic way. Because earthworms are important biological agents, not only in soil formation and redistribution of organic matter in the soil but also in organic litter decomposition and mineralization, toxicity tests using earthworms as test organisms in microcosms have a relatively high ecological relevance. Multi-species studies, in which interactions between the below- and above-ground communities also can be studied, should be considered to increase the ecological relevance of these tests. These tests allow not only longer exposure, more levels of biological organization, and more toxicity endpoints [1], but they offer a better integration with ecosystem structure and function. This higher ecological relevance, however, has its price, both in difficulty (e.g., discriminating between direct toxicity and indirect effects of toxicants) and in cost-benefit ratio.

In order to increase the ecological relevance of ecotoxicological risk assessment, more attention should be given in toxicity testing to the effects of toxicants on biotic interactions. Because multi-species testing requires thorough knowledge of the trophic interactions between the biota, using or developing ecosystem (foodweb) models in risk assessment is recommended.

The process of conducting environmental risk assessments requires carefully identifying assessment objectives, deciding the scale and level of biological organization, assembling multidisciplinary data collection and assessment teams, interpreting results using quantitative and qualitative methods, and finally communicating the results in a manner that facilitates risk assessment. The following schematic framework for environmental risk assessment is presented in this chapter; it contains unified principles for assessing the ecological risks of toxic chemicals and other stresses (Figure 29-1). This figure shows the 4 components of the suggested framework: 1) hazard identification, 2) exposure and exposure-response assessment, 3) risk characterization, and 4) risk management. Science is involved in the first 3 components. It has the task to identify the resources at risk and to define the desired or acceptable conditions of the resources. In this chapter, the focus is on the ecosystem as the environmental unit, which leads to the problem of the identification of the values of that ecosystem. Only after a satisfactory definition of a healthy ecosystem is formulated, and a standard is provided against which the current condition of the ecosystem can be evaluated, can the potential changes in that condition over time, in response to a stress, be predicted.

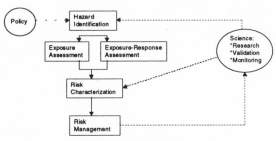

However, suggested ecosystem properties or attributes, e.g., homeostasis, absence of disease, diversity, stability, resilience, vigor,

Figure 29-1   Schematic framework for ecological risk assessment (from Linthurst et al. [2]).

or scope for growth [3–5], all have shortcomings that limit their utility. Still, it is possible to evaluate the condition of terrestrial ecosystems by studying their structural and functional characteristics. At a recent workshop on methods to assess the effects of chemicals on ecosystems organized by the Scientific Group on Methodologies for the Safety Evaluation of Chemicals, structural and functional characteristics of 5 different terrestrial ecosystems, ranging from forests to agro-ecosystems, were summarized (see Table 29-1), with a possible ranking of the importance of the specific characteristics of the different ecosystems [2]. A proposal to conduct exposure and exposure-response assessment in

Table 29-1   Structural and functional characteristics as indicators of the condition of terrestrial ecosystems

Characteristic	Forests	Grasslands	Deserts	Tundra	Agro-ecosystems[1]
Structural					
Species composition	M	M	M	M	L
Biodiversity	M	M	M	M	L
Food web complexity	M	M	L	L	L
Relative density	M	M	M	M	L
Genetic diversity	L	L	L	L	L
Functional					
Primary productivity	H	H	L	L	H
Secondary productivity	M	M	M	M	M
Decomposition	M	L	L	L	L
Mineralization	M	M	L	L	H

Importance of the characteristic: L = low; M = medium; H = high
[1]Conventional tillage system (after Linthurst et al. [2])

such a way that these ecosystem characteristics can be incorporated into the assessment is suggested. This leads to the following 2 strategies in the methods to assess risk of chemical contaminants: a) examination of biota from the ecosystem for contaminant effects and b) examination of the characteristics of the ecosystem under study [2].

Using a representative, minimally disturbed part of the ecosystem, including the native biota, and considering this micro-ecosystem or microcosm as an intermediate between studies of single species toward studies of true ecological complexity is recommended.

Sheppard [1] recently summarized the advantages of microcosm toxicity tests by comparing them with single-species tests, as follows: tests using microcosms allow longer exposures, more levels of biological organization, more species, and more toxicity endpoints. Overall, microcosms might be more realistic. But, what is even more important, microcosms allow integration between the effects of contaminants on biota and effects on ecosystem structure and function [6]. The scale of the microcosms is of importance [7]. Choosing volumes that allow, apart from micro- and mesofauna, the presence of earthworms is suggested. Earthworms can be considered as an important key group in most soils, based on their abundance and biomass. Due to their bioturbation activities and their relatively high consumption rate, earthworms influence both soil structure and decomposition of organic matter [8]. However, they should be included in the toxicity test only if they belong to the native biota of the ecosystem under study.

Ecological relevance of microcosm studies

In a recent special feature of *Ecology* entitled "Can you bottle nature; the roles of microcosms in ecological research," it was concluded that the value of microcosms depends on the questions asked and the ecosystem or community studied [9]. However, it was the general opinion that important species interactions and community and ecosystem questions can be studied using microcosms in a level of detail that has not been possible in the field [10]. Also, ecosystem functions such as mineralization and decomposition can be measured in a realistic way, as was shown by Verhoef [11]. Decomposition (as CO_2 production) and nutrient mobility from coniferous forest litter in microcosms were compared with the results from field cylinder experiments and field plots (Table 29-2). There were significant differences between these 3 systems. Still, the magnitudes of the 3 values were more or less the same for almost all variables. Furthermore, the relative ranking of the extractability of the different nutrients was comparable for the 3 different systems

Table 29-2   Values of soil-process variables ($X \pm 1$ SD) in coniferous forest litter in microcosms ($n = 60$), field cylinders ($n = 50$), and field plots ($n = 10$)

	Microcosms	Field cylinders	Field plots
Ratio: dry mass/fresh mass (g/g)	0.30 ± 0.03	0.26 ± 0.01	0.29 ± 0.02
CO_2 production (mg/g of dry mass)	27.7 ± 4.58	9.6 ± 1.43	10.8 ± 0.86
Nutrient extractability [1] (µmol/g of dry mass)			
K^+	7.0 ± 2.55	9.9 ± 2.13	12.2 ± 1.49
Ca^{2+}	30.9 ± 3.97	72.5 ± 6.68	61.8 ± 13.31
Mg^{2+}	14.0 ± 1.96	22.5 ± 1.93	19.2 ± 2.93
NO_3^-	1.3 ± 0.56	0.6 ± 0.32	2.6 ± 0.88
NH_4^+	29.1 ± 6.20	23.6 ± 4.26	15.6 ± 2.23
PO_4^{3-}	2.9 ± 0.44	1.3 ± 0.38	1.6 ± 0.35

[1]Availability of nutrients was measured by extracting the individual microcosms, field cylinders, and random soil samples with 1.0 mol/L ammonium acetate (for K^+, Ca^{2+}, Mg^{2+}) or with 1.0 mol/L potassium chloride (for NO_3^-, NH_4^+, PO_4^{3-}).
Source:Verhoef [11].

(Figure 29-2). In all systems, Ca^{2+} was the most extractable, and K^+, PO_4^{3-}, and NO_3^- were the least extractable. NH_4^+ and Mg^{2+} could be found at intermediate positions. These systems were relatively simple soil cores with a simple soil community, containing bacteria, fungi, microfauna (such as protozoa, nematodes) and mesofauna (such as micro-arthropods), and enchytraeids. Although these are the relevant groups for the coniferous forest soil studied by Verhoef [11], in many other soils, such macrofauna components as earthworms should be included.

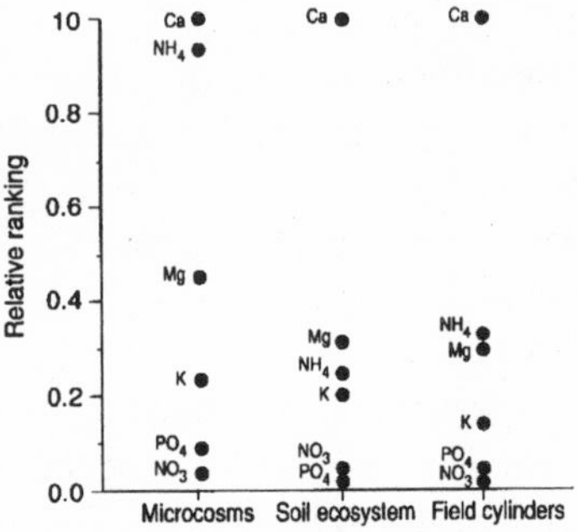

Figure 29-2   Relative ranking of the nutrient extractability spectrum of the microcosms, soil ecosystem, and field cylinders (from Verhoef [11]).

Each of these biotic groups play significant roles in affecting and influencing nutrient dynamics and soil structure (Table 29-3). Table 29-3 also shows that these soil processes are often the result of the interactions between the different biotic groups. Soil-pollution studies should for that reason consider not only the effects of contaminants on isolated species but also the changes in community structure due to interactions [13].

Vegetation is another ecologically important addition to the soil-core microcosms. The root system is an important sink and source of nutrients, influencing decomposition and nutrient flow, and there are important interactions between the below- and aboveground community [14]. The following example is the first Ecotron study [15]. In this microcosm

Table 29-3   Influences of soil biota on soil processes[1]

Soil biota	Nutrient dynamics	Soil structure
Microflora	Catabolize organic matter (OM)	Produce organic compounds that bind aggregates
	Mineralize and immobilize nutrients	Hyphae entangle particles onto aggregates
Microfauna	Regulate bacterial and fungal populations	May affect aggregate structure through interactions with microflora
	Alter nutrient turnover	
Mesofauna	Regulate fungal and microfaunal populations	Produce fecal pellets
	Alter nutrient turnover	Create biopores
	Fragment plant residues	Promote humification
Macrofauna	Fragment plant residues	Mix organic and mineral particles
	Stimulate microbial activity	Redistribute OM and microorganisms
		Create biopores
		Promote humification
		Produce fecal pellets

[1]Source: Hendrix et al. [12]

experiment, the belowground community consisted of micro-organisms, Collembola, and 2 earthworm species: *Lumbricus terrestris* and *Aporrectodea longa*. The aboveground community consisted of 3 plant species: *Trifolium dubium, Poa annua,* and *Senecio vulgaris,* including the herbivore snail *Helix aspersa.* The mean percentage of the total area of the Ecotron units occupied by *Trifolium* remained relatively low (up to 20%) in all animal treatments for the first 4 months (111 d), after which the units containing earthworms (and no snails) doubled the cover of *Trifolium,* compared to the other treatments (see Figure 29-3). This high cover was maintained for about 2 months. The influence of the earthworms on the performance of *Trifolium* operates both through the soil (due to increased availability of phosphates) and by their interactions with the other plant species: earthworms increased root nodulation and establishment of *Trifolium* by creating microsites by depositing casts. Long-term effects of earthworms included the burying of the seedbank of the other 2 plant species, thereby reducing the germination of new seedlings. This shows the important interactions between belowground organisms, especially earthworms, and plant community dynamics.

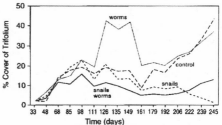

Toxicity testing using microcosms

Figure 29-3   Mean percentage of the total area of the Ecotron-unit occupied by *Trifolium dubium,* with no additional animals (control), worms, snails, and both worms and snails (from Thompson et al. [15]).

Two examples will be discussed here in which microcosms were used for toxicity testing. They differ, among other aspects, in the number of trophic interactions. The first example concerns a study of van Wensem et al. [16], in which microcosms contained poplar leaves as substrate, micro-organisms, and isopods. They studied the effects of the fungicide triphenyltin hydroxide (TPTH). In the presence of the isopods (*Porcellio scaber*), the content of leachable NH_4^+ was strongly increased, which is caused by the excretion products of the isopods and indirect effects on micro-organisms [17]. In systems without isopods, the toxicant had no effect on NH_4^+-mobilization, but in systems with isopods, the addition of TPTH inhibited ammonification. It can be concluded that addition of isopods made this system a more susceptible tool for demonstrating toxic effects than single-species tests using isopods only [18].

The second example concerns a study of Parmelee et al. [19]. The first author refers to this study in his contribution to *Linking Species and Ecosystems* [20], and it is notable that in the whole book this is the only ecotoxicological case [21], despite the importance of the subject for this discipline [19]. The microcosm contains soil from a mature oak-beech forest with natural nematode and micro-arthropod communities. Figure 29-5 depicts the numerical responses of the animal communities to different levels of Cu application. The different Cu sensitivity found for nematodes and micro-arthropods might have to do with the different niches these 2 animal groups inhabit in this soil, as nematodes live in pore water and the micro-arthropods live in air-filled pores [22]. It is clear that the relatively natural substrate in which these experiments are performed add to the ecological realism of the results. The importance of indirect effects of toxicants due to the presence of multi-trophic interactions is seen by the fact that because omnivore-predator nematodes have the highest sensitivity to Cu, there is a positive, indirect, effect of the toxicant on the herbivorous nematodes. Although the length of this test was relatively short (7 d at room temperature) to detect indirect effects on the soil-foodweb structure, more recent experiments [23] in which the microcosm technique was applied during longer periods (up to 40 d) show the advantages of the multi-species approach to link direct effects of a toxicant on soil organisms to its indirect effects on ecosystem-level processes. Because feeding relationships play an essential part in ecosystem processes such as nutrient and energy flows, integration between species interactions and ecosystem processes is possible by the application of foodweb models [24]. The foodweb approach enables predictions about the effects of toxic substances, in combination with other environmental factors, at a higher level of biological integration. The use of this approach in site-specific ecotoxicological risk assessment is advocated in a report from the Health Council of the Netherlands [25].

A similar conclusion has been drawn in the U.S. Environmental Protection Agency (USEPA) document "A Framework for Ecological Risk Assessment" [26], in which the use of food-web models is suggested as a useful tool. The number of suitable terrestrial ecosystem models appears to be scarce and outnumbered by aquatic ecosystem models [27].

However, as stated in [24], foodweb models addressing ecological effects of toxic substances have primarily been used for scientific purposes, and the use of such models in ecotoxicological risk assessment is still in its infancy.

Conclusions

Toxicity tests, in which the effects of toxicants on biota interactions are studied, greatly increase the ecological relevance of ecotoxicological risk assessment. The use of microcosms, in which the responses of organisms and system processes as well as abiotic factors of toxicants can be studied in an integrated way, makes extrapolation of the observed effects to the field more reliable [28]. Because multi-species testing requires thorough knowledge of the trophic interactions between the biota, the use (or development) of food-web models in site-specific risk assessment is highly recommended. In order to increase the ecological relevance of the test, the composition of the biota involved in the test should reflect the composition of the biota of the ecosystem under study. Because earthworms are a ubiquitous group with a great influence on soil structure and function, the scale of the microcosms should allow the presence of this important macrofauna component.

Acknowledgment - I thank Adriaan Reinecke, Steve Sheppard, and 2 anonymous reviewers for their valuable comments on the earlier version of the manuscript.

References

1. Sheppard SC. 1997. Toxicity testing using microcosms. In: Tarradellas J, Bitton G, Rossel D, editors. Soil ecotoxicology. Boca Raton FL: CRC. p 345–373.

2. Linthurst RA, Bourdeau Ph, Tardiff RG. 1995. Methods to assess the effects of chemicals on ecosystems. New York: Wiley. 416 p.

3. Calow P. 1992. Can ecosystems be healthy? Critical consideration of concepts. *J Aquatic Ecosys Health* 1:1–5.

4. Costanza R. 1992. Toward an operational definition of ecosystem health. In: Costanza R, Norton BG, Haskell BD, editors. Ecosystem health: new goals for environmental management. Washington DC: Island. p 236–253.

5. Costanza R, d'Arge R, de Groot R, Farber S, Grasso H, Hanon B, Limburg K, Naeem S, O'Neill RV, Parvelo J, Raskin RG, Sutton P, van den Belt M. 1997. The value of the world's ecosystem services and natural capital. *Nature* 387:253–260.

6. Giesy JP, Odum EP. 1980. Microcosmology: introductory comments. In: Giesy JP, editor. Microcosms in ecological research. DOE Symposium Series 52: CONF-781101.

7. Teuben A, Verhoef HA. 1992. Relevance of micro- and mesocosm experiments for studying soil ecosystem processes. *Soil Biol Biochem* 24:1179–1183.

8. Coleman DC, Crossley Jr DA. 1996. Fundamentals of soil ecology. New York: Academic. 205 p.

9. Daehler CC, Strong DR. 1996. Can you bottle nature? The role of microcosms in ecological research. *Ecology* 77:663–664.

10. Lawton JH. 1996. The Ecotron facility at Silwood Park: the value of "big bottle" experiments. *Ecology* 77:665–669.

11. Verhoef HA. 1996. The role of soil microcosms in the study of ecosystem processes. *Ecology* 77:685–690.

12. Hendrix PF, Crossley Jr DA, Blair JM, Coleman DC. 1990. Soil biota as components of sustainable agroecosystems. In: Edwards CA, Lal R, Madden P, Miller RH, House G, editors. Sustainable agricultural systems. Ankeny IA: Soil and Water Conservation Society. p 637–654.

13. Van Gestel CAM, van Straalen NM. 1994. Ecotoxicological test systems for terrestrial invertebrates. In: Donker MH, Eijsackers H, Heimbach F, editors. Ecotoxicology of soil organisms. Boca Raton FL: CRC Lewis. p 205–228.

14. Wardle DA, Verhoef HA, Clarholm M. 1997. Trophic relationships in the soil microfoodweb: predicting the responses to a changing global environment. *Global Change Biology* (In press).

15. Thompson L, Thomas CD, Radley JMA, Williamson S, Lawton JH. 1993. The effect of earthworms and snails in a simple plant community. *Oecologia* 95:171–178.

16. van Wensem J, Jagers GAJM, Akkerhuis OP, van Straalen NM. 1991. Effects of the fungicide triphenyltin hydroxide on soil fauna mediated litter decomposition. *Pestic Sci* 32:307–316.

17. Teuben A, Verhoef HA. 1992. Direct contribution by soil arthropods to nutrient availability through body and faecal nutrient content. *Biol Fertil Soils* 14:71–75.

18. van Straalen NM, van Gestel CAM. 1993. Soil invertebrates and micro-organisms. In: Calow P, editor. Handbook of ecotoxicology, Vol. 1. Oxford UK: Blackwell Scientific. p 249–251.

19. Parmelee RW. 1995. Soil fauna: linking different levels of the ecological hierarchy. In: Jones CG, Lawton JH, editors. Linking species and ecosystems. New York: Chapman and Hall. p 107–116.

20. Jones CG, Lawton JH. 1995. Linking species and ecosystems. New York: Chapman and Hall.387 p.

21. Parmelee RW, Wentsel RS, Phillips CT, Simini M, Checkai RT. 1993. Soil microcosm for testing the effects of chemical pollutants on soil fauna communities and trophic structure. *Environ Toxicol Chem* 12:1477–1487.

22. Verhoef HA, Brussaard L. 1990. Decomposition and nitrogen mineralization in natural and agroecosystems: the contribution of soil animals. *Biogeochem* 11:175–211.

23. Bogomolov DM, Chen S-K, Parmelee RW, Subler S, Edwards CA. 1996. An ecosystem approach to soil toxicity testing: a study of copper contamination in laboratory soil microcosms. *Appl Soil Ecol* 4:95–105.

24. Berg MP. 1997. Decomposition, nutrient flow and food web dynamics in a stratified pine forest soil [Ph.D. thesis]. Amsterdam, the Netherlands: Vrije Universiteit. 310 p.

25. Health Council of the Netherlands: Standing Committee on Ecotoxicology. Publication #1997/14.1997. The food-web approach for ecotoxicological risk assessment. Rijswijk: Health Council of the Netherlands.

26. [USEPA] U.S. Environmental Protection Agency. 1992. Framework for Ecological Risk Assessment. Washington DC: USEPA, Risk Assessment Forum. EPA/630/R-92/001.

27. Suter II GW. 1993. Ecological risk assessment. Boca Raton FL: Lewis. 505 p. 28. Verhoef HA, van Gestel CAM. 1995. Methods to assess the effects of chemicals on soils. In: Linthurst RA, Bourdeau Ph, Tardiff RG, editors. Methods to assess the effects of chemicals on ecosystems. New York: Wiley. p 223–257.

Chapter 30

Use of a model ecosystem for ecotoxicological studies with earthworms

Tonny M. E. Olesen and Jason M. Weeks

A number of experiments have successfully linked different levels of biological organization with exposure to chemicals between field and laboratory experiments. These have shown that an ecosystem approach for soil ecotoxicological studies can provide a more complex understanding of the overall effects of a pollutant on the soil system by providing information on both ecosystem structure (organisms) and function (processes). Mesocosm and microcosm designs enable variations of field experiment from purely ecological in situ studies to totally manipulated studies, ranging, for example, from individual species to ecosystem-level models using standard soils, introduced species, pollution compounds, and changed physicochemical soil parameters. This chapter highlights the current status of earthworm mesocosm and microcosm research.

The importance of earthworms in testing the undesirable effects of chemicals in the soil environment has been recognized by various environmental organizations and has resulted in standard test guidelines such as the Organization for Economic Cooperation and Development (OECD) guidelines [1]. Such laboratory tests tend to give good reproducibility, but only of worst-case results due to the intimate contact between the earthworms and the chemicals. Also, these laboratory single-species tests are carried out under artificial conditions, thereby disregarding any ecological interaction and making realistic extrapolations to the field unpredictable. Thus, field testing at an ecosystem level is necessary to confirm the validity of the predictions made on the basis of standard laboratory tests and to recognize fully the ecological significance of chemicals on earthworms [2]. Comparison of OECD artificial-soil test results and effects on earthworm populations in the field can be made [3,4], but often the differences in bioavailabilities of a given chemical can complicate such comparisons [5].

One of the major difficulties in comparing results from similar field tests that assess the toxicity of chemicals to earthworms is the wide range of inherent variability in parameters such as earthworm species, population sizes and age classes, competition, migration, and predation, which combined with the existence of strong heterogeneity of soils [6] makes it more difficult to design reproducible field tests [7]. Furthermore, chemicals may not be distributed evenly within test soils, may have reduced bioavailability due to various binding processes in the soil, and are all subject to various biotransformation and biodegradation processes once they are released into the environment. These factors make field tests adequate only to identify chemicals that are extremely toxic to earthworms but insufficient to identify moderately toxic compounds [8].

Many functional properties of ecosystems can be monitored to evaluate ecosystem conditions, e.g., primary production, biotic factors and trophic interactions, rates of decomposition and mineralization. However, assessment of the impacts of chemicals on ecosystems at the ecosystem level raises considerable theoretical and methodological problems. Model ecosystems have been designed as a compromise between these 2 extremes: the highly standardized laboratory single-species tests and the complex ecosystem level of field tests.

The use of such model ecosystems provides scope for experiments ranging from pure field studies to completely standardized experiments running under field conditions or possibly in a climate room, and it has enabled the ecotoxicological objectives of a field study to include measurements (changes in bodyweight, mortality, cocoon production) that would otherwise be difficult, if not impossible, to measure in an open-field experiment. Holmstrup et al. (Chapter 27) presented results on how to estimate the production time of each cocoon, the density of adult earthworms, and their reproductive rates through given periods of time for earthworms collected in the field. Still, in an open-field experiment, the question remains whether migration of earthworms between exposed and nonexposed plots has taken place and to what degree sampled earthworms have actually been exposed to the chemicals in question.

Designs and approaches

Microcosm and mesocosm tests are useful intermediates between bioassays and ecosystem experiments, capturing certain aspects of real ecosystems, but at the same time are simple enough to enable plausible interpretation of the results. They provide controlled experimental conditions in the laboratory or field with the potential for studying changes at any organizational level— individual, population, and community—as well as different aspects of soil-foodweb (system structure) and ecosystem-level processes (system functions) resulting from exposure of a chemical or other stressor.

Several different terrestrial model ecosystems have been described. Usually the system takes the form of a cylinder (e.g., polyethylene or polyvinyl chloride) with an encased soil column, and there often are various devices serving to collect leachate from the column and devices to isolate the whole unit or simply to entrap test species (e.g., Figure 30-1, from [9]). Microcosm studies are usually small, contain a single or a few species, and are generally conducted indoors under highly controlled conditions (e.g., temperature, light, humidity, and fluxes of air and water) (see reviews by [10–13]). A mesocosm can be defined as a bounded and partially enclosed outdoor experimental unit that closely simulates the natural environment [14]. Mesocosm studies are normally relatively large and contain more species or ecosystem levels with a lower degree of control over environmental conditions than for laboratory-based systems. The experiments generally attempt to isolate a small part of an ecosystem and thereby control some of the physicochemical factors and conditions and composition of the biota under near-natural conditions for prolonged periods [9].

Approaches to implement effect endpoints

Depending on the characteristics and uses of a test chemical, the set of endpoints chosen can vary considerably. Soil used in the various ecocosms may either be extracted as intact cores [13,15,16], disturbing the micro-organisms, mesofauna, and physical soil structure as little as possible, or extracted and homogenized [9,12] to reduce natural variability. Also, the experimental containers may be designed as closed (sealed) systems or as open systems with contact to the surrounding atmosphere under strict environmental control, or can be placed in the field under realistic but highly variable environmental conditions.

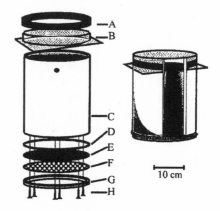

Figure 30-1   Schematic representation of an exploded and assembled mesocosm unit: (A) broad rubber band, (B) fine meshed 60% shading top-netting, (C) medium density polyethylene (MDPE) tube section, (D) PE tubing gasket, (E) Phormisol membrane, (F) strong MDPE netting, (G) MDPE bottom rim, (H) stainless steel screws (from Svendsen and Weeks [9]).

Bogomolov et al. [12] used small laboratory microcosms with a homogenous soil treated with Cu in an open system. Several organism-level measurements were made (microbial biomass N, CO_2 evolution, soil urease activity) as well as nematode and micro-arthropod counts and earthworm growth and mortality. By examining both the structure and function of the soil system, they were able to link the direct effects of Cu on a number of organisms to indirect effects on ecosystem-level processes. Svendsen and Weeks [9,17] measured tissue residues and biomarker responses in earthworms exposed to Cu and linked results at different biological levels (cellular, individual, and population) from a modified OECD test with results from a semi-field mesocosm setup (homogenous soil, open system). Römbke et al. [13] used intact cores in an open laboratory setting, comparing effects of 2 pesticides on earthworm and enchytraeid biomass and abundance. These results were compared with those obtained from an OECD guideline laboratory test and an open field site. Results from microcosms compared well with those obtained from the laboratory, although the field results were not reflected in the microcosm data.

The use of intact soil cores offers the advantage of being close to a real situation, having minimally disrupted microorganisms and mesofauna, but suffers from a high variability due to the inherent differences in soil structure. Open systems cannot be used with volatile substances but are technically less complex than the closed systems and therefore less expensive. Application of the test substance must also be considered with some importance. Surface-applied compounds may be accomplished in such a fashion as to mimic the realistic application method (e.g., pesticides). However, the real or direct exposure of the test animals or system is then largely unknown, and a big part of the advantage of the technique may be lost. When using systems with homogeneously mixed soils and compounds, the test organisms will be in intimate contact with the chemical. This, however, might not be a realistic approach. Accurate estimates of exposure in the environment are difficult to obtain because they depend on a number of interacting factors, e.g., soil-binding properties, meteorological conditions, and a range of other ecological parameters. Careful consideration of available laboratory data will be essential in order to perform a valuable field or semi-field mesocosm test and to link the results from these to the real world. The size of the system will also have some importance. If the test organisms only include, for example, microorganisms and micro-arthropods, then a relatively small containment system will suffice, whereas if earthworms or other mesofauna are included, larger systems allowing for movement of the organisms (on the soil surface or in the soil column) will have to be considered.

The major disadvantage in using any type of mesocosm is that normally they are very costly, both in true economic terms and in relation to manpower. Previously, this has meant that few mesocosm experiments have been undertaken, and, when adopted, they were often poorly designed statistically and underreplicated. Together with the large inherent variability of these more complex systems, several large, expensive studies have rendered inconclusive results [7,18]. This problem can be addressed on several fronts: first, by decreasing the variation caused by nonexperimental factors (i.e., various soil gradients) by the provision of more standardized conditions, and second, by making the experimental setup more economically achievable to enable greater replication. Finally, the statistical design of such experiments also requires a priori consideration. In many cases, statistics are considered and statisticians consulted only at the final stages of analyzing the datasets, and the statistician is then unfortunately faced with the task of making the best of a bad job [19]. The ideal mesocosm setup would, therefore, be one that could provide the necessary degree of standardization, while being inexpensive to construct, maintain, and operate. This would enable more experiments to be undertaken with the appropriate design and replication required for more satisfactory results.

In terms of risk assessment, effects on earthworms are considered chiefly because of the potential risks of secondary poisoning for birds and mammals. In principle, risk to predators can be judged from residue measurements in field trials. However, routes of exposure are often more complex (e.g., birds may eat both worms and the soil contained within their guts and granular pesticides directly), and a better understanding of such mechanisms is desirable. It is also of importance to understand the cause of adverse effects of pesticides on earthworm populations, which if lowered may adversely affect crop performance [20]. Mesocosm studies will aid in the interpretation of the potential risk of chemicals, in particular pesticides, as more control on variables such as initial earthworm density, dose of compound, worm-burrowing depth, soil moisture, and many other characteristics are known both at the start and end of a semi-field trial. Thus, information may be more easily interpreted and incorporated into a risk assessment procedure. It is imagined that ultimately such semi-field testing will eliminate the need for large-scale, real field trials.

Tests are yet to be developed to the stage of routine and regulatory use. This has been attempted by the UBA (Umweltsbundesamt), Germany, in a draft guideline on using earthworms in "Terrestrial Model Ecosystems" [21].

References

1. [OECD] Organization for Economic Cooperation and Development. 1984. Organization for Economic Cooperation and Development guidelines for testing of chemicals: earthworm acute toxicity test. OECD Guideline No. 207. Paris, France: OECD.

2. Christensen OM, Mather JG. 1994. Earthworms as ecotoxicological test-organisms. 5. Bekæmpelsesmiddelforskning fra Miljøstyrelsen. Danish Environmental Protection Agency.

3. Heimbach F. 1992. Correlation between data from laboratory and field tests for investigating the toxicity of pesticides to earthworms. *Soil Biol Biochem* 24:1749–1753.

4. Van Gestel CAM. 1992. Validation of earthworm toxicity tests by comparison with field studies: A review on benomyl, carbendazim, carbofuran and carbaryl. *Ecotox Environ Safety* 23:221–236.

5. Spurgeon DJ, Hopkin SP. 1995. Extrapolation of the laboratory-based OECD earthworm toxicity test to metal-contaminated field sites. *Ecotoxicol* 4:190–205.

6. Becher HH. 1995. On the importance of soil homogeneity when evaluating field trials. *J Agron Crop Sci* 74:33–40.

7. Lofs-Holmin A, Bostrom U. 1988. The use of earthworms and other soil animals in pesticide testing. In: Edwards CA, Neuhauser EF, editors. Earthworms in waste and environmental management. The Hague, the Netherlands: SPB Academic. p 303–313.

8. Goats GC, Edwards CA. 1988. The prediction of field toxicity of chemicals to earthworms by laboratory methods. In: Edwards CA, Neuhauser EF, editors. Earthworms in waste and environmental management. The Hague, the Netherlands: SPB Academic. p 283–294.

9. Svendsen C, Weeks JM. 1997. A simple low-cost field mesocosm for ecotoxicological studies on earthworms. *Comp Biochem Physiol* 117C:31–40.

10. Morgan E. 1994. A review of the use of terrestrial model ecosystems in the ecotoxicological testing of chemicals and an outlining of a proposed test guideline. In: UBA-Workshop on terrestrial Model Ecosystems. UBA-texte 54/94: 7-19. UBA (Umweltbundesamt).

11. Sheppard SC. 1997. Toxicity testing using microcosms. In: Tarradellas J, Bitton G, Rossel D, editors. Soil ecotoxicology. Boca Raton FL: Lewis. p 345–373.

12. Bogomolov DM, Chen S-K, Parmalee RW, Subler S, Edwards CA. 1996. An ecosystem approach to soil toxicity testing: a study of copper contamination in laboratory soil microcosms. *Appl Soil Ecol* 4:95–105.

13. Römbke J, Knacker T, Forster B, Marcinkowski A. 1994. Comparison of effects of two pesticides on soil organisms in laboratory tests, microcosms, and in the field. In: Donker M, Eijsackers H, Heimbach F, editors. Ecotoxicology of soil organisms. Boca Raton FL: CRC. p 229–240.

14. Odum EP. 1984. The mesocosm. *Bioscience* 34:558–562.

15. Baker GH, Barrett VJ, Carter PJ, Woods JP. 1996. Method for caging earthworms for use in field experiments. *Soil Biol Biochem* 28:331–339.

16. Tolle DA, Frye CL, Lehmann RG, Zwick TC. 1995. Ecological effects of PDMS-augmented sludge amended to agricultural microcosms. *Sci Total Environ* 162:193–207.

17. Svendsen C, Weeks JM. 1997. Relevance and applicability of a simple biomarker of copper exposure. II Validation and applicability under field conditions in an esocosm experiment with *Lumbricus rubellus*. *Ecotoxicol Environ Safety* 36:80–88.

18. Steel JH. 1979. The use of experimental ecosystems. *Philos Trans R Soc Lond B Biol Sci* 286:583–595.

19. Sparks TH, Gadsden RJ. 1997. The trials and tribulations of experimental design in ecotoxicology. *Toxicol Ecotoxicol News* 4:55–59.

20. Greig-Smith PW. 1992. Risk assessment approaches in the UK for the side-effects of pesticides on earthworms. In: Greig-Smith PW, Becker H, Edwards PJ, Heimbach F, editors. Ecotoxicology of earthworms Hants, Andover UK: Intercept. p 159–168.

21. Knacker T, Morgan E. 1994. Test guideline—terrestrial model ecosystem. In: UBA-Workshop on Terrestrial Model Ecosystems. Appendix UBA-Texte 54/94: 1-35. UBA (Umweltbundesamt), Germany.

Section 7
Recommendations

Chapter 31

Recommendations from the Second International Workshop on Earthworm Ecotoxicology, Amsterdam, Netherlands (April 1997)

John D. Bembridge

The Second International Workshop on Earthworm Ecotoxicology was held in April 1997 at the Vrije Universiteit, Amsterdam. The aim of the workshop was to build upon the success of the first workshop in Sheffield in 1991, evaluate progress made against the recommendations published at that time [1], and develop further improvements in earthworm ecotoxicology. Seventy-eight delegates from Europe, North America, Africa, Asia, and Australasia participated in the workshop, which took the form of platform presentations, poster sessions, and discussion. Participants, although linked by an interest in earthworms, came from a variety of roles in academia, industry, and regulatory bodies. The workshop aimed to produce clear recommendations to optimize the use of earthworms in the assessment of effects for regulatory and policy-setting purposes. Where research needs were identified, the aim is for researchers to take up these challenges and set up collaborations to advance knowledge.

The workshop focused on 5 main topics formulated from a questionnaire circulated in 1996 representing the major areas of interest of workers in the earthworm field. The areas selected were the following: the significance of earthworm populations and how to identify what constitutes a significant effect, the refinement of laboratory tests so that they are relevant to field conditions, exposure quantification, flexible field test design, and risk assessment. Sessions addressing these topics culminated in general discussions and from these and the conclusions of the presentations and posters, the session chairmen (PJ Edwards, CAM Van Gestel, S Sheppard, and A Reinecke) drafted recommendations that were presented to the whole workshop in a final session chaired by H Eijsackers. The recommendations presented here are based this discussion and on further comments from the delegates when the formulated recommendations were circulated prior to publication.

The recommendations can be divided into 4 main areas:
1) laboratory testing,
2) semi-field and field testing,
3) risk assessment, and
4) modeling.

Laboratory testing

Laboratory testing methods are well established, and guidelines exist for acute and sublethal studies. Further optimization of these study guidelines may be possible in some areas, but large-scale changes are not considered to be necessary or useful. However, a word of caution: in regulatory ecotoxicology, we should not lose sight of what information is actually required by the regulators and the way in which they use these data.

The acute laboratory study that follows the Organization for Economic Cooperation and Development (OECD) Guideline #207 [2] appears to be the least contentious of all studies and is generally accepted by the majority of workers.

Recommendation 1

For the first screening of new chemicals, a simple acute toxicity test with *Eisenia fetida* or *Eisenia andrei*, such as the OECD artificial-soil test, using mortality as the endpoint, should be considered as sufficient.

The acute toxicity study is principally designed to measure toxicity with mortality as the major endpoint, however bodyweight data are also recorded, and the need for or most appropriate use of these data needs to be established. Bodyweight information can be regarded as a general indication of the health of the earthworms during the study and therefore changes in the control worms can be used as an assessment of the quality and validity of the study. Changes in bodyweight in the treated vessels may be an indication of potential sublethal effects. For both these situations, it is important that there is agreement as to the scale of change in bodyweight that is regarded as significant.

Recommendation 2

Determination of bodyweight before and after the test is recommended in acute studies as a quality control. A reduction in fresh weight of less than 15 to 20% in the control should be evaluated as a suitable validity criterion. In treated vessels, a significant reduction beyond that of the control should be considered as an indication of sublethal effects.

Bodyweight measurements may also be used to improve the design and accuracy of laboratory studies. The weight of an earthworm at the beginning of the study will have an impact on its subsequent growth and therefore will directly influence the final body weight. If the earthworms are not uniform at the beginning of the study, then the final bodyweight measurements will be a factor of any treatment effects and the initial bodyweight. Efforts should therefore be taken to negate this potential influence by using the initial body weight measurements to assign earthworms to treatments.

Recommendation 3

Initial bodyweight measurements should be used to guarantee a uniform assignment of earthworms to treatments in laboratory studies. The method used to assign earthworms to treatments should be included in study reports to facilitate interpretation of the results.

Sublethal testing has now been accepted as part of the routine regulatory testing scheme in Europe. Although there have been problems with its introduction and in how the resulting data are interpreted, it is generally thought that the existing guidelines are sufficient. The only guideline in its final form is issued by the Biologische Bundesanstalt Für Land-Und Forstwirtschaft (BBA) [3], but this is soon to be replaced by the International Standards Organizations (ISO) guideline, which currently is in draft form [4].

Recommendation 4

For a second-stage assessment, a sublethal endpoint is needed. The ISO draft guideline for a reproduction study using *E. fetida* or *E. andrei* should be suitable for this purpose. Because there is an interrelationship between growth of adult earthworms and reproduction, both these parameters should be measured as part of the test.

There is still some discussion as to what is the most appropriate endpoint for assessing the potential effects of a chemical on earthworms in the field. The rate of growth of immature worms has been suggested as a more realistic endpoint because it gives some indication of the dynamics of the field population and its ability to grow and recover from perturbations. At present there are no standard test methods to measure the growth of immature earthworms, and these would need to be produced before the current system based on reproduction could be changed. Workshop delegates were reluctant to introduce the need for further studies unless it was clearly proved that such studies provided useful and relevant information.

Recommendation 5

The rate of growth of immature earthworms is an important endpoint in terms of population growth. However, some chemicals that have a direct impact on reproduction may not affect growth. Additional research is needed to evaluate whether a standardized toxicity test using growth of immature earthworms should be developed.

Existing test guidelines are aimed primarily at evaluating the effects of pesticides on earthworms. Special consideration needs to be given to other types of substances that require testing, as their properties may require the modification of existing test methods.

Recommendation 6

Consideration should be given to the problems of using existing methods for organic toxicants, particularly those with short environmental half-lives and those that are volatile.

Measurement of exposure may be useful in some circumstances. While it was not considered necessary to include such measurements in the standard laboratory tests, there are a number of variables that could be measured in other types of study.

Recommendation 7

Some quantification of exposure specific to the contaminant and the bioassay method should be included in nonstandard testing. Suitable methods may be the measurement of the dissociation constant (Kd), the partition coefficient, and tissue residues or the use of biomarkers or behavioral observations. Toxicity should be related to both solid- and liquid-phase concentrations in soil in such studies.

Semi-field and field testing

While the guidelines for acute and sublethal laboratory studies are now becoming established, it was felt that they may not be flexible enough to deal with all situations. Some specialized application methods are not accurately represented in the laboratory studies, leading to unrealistic exposure. In other cases, it is felt that widening the species range in the studies could provide important information and that groups other than earthworms should be considered. One of the key groups mentioned were the enchytraeids, which are dominant in more acidic soils, and test guidelines are currently being developed for these animals. Other suggestions were Collembola and Oribatid mites.

Recommendation 8

At higher tiers of assessment, tailor-made testing may be more appropriate, particularly in the case of specialized application techniques (e.g., seed dressing, band applications, slow-release formulations, etc.). As well as semi-field and field studies, tests using other appropriate earthworm species should be considered. In ecosystems with more specialized conditions (highly organic, acidic, or water-saturated soil conditions) or in soils where earthworms are absent or at low densities, then tests on other species fulfilling a similar ecological niche might be considered.

Field testing methods for earthworms are still much discussed. New field testing guidelines have been published in Germany [5] and are currently being developed by the ISO [6]; it was felt that these were adequate and should provide the basis for routine field testing in the future. It was generally thought that effort should be centered on reviewing the data produced from these guidelines and using these data to suggest refinements, rather than starting at the beginning again.

There is currently a large amount of interest in mesocosm/semi-field/model ecosystem-type studies that could provide an interim step between the laboratory and the field. A number of different research groups have been working with a range of study designs, from small to large scale. However, at present there are still considerable concerns about the validity of these approaches and the methodology used. It was not thought that there was yet an acceptable method for general use of these interim studies. This is a controversial area that is likely to expand in the future, and therefore, a close watch will need to be given to developments and how these could be used for improving understanding of effects on earthworms.

Recommendation 9

A working group should be formed to produce guidance for the design and interpretation of flexible field, and particularly semi-field, tests. This group should build on experience with the existing guidelines by reviewing the data produced and should use that experience to promote guideline refinement and to issue guidance on how to use such data in risk assessment.

If tailored studies are used as part of a risk assessment, it is important that they can be linked to the information required by regulators. All such studies should be carefully designed to ensure that they provide data that add value to available information. The design of studies should reflect the use of the compound, the application methods, the environmental conditions, the species that will be exposed, and the potential effects on earthworm communities.

Recommendation 10

Guidelines should be developed to allow the linkage of appropriate testing to the level and type of regulatory information required according to the substance involved. For example, semi-field or mesocosm-type studies may be more appropriate for the secondary or tertiary testing of some pesticides. Potential genotoxins should be tested in the laboratory or field for a period that covers more than 1 life cycle of the species involved.

A number of areas of basic earthworm behavior and physiology still are not fully understood. Particular interest has been shown in the area of earthworm migration or surface movement. This may have significant effects on the results of field trials, and further work is required to identify the scale and cause of this movement. If earthworms are moving considerable distances, then this will have implications not only for the design of field studies but also for risk assessment.

Equally, there is a need to understand more about what happens to a compound or contaminant when it is taken into an earthworm and how measured concentrations in various tissues relate to effects on the earthworm as a living unit.

Recommendation 11

Some fraction of earthworm ecotoxicology effort should address earthworm ecology and physiology to help understand earthworm responses to chemicals. One possible area of research is how and why earthworms move on the soil surface. A better understanding of the concentrations of compounds that have a physiological effect on tissues and how these effects relate to measured residue levels is another potential area. The results of this research should be used to help design and interpret ecotoxicological studies.

The use of earthworm biomarkers is another area in which there is considerable interest. Most of the work so far has been centered around the use of biomarkers to identify heavy-metal pollution. While the expression of biomarker genes can be demonstrated in the laboratory, there remains the need for the ecological relevance of this response to be quantified.

Recommendation 12

Biomarkers may be useful tools for post-registration monitoring. More research is required to develop an understanding of their ecological relevance and of the impact of confounding factors on their response.

Risk assessment

In the regulatory area, tests on earthworms aim to provide information for regulatory authorities to carry out risk assessments. Regulators wish to evaluate whether a substance will have any effects on earthworm populations that may subsequently have effects on soil fertility or on earthworm predators. To be able to do this effectively, they must be provided with clear data on the ecologically relevant endpoints, which should allow them to make the appropriate judgment as to the level of risk. Efforts should therefore be made to ensure that the most relevant data are provided, and the experience of earthworm workers should provide guidance in how data should be interpreted.

Recommendation 13

Regulators must be provided with information and guidance to allow them to interpret data effectively and make informed decisions. This information may include how the structure and composition of earthworm populations may affect soil functions, their importance as a food resource for vertebrates, and their associated role in transferring chemical residues to vertebrates by this route. The identification of appropriate regulatory endpoints will allow their use in risk management and labeling schemes.

To assess the importance of an effect on an earthworm population, it is vital to relate this effect to other pressures on the population. We know that earthworms exhibit considerable seasonal variation in their population size and makeup; therefore, this should be used as a benchmark against which to judge the ecological relevance of an imposed perturbation.

Recommendation 14

It must be recognized that earthworm populations fluctuate widely in response to biotic and abiotic factors. Chemical perturbations may be only 1 factor in agricultural habitats, and field assessments should attempt to identify the other factors. Methods should be developed to improve the interpretation of population effects caused by common abiotic factors and those that are chemical effects.

One of the key topics addressed at the first workshop was what should be considered an ecologically relevant effect, and in many respects this question still remains unanswered. The suggestion of a 30 to 50% reduction in population density still appears to be a good one, and the workshop was unable to improve upon it. The concept of recovery was again identified as a key factor, and field trials should be designed appropriately so that this can be measured by, for example, considering the length of the life cycle of the earthworm species.

While triggers can be set as to what should be considered to be a significant effect, there is some concern for the ecosystem about effects that are below this trigger. If recovery is regarded as a key factor, then the effects on the ecosystem while earthworm numbers are reduced during recovery must be considered.

Recommendation 15
The definition of what constitutes an ecologically relevant effect on earthworm communities must be established. Failure to recover to the level of untreated control plots within a period relevant to the life cycle of the earthworm species involved, but not normally less than 12 months, should be considered a useful measure. A 30 to 50% reduction in population density may also be a useful definition, but the significance of this effect on ecosystems of these and smaller reductions should be considered.

Part of the difficulty in determining ecologically relevant effects from field trial data is the complexity of the data produced. Traditional field trials produce a matrix of univariate statistical operations, with some species appearing to be significantly affected while others are not. The overall ecological effect on the earthworm community may be unclear. Similar problems have been dealt with in the area of aquatic ecotoxicology by using multivariate approaches that aim to look at the data set as a whole. This technique may prove useful in the assessment of earthworm trials in which measurements are taken on a number of different species and life stages.

However, when new statistical approaches are used, it is important that regulators are able to interpret the outputs; otherwise, nothing is added to the improvement of the regulatory process.

Recommendation 16
Multivariate statistical methods provide a powerful way of investigating effects at the population and community levels. Further work is required to evaluate the use of these methods with earthworm data. Information should be sought from their use in aquatic ecotoxicology.

Within the European Union (EU), earthworm testing and risk assessment is carried out according to a tiered system that involves a number of triggers. There has been considerable discussion about the validity of these triggers, especially those concerning sublethal testing. The purpose of the tiered approach is to ensure that, if there is concern from a study in which there may be an effect on earthworms, a higher-tier study is triggered to address this concern. However, the success of such a scheme must rely not only on identifying areas of concern but equally in not triggering extra studies when there is no need for concern.

A point of concern has been the high sensitivity of the sublethal effects test in detecting effects. Work carried out by Heimbach (Chapter 19) has shown that for some compounds, tested at the top rate of 5× the field rate in the sublethal effects test, there are indications of potential effects. However, when these compounds have subsequently been tested in the field, no effects have been seen. Therefore, a group from the European Crop Protection Association (ECPA) has proposed a revised set of triggers to take this into account. Regulators have been interested in this approach, but it is likely that further discussions will need to take place before such a scheme could be adopted.

Recommendation 17

The alterations proposed by the ECPA earthworm group (Chapter 21) to the triggers currently used within EU registration are recommended as a basis for future discussions.

Modeling

Modeling is seen by many as the key activity in the future for earthworm risk assessment. At present, however, only a few models are available for earthworms, and these deal with only a small number of species. Equally, there is still no full agreement as to the key parameters that should be considered in models. Even when parameters have been identified, there is often a lack of data to allow the models to be calibrated against real conditions. However, as in other areas of ecotoxicology, models present an opportunity to look more closely at the impact of perturbations on earthworms, and they will be invaluable tools. To reach this stage, much work is required in the design and calibration of such models, and modelers were urged to form collaborations to take this forward.

Recommendation 18

Models should be regarded as valuable tools for developing the understanding of the effects of toxicants on earthworm population dynamics and in identifying key parameters and endpoints for risk assessment. Food availability, food quality, temperature, and soil properties are key parameters affecting earthworm populations and should be considered in models. Appropriate endpoints to measure may be survival, growth, and reproduction rate. A working group should be set up to work toward developing and validating a generally applicable population-dynamics model to improve understanding of the processes involved in regulating earthworm populations.

It must be recognized that there are still many gaps in our understanding of the population dynamics of earthworms and how they respond to perturbations. Until this information is provided, it will be impossible to accurately model the system. Therefore, a number of key research areas need addressing to provide input to the modelers.

Recommendation 19

A full understanding of the life cycle of relevant earthworm species is required to support population modeling. Research is required to provide information on those parts of the life cycle that are still insufficiently known to provide proper model input parameters. It is recommended that the effect of a number of chemicals, with different modes of action, throughout the entire life cycle of earthworms should be evaluated.

Recommendation 20
Further work to investigate the interaction of the impact of specific toxicants with other stressors, e.g., temperature, soil moisture content, food availability, predation, and the importance of indirectly driven effects, should be promoted.

Recommendation 21
Bioconcentration-factor, toxicokinetic, and coefficient data may be useful for modelers and ecotoxicologists. The development of a database to help link data from the laboratory and the field, plus its potential structure and use, should be evaluated. Standardized methods should be recommended for the measurement of these values.

Recommendation 22
Another workshop should be convened to assess advances in earthworm ecotoxicology. All those involved in working with earthworms are encouraged to address the topics arising from this workshop in preparation for the next one.

Conclusion
It is hoped that participants of the Second International Workshop on Earthworm Ecotoxicology will these recommendations to help develop their work programs for the coming years. The workshop organizers will present the outcomes to the appropriate authorities, e.g., EU, OECD, European Plant Protection Organization (EPPO), and national regulatory bodies. It was encouraging to see the number of people who were addressing the issues that had arisen from the first workshop, and the aim is to maintain this momentum. Plans are already underway for a future workshop, and all earthworm workers are set the challenge of addressing the issues of the second workshop so that further advances can be made in our understanding of this interesting and ecologically important group of animals.

References

1. Greig-Smith PW. 1992. Recommendations of an international workshop on ecotoxicology of earthworms, Sheffield, UK (April 1991). In: Greig-Smith PW, Becker H, Edwards PJ, Heimbach F, editors. Ecotoxicology of earthworms. Andover, Hampshire UK: Intercept.

2. [OECD] Organization for Economic Cooperation and Development. 1984. Organization for Economic Cooperation and Development guidelines for testing of chemicals: earthworm acute toxicity test. OECD Guideline No. 207. Paris, France: OECD.

3. Kula C, Bauer C, Dohmen P, Heimbach F, Kampmann T, Kula H, Riepert F, Römbke J, van Gestel CAM. 1994. Guidelines for pesticide testing in the registration procedure, Part VI, 2-2. Effects of pesticides on the reproduction and growth of *Eisenia fetida/Eisenia andrei*. Federal Biological Research Centre for Agriculture and Forestry (BBA), Germany.

4. [ISO] International Standards Organization. 1993. DIS 11268-2 (Draft) Soil quality—effects of pollutants on earthworms (*Eisenia fetida*) Part 2: Methods for the determination of effects on reproduction. Geneva, Switzerland: ISO.

5. Kula C, Bachhenß J, Dohmen P, Heimbach F, Kula H, Kampmann T, Riepert F, Römbke J. 1994. Guidelines for pesticide testing in the registration procedure, Part VI, 2-3. Effects of pesticides on earthworms in the field. Federal Biological Research Centre for Agriculture and Forestry (BBA), Germany.

6. [ISO] International Standards Organization. 1996. DIS 11268-3. (Draft) Soil quality - effects of pollutants on earthworms. Part 3: Field method. Geneva, Switzerland: ISO.

Section 8
Acknowledgments

Acknowledgments

The Second International Workshop on Earthworm Ecotoxicology and this book are the result of an impressive international cooperation among specialists from academia and industry. Great care was taken at all stages to ensure high quality. Prior to the workshop, the proposed papers were screened for relevance by the scientific committee. Once the manuscripts were received, each was directed to review by 1) a peer who was present at the workshop, 2) an expert peer who did not attend the workshop, and 3) the Chair of the corresponding workshop session, who made sure that discussions during the Workshop were reflected in the papers. The reviews were then gathered by the 4 editors, who reviewed the papers again and negotiated revisions with the authors. The full book was read by the senior editor and by reviewers selected by SETAC. Throughout, the process was overseen by the SETAC liaison, Roman Lanno.

The Session Chairs and discussion leaders were, in alphabetic order: Peter Edwards, Herman Eijsackers, Jari Haimi, Martin Holmstrup, Steve Sheppard, Kees van Gestel, and Adriaan Reinecke. They played a key role in integrating the diverse papers into a coherent assemblage.

The workshop was chaired by Herman Eijsackers, with Adriaan Reinecke as Chair of the Scientific Committee. Leo Posthuma and Kees van Gestel skillfully directed the organizational and local arrangements. Karin van Ginkel deserves exceptional credit for smooth operation of the full process, from start to finish. Kees Verhoef and Desiree Hoonhout also contributed substantially. Eline Kruger was instrumental in drafting the workshop program.

It is most important to acknowledge the thoughtful contributions of the reviewers. They were anonymous to the authors, but most agreed to be listed here. A few reviewers are not listed for various reasons. All the reviewers were pressed into service with demands for quick replies by the Editors, and the response was excellent. Many authors wrote special congratulations for the insights they obtained from the reviews. Clearly, this process was an international collaboration. The Editors are indebted to the following for their timely review comments:

Jorgen Axelsen (Denmark)
John D. Bembridge (UK)
Göran Bengtsson (Sweden)
W. Nelson Beyer (USA)
Kevin C. Brown (UK)
Kevin Butts (UK)
Peter M. Chapman (Canada)
Marianne Donker (The Netherlands)
Charles T. Eason (New Zealand)
Clive A. Edwards (USA)
M.C. Fossi (Italy)

Peter Grieg-Smith (UK)
John W. Hall (Canada)
Andrew D.M. Hart (UK)
Fred Heimbach (Germany)
Martin Holmstrup (Denmark)
Steve Hopkin (UK)
Tjalling Jager (The Netherlands)
John Jensen (Denmark)
Tim J. Kedwards (UK)
Chris Klok (The Netherlands)
Christine Kula (Germany)
Hartmut Kula (Germany)
Otto Larink (Germany)
Johan Lembrechts (The Netherlands)
D.M. Lowe (UK)
Colin Macdonald (Canada)
Janice G. Mather (Denmark)
A. John Morgan (UK)
Jos Notenboom (The Netherlands)
Robert W. Parmelee (USA)
Leo Posthuma (The Netherlands)
Poul Prentö (Denmark)
Richard Protz (Canada)
Hans Ramlöv (Denmark)
Jörg Römbke (Germany)
Kees Romijn (France)
Sten Rundgren (Sweden)
Janne Salminen (Finland)
Janek Scott-Fordsman (Denmark)
Marsha I. Sheppard (Canada)
Steve Sheppard (Canada)
Vibeke Simonsen (Denmark)
Tim Sparks (UK)
Jorgen Stenersen (Norway)
Gladys Stephenson (Canada)
Geoffrey Sunahara (Canada)
Kim Z. Travis (UK)
Kees van Gestel (The Netherlands)
Nico van Straalen (The Netherlands)
Joke van Wensem (The Netherlands)
H. Verhoef (The Netherlands)
Nadja Wagman (Sweden)
Jason Weeks (UK)

Index

A